AF812847

Progress on Ecosystem Restoration of Tropical Inland Waters

Adelina C. Santos-Borja · Luki Subehi ·
Miratul Maghfiroh · Ali Rahmat
Editors

Progress on Ecosystem Restoration of Tropical Inland Waters

On the Path of the UN Decade on Ecosystem Restoration

Springer

Editors
Adelina C. Santos-Borja
Southeast Asian Limnological Network (SEALNet)
Bogor, Jawa Barat, Indonesia

Miratul Maghfiroh
Research Center for Limnology and Water Resources
National Research and Innovation Agency (BRIN)
Bogor, Jawa Barat, Indonesia

Luki Subehi
Southeast Asian Limnological Network (SEALNet)
Bogor, Jawa Barat, Indonesia

Research Center for Limnology and Water Resources
National Research and Innovation Agency (BRIN)
Bogor, Jawa Barat, Indonesia

Ali Rahmat
Research Center for Limnology and Water Resources
National Research and Innovation Agency (BRIN)
Bogor, Jawa Barat, Indonesia

ISBN 978-981-96-2283-2 ISBN 978-981-96-2284-9 (eBook)
https://doi.org/10.1007/978-981-96-2284-9

© The Editor(s) (if applicable) and The Author(s), under exclusive license to Springer Nature Singapore Pte Ltd. 2025

This work is subject to copyright. All rights are solely and exclusively licensed by the Publisher, whether the whole or part of the material is concerned, specifically the rights of translation, reprinting, reuse of illustrations, recitation, broadcasting, reproduction on microfilms or in any other physical way, and transmission or information storage and retrieval, electronic adaptation, computer software, or by similar or dissimilar methodology now known or hereafter developed.
The use of general descriptive names, registered names, trademarks, service marks, etc. in this publication does not imply, even in the absence of a specific statement, that such names are exempt from the relevant protective laws and regulations and therefore free for general use.
The publisher, the authors and the editors are safe to assume that the advice and information in this book are believed to be true and accurate at the date of publication. Neither the publisher nor the authors or the editors give a warranty, expressed or implied, with respect to the material contained herein or for any errors or omissions that may have been made. The publisher remains neutral with regard to jurisdictional claims in published maps and institutional affiliations.

This Springer imprint is published by the registered company Springer Nature Singapore Pte Ltd.
The registered company address is: 152 Beach Road, #21-01/04 Gateway East, Singapore 189721, Singapore

If disposing of this product, please recycle the paper.

Preface

In the era where human interference has severely impacted the environment, ecosystem has to endure such an immense pressure that leads to ecosystem degradation. In the tropical region with populous inhabitants, the increase in human population and rapid urbanization often correspondingly result in urban problems such as inadequate water supply, environmental pollution, and various wastes. The impacts of those problems are reflected in the obvious signs of environmental changes. The changes of physicochemical properties of the water bodies due to habitat degradation and destruction trigger elimination of vulnerable taxa, thereby, loss of important niches in ecosystems impacts negatively on the species diversity and significantly damages the sustainability of ecosystem in the future.

Inland water ecosystems provide humans with numerous services such as food, drinking water supply, power, transportation, recreation, flood prevention. Degradation and destruction upon inland water ecosystem induce threats to humans, especially those who rely their lives on inland waters, i.e., lakes, rivers. Eventually, ecosystem restoration is then necessary to ensure that the ecosystem degradation is reduced or does not continue to occur.

Initiatives among communities regionally or globally have been established to encourage the practice of ecosystem restoration. Academia, researchers, the government, communities and other stakeholders need to be aware that the goals of ecosystem restoration can only be achieved through synergistic efforts and relentless participations. The efforts on ecosystem restoration in tropical countries such as in Southeast Asia seem to be complex since there are limitations in terms of regulations, policies, government support, commitments, adaptable technologies and approaches. Successful ecosystem restoration is not impossible but the journey is surely winding.

We appreciate UNEP (UN Environment Programme) representatives for their contribution in presenting the ideas, insights, and vision of the UN Decade on Ecosystem Restoration. The first part of this book allows us to revisit the retrospective view on ecosystem restoration and envision possibilities, challenges, and strategies in the future. Threats, problems and impacts of abiotic and biotic constituents are discussed in the second part to lucidly remind us of what has happened in our tropical inland water ecosystems such as algal blooms, the occurrence of invasive species and the impacts of emerging pollutants. Moving forward to the next part, the discussion subsequently continues with the compilation of trending research and recent studies which can offer opportunities for adoption and inspiration. A separate chapter is dedicated to addressing the monitoring and measurement tool involving the concepts and their implementation such as flood warning system in Indonesia and the Philippines using satellite images and the monitoring of water quality through microbial-source tracking. Lastly, the final chapter underlines the socio-cultural aspects which describe how the local communities in the Philippines and Indonesia have established values and wisdom to contribute to protecting the ecosystem. After all, this book denotes the collation of updates and current states of tropical inland water restoration and optimistically encourages the continuation and sustainability of restoration efforts.

Bogor, Indonesia

Adelina C. Santos-Borja
(Editor in Chief)

Luki Subehi
Ali Rahmat
Miratul Maghfiroh

Contents

Part I
Backgrounds

Chapter 1
The UN Decade on Ecosystem Restoration: The Opportunity for Tropical Inland Waters

Susan C. Gardner, Ann-Kathrin Neureuther, and Tiffany Njoki

By restoring ecosystems, we can drive a transformation that will contribute to the achievement of all the Sustainable Development Goals. Everyone has a critical role to play—Governments, international organizations, cities, businesses, civil society and individuals.
UN Secretary General António Guterres, at the start of the UN Decade on Ecosystem Restoration.

The United Nations General Assembly declared the years 2021 through 2030 as the UN Decade on Ecosystem Restoration with the aim to galvanize action across all ecosystem types including tropical inland waters. This reflects the growing global understanding of the importance of these ecosystems as hosts of remarkably high biodiversity and important sources of nutrition and livelihoods. However high rates of global wetlands loss have had far-reaching impacts for food security, climate resilience and sustainable development. Halting and reversing ecosystem degradation through restoration can improve ecosystem health, recover biodiversity and regain valuable ecosystem services. Despite the relatively low number of restoration efforts in freshwater ecosystems, there is a growing evidence-base of successful restoration efforts. The UN Decade has generated principles and best practices for inclusive restoration that achieves multiple benefits for nature, climate and communities. Moving forward into the second half of the decade, the benefits of restoration interventions can be better understood through increased monitoring and technology innovations that document the economic value of restoration investments. Strengthening engagement of youth and local communities can help prepare the next generation of restoration practitioners to build a future where nature, prosperity and human wellbeing are mutually secured.

S. C. Gardner (✉) · A.-K. Neureuther · T. Njoki
Ecosystems Division, United Nations Environment Programme, Nairobi, Kenya
e-mail: susan.gardner@un.org

© The Author(s), under exclusive license to Springer Nature Singapore Pte Ltd. 2025
A. C. Santos-Borja et al. (eds.), *Progress on Ecosystem Restoration of Tropical Inland Waters*, https://doi.org/10.1007/978-981-96-2284-9_1

1.1 The Opportunity of Ecosystem Restoration

In 2022, the 15th Conference of Parties of the Convention on Biological Diversity adopted the Kunming-Montreal Global Biodiversity Framework. The Framework includes, among 23 targets, a call to ensure by 2030 at least 30% of areas of degraded terrestrial, inland water, coastal and marine ecosystems are under effective restoration in order to enhance biodiversity and ecosystem functions and services, ecological integrity and connectivity. Freshwater ecosystems and other wetlands are also protected by the Ramsar Convention on wetlands, which has 168 contracting parties who have designated 2186 protected Ramsar sites, covering over two million square kilometres worldwide. This reflects the global understanding that healthy wetland ecosystems are essential for sustainable development (Intergovernmental Science-Policy Platform on Biodiversity and Ecosystem Services 2019). The Convention on Biological Diversity defines ecosystems as “a dynamic complex of plant, animal and micro-organism communities and their non-living environment interacting as a functional unit”. The results of these functional units are ecosystem services that provide essential benefits for people. Freshwater ecosystems in particular—including rivers, lakes and wetlands—provide essential functions that benefit billions of people and are especially critical for impoverished and vulnerable communities (Lynch et al. 2016). These include stabilizing water for drinking and food production including for fisheries, agriculture and livestock, while they also enhance resilience to climate change by absorbing the worst impacts of both droughts and floods.

Marshlands of Mesopotamia UNEP works with partners to ensure that all ecosystem types are covered by restoration efforts, including inland waters. One example is a number of projects to restore the Marshlands of Mesopotamia (Al-Ahwar) in southern Iraq, once the largest wetland ecosystem in the Middle East. The Iraqi Marshlands have traditionally provided habitat for the Marsh Arabs and critical populations of wildlife but have been degraded significantly since the 1970s due to a combination of human activities, conflict, drought, desertification and climate change. Since 2004, UNEP has supported the Iraqi government in consultation with relevant stakeholders, to develop a sustainable management plan for the Marshlands. Those efforts have led to the strengthening of transboundary cooperation to sustain upstream water flows, and have curtailed interference from human activities, resulting in the recovery of natural wetlands and providing communities with a sustainable source of food, water, fiber and health materials.

Freshwater ecosystems host remarkably high biodiversity relative to their areas. Although they only cover less than 1% of Earth’s surface, these habitats host approximately 10% of all species including one-third of vertebrate species (Strayer and Dudgeon 2010). The greatest species richness of freshwater fishes is found in the tropics, especially in South America, East Africa and South-East Asia (Tisseuil et al.

2012; Jézéquel et al. 2020). Tropical inland fisheries catch is estimated to reach up to 5.5 million tons with an estimated economic value of US$6 billion, but the social and food security value for rural communities may be substantially underestimated (Neiland and Bene 2008; Lymer et al. 2016). For example, inland fisheries in Africa and parts of Asia provide an important source of nutrition and livelihoods for an estimated 600 million people (Dugan et al. 2010). In addition, over 5000 species of tropical freshwater fish are traded every year as ornamental fish in 125 countries, at a retail value of up to US$30 billion (Evers et al. 2019).

South Africa Peatlands Restoration Project Carbon-rich peatlands provide vital ecosystem services and benefits to local communities, including water regulation and a rich habitat for biodiversity. Water scarcity and frequent droughts and floods, in addition to impacts from human activities make peatlands one of the most threatened ecosystems in South Africa. With Global Environment Facility funding, UNEP assists with restoration across three peatland ecosystems, aiming to restore ecosystem functions across 1860 ha of South African peatland ecosystems by ensuring sufficient water supply and allowing ecosystem function to recover. The South Africa Peatlands Project aims to sustainably protect these threatened ecosystems for the future, through a combination of improved governance and management of the ecosystems at the national and local level, inclusive stakeholder engagement and evidence-based decision-making. Restoration of the peatlands is expected to kick start the carbon sequestration function of the systems, which will contribute to improved climate regulation, as well as improving water purification and regulation, and soil retention, while protecting livelihoods of local communities that are dependent on biodiversity.

Globally, the rate of wetlands loss is occurring three times faster than forests (Gardner and Finlayson 2018). More than 50% of freshwater wetlands have been lost globally since 1900 (Almond et al. 2022). Freshwater ecosystems in particular are under increasing pressure from human-induced environmental degradation including due to pollution, urbanization, over-exploitation, climate change and invasive species (Gardner et al. 2001; Acreman et al. 2019; Almond et al. 2022). As a result, global freshwater biodiversity is declining at startling rates, and there is clear understanding that biodiversity loss will decrease ecosystem functioning and nature's contributions to people (Isbell et al. 2022). From 1970 to 2018, the decline in freshwater species was faster than any other wildlife population globally (Almond et al. 2022). Currently approximately one-third of freshwater species are threatened with extinction and declines of freshwater vertebrate species have been more than twice that of terrestrial or marine populations (Grooten and Almond 2018; Tickner et al. 2020). Tropical habitats in particular are estimated to have the greatest percentage of species threatened or driven to extinction (Isbell et al. 2022), and because tropical freshwater ecosystems have been historically understudied, these rates may be underestimated (Isbell et al. 2022).

Regreening Africa Africa has faced severe and recurrent droughts, and desertification now affects 45% of the continent. To lessen the impacts of droughts and help increase water retention, UN World Restoration Flagship Regreening Africa is starting with agriculture. So far, this winning initiative has implemented sustainable farming practices to restore over 300,000 ha. A total of five million hectares are planned to be under restoration by 2030. In addition to water conservation practices, restoration techniques implemented by Regreening Africa include tree planting, home gardening with trees, pastoral-managed natural regeneration, farmer-managed natural regeneration and diversifying crops.

Because of the extent of historical ecosystem destruction, conservation of nature, while essential, is alone not sufficient to achieve the goals of the Paris Agreement, the Global Biodiversity Framework and the Sustainable Development Goals (United Nations Environment Programme [UNEP] and Food and Agriculture Organization of the United Nations [FAO] 2021) since the losses already incurred of wetland species and habitats have resulted in corresponding losses of critical ecosystem services (Almond et al. 2022). Fortunately, there is hope. Just as human activities have driven the degradation of ecosystems, action can be taken to aid their recovery. Halting and reversing ecosystem degradation through restoration can improve ecosystem health, recover biodiversity and as a result, regain valuable ecosystem services. Restoration initiatives present an enormous opportunity to create livelihoods and support resilient, sustainable economies including for vulnerable rural communities that depend heavily on natural resources. For instance, it is estimated that every dollar spent on restoration generates around 30 dollars of economic benefits. Between now and 2030, the restoration of 350 million hectares of degraded terrestrial and aquatic ecosystems could generate US$9 trillion in ecosystem services (UNEP and FAO 2021).

Building with Nature in Indonesia The coast of Java, Indonesia's most populated island, has been plagued by flooding and erosion caused by the loss of mangroves. Millions of people suffer from these impacts. Rather than replanting mangrove trees, this innovative UN World Restoration Flagship has built fence-like structures with natural materials along the shore to calm waves and trap sediment, creating conditions for mangroves to rebound naturally. In addition, farmers have been trained in sustainable techniques to increase shrimp production. With mangroves providing habitat for a host of marine organisms, fishers have seen improved near-shore harvest.

In recognition of the enormous benefits of restoration for economies and development, as well as human wellbeing, countries have already committed to restore over one billion hectares—an area larger than China—as part of international climate, nature and land goals. These commitments come from over 100 governments eager to utilize restoration as part of their response to the Bonn Challenge as well as global climate, biodiversity and land degradation neutrality pledges (Sewell et al. 2020). Therefore, rather than more commitments on action, the world needs urgent action on existing commitments. In order to effectively reverse the rapid decline in global freshwater ecosystems, scientists have called for restoration as a central aspect of recovery plans (Tickner et al. 2020). Despite the relatively low number of restoration efforts in freshwater and other aquatic ecosystems (Sewell et al. 2020), meaningful examples do exist of successful inland water restoration. Recently, the commitment by 47 countries to join the *Freshwater Challenge*—a global pledge to protect and restore 300,000 km of rivers and 350 million hectares of wetlands to reach biodiversity and climate goals—is a positive sign of momentum and attention on this issue. Building upon those restoration solutions and successes can have an immediate and far-reaching impact, for resilience, for communities and for livelihoods; particularly if they are scaled across regions and focused on underrepresented areas such as the tropics.

Mangrove Regeneration in Sri Lanka Following the tsunami, Sri Lanka led a mass-drive to plant mangroves. However, noticing that only approximately three per cent of the planted saplings survived, the country changed tactics: from planting mangroves to nurturing them. By creating the right conditions for mangroves to thrive, the plants started regenerating naturally. The survival rate is now higher than 90%, and the island nation is on track to increase their mangrove cover by 50% by 2030. They are doing so following a uniquely inclusive approach involving communities, young scientists, policy makers, national private sector and international partners.

1.2 The UN Decade on Ecosystem Restoration

In response to the staggering rate of ecosystem degradation and loss, and its negative effects on the global climate, biodiversity and human development, the UN General Assembly declared the years 2021 through 2030 as the UN Decade on Ecosystem

Restoration in resolution 73/284 (UNGA 2019). The timeline is critical since scientists have identified this ten-year period as the last opportunity to avoid catastrophic climate change and mass biodiversity loss (UNEP and FAO 2021). The UN General Assembly called for the UN Decade on Ecosystem Restoration (UN Decade) to prevent, halt and reverse the loss of nature worldwide, with UNEP and FAO leading its implementation in collaboration with the Secretariats of the three Rio Conventions (the UN Convention on Biological Diversity, the UN Convention to Combat Desertification, and the UN Framework Convention on Climate Change), as well as other relevant multilateral environmental agreements and entities of the United Nations system (UNGA Resolution 73/284).

The UN Decade aims to galvanize restoration action across all ecosystem types including terrestrial and aquatic ecosystems, through three main pathways for change: (i) building a global movement; (ii) generating political support; and (iii) building technical capacity (UN Decade on Ecosystem Restoration 2021). Bringing back ecosystems at the massive scale necessary to address global challenges is a task too daunting for any single organization to shoulder alone. This is why the strategy of the UN Decade established a partnership framework that lays out different partner types tailored to each organization's interest and capacities. To date, more than 260 entities across all regions have pledged their support, with many partners amplifying the movement by bringing their vast networks in support of the UN Decade's goals.

In a consultative process with hundreds of restoration experts, practitioners, researchers, non-governmental organizations, businesses and indigenous peoples, the UN Decade's Task Force on Best Practices established 10 Principles for Ecosystem Restoration to guide partners, governments and other stakeholders in effective, inclusive ecosystem restoration. The open consultation enabled the taskforce to draw upon diverse expertise and perspectives for creating a shared vision that can be applied across all sectors of society, all ecosystems and all activities under the decade (FAO, Society for Ecological Restoration, International Union for Conservation of Nature, the Commission on Ecosystem Management 2023).

The 10 Principles are:

1. Global Contribution; that restoration contributes to the UN Sustainable Development Goals and the goals of the Rio Conventions.
2. Broad Engagement; that restoration promotes inclusive and participatory governance, social fairness and equity.
3. Many Types of Activities; that ecosystem restoration encompasses a wide range of activities such as pollution reduction, rehabilitation of highly modified areas such as industrial sites and ecological restoration.
4. Benefits to Nature and People; that restoration aims to achieve the highest level of recovery for biodiversity, ecosystem health and integrity and human wellbeing.
5. Addressing Causes of Degradation; that restoration addresses the direct and indirect causes of ecosystem degradation.

6. Knowledge Integration; that restoration incorporates all types of knowledge; including indigenous, traditional, local and scientific knowledge, and promotes knowledge exchange and integration throughout the process.
7. Measurable Goals; that restoration is based on well-defined ecological, cultural and socio-economic objectives and goals.
8. Local and Land/Seascape Contexts; that restoration is tailored to the local ecological, cultural and socio-economic contexts, while considering the larger landscape or seascape.
9. Monitoring and Management; that restoration includes monitoring, evaluation and adaptive management throughout and beyond the lifetime of the project.
10. Policy Integration; that restoration is enabled through intersectoral policy coordination.

Through the establishment of these 10 Principles, which have been complemented by the Standards of Practice to facilitate implementation, all stakeholders now have a common language for the planning, implementation and monitoring of restoration activities, whilst ensuring that restoration efforts contribute to broader environmental and societal objectives.

In addition to the 10 Principles and Standards of Practice, the UN Decade issued an "Action Plan" for partners to take leadership on delivering across 12 critical "Restoration Challenges" to achieve global restoration goals by 2030. Among them are goals for biodiversity and climate, but also key deliverables for cities and freshwater ecosystems—two areas historically neglected in restoration action (UN Decade on Ecosystem Restoration Action Plan 2023). Under the Action Plan, the Freshwater Challenge brings together countries for a global goal to restore an area equivalent to India, while aiming to increase socio-economic returns on restoration investments (Freshwater Challenge 2023).

1.3 Scaling Implementation: From Commitments to Action

To provide meaningful examples of successful large-scale ecosystem restoration, the UN Decade calls on member states and partners to share experiences that "show where they restore". Under the lead of the UN Decade's Task Force on Best Practices, a review panel evaluates all proposals against an established set of criteria. This includes hectares restored to date, species reintroduced, greenhouse gases sequestered, jobs created and other factors in adherence with the 10 Principles of Ecosystem Restoration. Submissions also include details on geospatial locations of project areas to help gauge the impact on the ground with the use of satellite imagery. The selected "UN World Restoration Flagships" represent some of the most ambitious, effective and inspiring examples of ecosystem restoration. Eligible Flagships receive technical, financial and advocacy support. Through these Flagships, the UN honours large-scale and long-term ecosystem restoration and shares practical examples that other countries can replicate.

The first 10 UN World Restoration Flagships were announced during the Convention on Biological Diversity's 15th Conference of Parties and aim to restore over 60 million hectares and create nearly 15 million jobs. A second group of seven Flagships were announced at the sixth UN Environment Assembly, which are expected to restore 40 million hectares and create 500,000 jobs. Together, the 17 Flagships represent a diverse range of ecosystem restoration success stories—from increasing Sri Lanka's mangrove cover by 50% in the wake of the tsunami, to reversing 500 years of deforestation in South America, bringing back iconic species like saiga antelope, sea turtles, dugong and gorilla from the brink of extinction, to fighting fires with forests in the Mediterranean Region (visit the UN World Restoration Flagships' for films on each Flagship: https://www.decadeonrestoration.org/world-restoration-flagships).

1.4 Restoring Inland Water Ecosystems—UN Decade on Ecosystem Restoration Success Stories

While freshwater and other aquatic ecosystems are underrepresented in restoration pledges globally (Sewell et al. 2020), impactful and scalable successes of ecosystem restoration that brings back healthy inland water sources do exist. The current 17 UN World Restoration Flagships present a number of restoration success stories across almost all ecosystems including those linked to tropical inland waters.

Pakistan's Living Indus In response to the severe flooding that struck in 2022, this initiative targets the restoration of 30% of the Indus River Basin, a critical resource sustaining Pakistan's agriculture and economy. This vast basin supports an impressive range of biodiversity, including 195 mammal species, at least 668 bird species, over 150 fish species, and the endangered Indus Blind Dolphin, one of the world's rarest mammals. Recent years have brought Pakistan unprecedented floods and extreme heatwaves, drastically impacting the well-being and livelihoods of millions. Already, over one million hectares have been restored through this basin-wide effort. Now recognized as a World Restoration Flagship, the Living Indus initiative is positioned to receive additional technical and financial backing to support its ambitious target of restoring 25 million hectares by 2030—equivalent to more than 30% of Pakistan's territory. By advancing sustainable water management, reducing pollution, preserving biodiversity, and fostering community engagement, the initiative is shaping a climate-resilient future for Pakistan (Fig. 1.1).

Namami Gange The Ganges River, spanning 2700 km from the Himalayas to the Bay of Bengal, has faced significant degradation due to population growth, industrialization, and extensive irrigation, with climate change adding a grave additional threat. Initiated by the government in 2014, the Namami Gange programme focused on waste management, water treatment, public awareness, and community engagement to protect and restore the ecosystem. Recognized as a World Restoration Flagship,

Fig. 1.1 Efforts by local communities to protect and restore Pakistan's Indus Basin. *Photo credits* UNEP/Todd Brown

this initiative is reforesting parts of the Ganges basin and promoting sustainable agriculture. It is working to revive key wildlife species, such as river dolphins, softshell turtles, otters, and the hilsa shad, an important food fish (Fig. 1.2).

Fig. 1.2 Local communities cleaning up the Ganga in restoration efforts to conserve the Holy River. *Photo credits* AFP/Sanjay Kanojia

Fig. 1.3 Restoration activities provide livelihoods for local residents as part of the Shan-Shui initiative. *Photo credits* UNEP/Braunosarus Studios

Shan-Shui Initiative China's Shan-Shui Initiative, launched in 2016, represents an ambitious plan with 75 large-scale ecosystem restoration projects, covering diverse landscapes from mountain regions to coastal estuaries across the country. Developed through a systematic approach to restoration, these projects are integrated with national land use plans, work at the landscape or watershed scale, and incorporate agricultural and urban areas alongside natural ecosystems. Each project includes biodiversity goals and supports local industries. An example is the Oujiang River Headwaters Project in Zhejiang province, which merges scientific insight with traditional farming techniques, such as slope terracing and mixing crops with fish-and duck-rearing for more sustainable land use (Fig. 1.3).

Trinational Atlantic Forest Pact Along South America's coast, over 300 partners have joined forces to restore the Atlantic Forest across Brazil, Argentina and Paraguay. Centuries of deforestation has created extensive fragmentation of the Atlantic Forest putting at risk the function of important watersheds for highly populated cities. This became especially clear during the recurrent extreme droughts in Sao Paulo in 2014 when the city came close to having its reservoirs run dry. The Atlantic Forest is also home to the jaguar, the margay and the black and gold howler monkey, all endangered species. This restoration effort engaged local communities in the restoration of important watersheds surrounding Sao Paulo, with the vision that by 2050, the three countries will restore 15 million hectares of degraded forest. The results of this cooperation have already helped over 150 million people by increasing job opportunities, food security and water provision as well as helping communities contend with climate change (Fig. 1.4).

Fig. 1.4 Several endangered species, including the golden lion tamarin, call the Atlantic Forest home. *Photo credits* Getty/Mark Castiglia

1.5 What is Next for the Second Half of the UN Decade on Ecosystems Restoration?

The Decade's Action Plan will be updated to scale up implementation activities and pave the road for the second half of the decade. Ensuring a coordinated and synergetic approach, the updated plan will allow integration of lessons learnt during the first five years of the UN Decade and reflect all relevant developments across different agendas and areas. Meanwhile, the call for more examples of UN Decade Flagships continues, as a means of growing the evidence-base of successful restoration efforts, with a special focus on underrepresented regions and types of ecosystems, including tropical inland waters.

For the second half of the Decade, more robust monitoring and science are needed to better document the range and value of benefits achieved from restoration investments. The UN Decade's Framework on Ecosystem Restoration Monitoring is a publicly available geospatial tool to help keep track of restoration progress towards the UN Decade on Ecosystem Restoration's goal. Another tool that is contributing to our understanding of changes to water-related ecosystems is the Freshwater Ecosystems Explorer. This real-time, global tracking system uses satellite data and other global data sets to track changes to lakes, rivers, wetlands (including vegetated wetlands and mangroves) and groundwater aquifers worldwide, including river flow and changes in water quality of large lakes. This data will be refined and improved through the Global Wetlands Watch, an initiative that will provide additional data on specific wetland types by early 2025.

Additional technological innovations that close the information gap will enable more accurate understand of the value of restoration interventions. Improvements in water quality can be observed in a limited capacity from space by monitoring colour changes such as from turbidity or eutrophication due to excessive chlorophyll or plant growth. However, despite important technological innovations, there remains the need for in situ monitoring systems in order to gain a comprehensive understanding of the complex dynamics in freshwater ecosystems. At present, there is no global scientific assessment to measure restoration progress across all ecosystems, which is pivotal for taking stock of progress during the second half of the UN Decade.

It is critically important that we continue to build our understanding of the value of natural capital through natural capital accounting principles. Countries, communities and businesses alike will benefit from a full understanding of the economic value of the wide range of benefits provided by healthy, functioning ecosystems. This knowledge will provide better recognition of the economic benefits connected with ecosystems restoration along with the many economic risks arising from ecosystem degradation. The development of this information can assist both the public and private sector to undertake informed decision-making for financial commitments and policy development. UNEP's Economic of Ecosystems and Biodiversity (TEEB) initiative is working to develop natural capital accounting principles with the aim of mainstreaming the values of biodiversity and ecosystem services into decision-making at all levels. Through the production of reports and direct support to countries' restoration efforts, the work of TEEB is resulting in evidence-based policy development informed by a true understanding of the value of nature. By further developing such principles across all sectors of society, we can unlock finance for ecosystem restoration.

To ensure the widest possible uptake and participation in ecosystem restoration, the UN Decade is investing in strengthening the next generation of restoration practitioners through in-country education initiatives, collaborations with social media creators, a green jobs report and the UN Young Champions of the Earth programme. This is an area in which the UN Decade offers great hope. We have seen, through the Flagships, that everyone can contribute to ecosystem restoration. Indeed, restoration efforts are far more effective and durable in the long-term where local communities have a stake. By working together, by valuing nature in our policies, decision-making and financial commitments, we can build a future where nature, prosperity and human wellbeing are mutually secured and reinforced through nature-positive actions. As we chart out that path towards a brighter, greener future, ecosystem restoration offers the opportunity to leave the planet better for the next generation.

Acknowledgements The authors wish to express appreciation to Natalia Alekseeva, Elisabeth Bernhardt, Kilian Christ, Samuel Chavez-Rosales and Helene Van Rossum for their review and comments that contributed to the quality of this manuscript.

References

Acreman M, Hughes KA, Arthington AH, Tickner D, Dueñas M (2019) Protected areas and freshwater biodiversity: a novel systematic review distils eight lessons for effective conservation. Conserv Lett (art. e12684). https://doi.org/10.1111/conl.12684

Almond REA, Grooten M, Bignoli DJ, Petersen T (2022) Living planet report. Building a nature-positive society embargo 10-13-2022_LPR 2022 full 2 (panda.org)

Dugan P, Delaporte A, Andrew N, O'Keefe M, Welcomme R (2010) Blue harvest: inland fisheries as an ecosystem service. WorldFish Center, Penang, Malaysia

Evers H, Pinnegar JK, Taylor MI (2019) Where are they all from?—sources and sustainability in the ornamental freshwater fish trade. Fish Biol 94(6):909–916

Food and Agriculture Organization of the United Nations, Society for Ecological Restoration, International Union for Conservation of Nature, the Commission on Ecosystem Management (2023) Standards of practice to guide ecosystem restoration. A contribution to the United Nations Decade on Ecosystem Restoration. Summary report. Rome, FAO. https://doi.org/10.4060/cc5223en

Gardner RC, Finlayson C (2018) Global wetland outlook: state of the world's wetlands and their services to people. Ramsar Convention Secretariat

Gardner S, Grue C, Major III WW, Conquest LL (2001) Aquatic invertebrate communities associated with purple loosestrife (*Lythrum salicaria*), cattail (*Typha latifolia*), and bulrush (*Scirpus acutus*) in Central Washington, USA. Wetlands 21(4):593–601

Grooten M, Almond REA (2018) Living planet report 2018: aiming higher. World Wildlife Fund

Intergovernmental Science-Policy Platform on Biodiversity and Ecosystem Services (2019) Report of the plenary of the intergovernmental science-policy platform on biodiversity and ecosystem services on the work of its seventh session 2019

Isbell F, Balvanera P, Mori AS, He J-S, Bullock JM, Regmi GR, Seabloom EW, Ferrier S et al (2022) Expert perspectives on global biodiversity loss and its drivers and impacts on people. Front Ecol Environ. https://doi.org/10.1002/fee.2536

Jézéquel C, Tedesco PA, Bigorne R, Maldonado-Ocampo J, Ortega H, Hidalgo M, Martens K, Torrente-Vilara G et al (2020) A database of freshwater fish species of the Amazon basin. Sci Data 7(96). https://doi.org/10.1038/s41597-020-0436-4

Lymer D, Marttin F, Marmlla G, Bartley DM (2016) Global estimate of theoretical annual inland capture fisheries harvest. American Fisheries Society

Lynch AJ, Cooke SJ, Deines AM, Bower SD, Bunnell DB, Cowx IG, Nguyen VM, Nohner J et al (2016) The social, economic, and environmental importance of inland fish and fisheries. Environ Rev 24:115–121

Neiland AE, Béné C (eds) (2008) Tropical river fisheries valuation: background papers to a global synthesis. Studies and reviews No. 1836. The WorldFish Center, Penang, Malaysia, p 290

Sewell A, Esch S, Löwenhardt H (2020) Goals and commitments for the restoration decade: a global overview of countries' restoration commitments under the Rio conventions and other pledges, 6(2). Retrieved from pbl-2020-goals-and-commitments-for-the-restoration-decade-3906.pdf. Accessed on 26 Mar 2024

Strayer DL, Dudgeon D (2010) Freshwater biodiversity conservation: recent progress and future challenges. J N Am Benthol Soc 29:344–358

The Freshwater Challenge (2023) Rivers and wetlands are the frontlines of the climate and nature crises retrieved from Freshwater Challenge | Frontlines of the climate and nature crises. Accessed on 24 May 2024

Tickner D, Opperman JJ, Abell R, Acreman M, Arthington AH, Bunn SE, Cooke SJ, Dalton J et al (2020) Bending the curve of global freshwater biodiversity loss: an emergency recovery plan. Bioscience 70(4):330–342

Tisseuil C, Cornu JF, Beauchard O, Brosse S, Darwall W, Holland R, Hugueny B, Tedesco PA et al (2012) Global diversity patterns and cross-taxa convergence in freshwater systems. J Anim Ecol 82:365–376. https://doi.org/10.1111/1365-2656.12018

United Nations Decade on Ecosystem Restoration (2021) Strategy. https://www.decadeonrestoration.org/strategy

United Nations Decade on Ecosystem Restoration (2023) Action plan. https://wedocs.unep.org/bitstream/handle/20.500.11822/42095/UNDecade_ActionPlan.pdf?sequence=3&isAllowed=y.

United Nations Environment Programme and Food and Agriculture Organization of the United Nations (2021) Becoming #GenerationRestoration: ecosystem restoration for people, nature and climate

United Nations General Assembly, Resolution 73/284: United Nations Decade on Ecosystem Restoration (2021–2030). Accessed on 11 June 2024

Chapter 2
Primary Indicators of Degrading Ecosystem and Restoration Process

Miratul Maghfiroh, Adelina C. Santos-Borja, Luki Subehi, Awalina Satya, and Ali Rahmat

Inland water ecosystems carry special characteristics compared to other ecosystems, i.e., marine or coastal. The main dissimilarity between inland water and marine ecosystem is highlighted by the connectivity and greater openness in ocean system, while freshwater is much impacted by the surrounding land ecosystems such as vegetation, soil (Geist and Hawkins 2016). Freshwater lakes and rivers have shorter renewal times than those of the oceans (Wetzel 2001) with salt concentrations ranging from 0.02 to 1.0 g/L in freshwater (Saccò et al. 2021) and higher in seawater (Baseggio 1974). Other inland water ecosystems may have varying salt concentrations. Moderately saline, brackish and saline water bodies contain 1.1–3 g/L, 3.1–10 g/L and 10.1–35 g/L, respectively (Saccò et al. 2021). It is also recognized that inland aquatic organisms especially freshwater ones are distinguished from those in seawater in a way that the mechanism to cope with their surrounding water involves a restricted and selective interchange of dissolved matters between their bodies and the water (Adolph 1925). In addition, biodiversity in inland waters is often underrated. For example, 126,000 species of freshwater animals and 2614 species of freshwater macrophytes signify the representation of 9.5% and 1% of global animals and macrophytes respectively (Balian et al. 2008). With 0.5% portion of inland water from the total global water (Hahn 2006) and only 0.01% available in lakes, rivers, and others (Reid et al. 2019), the biodiversity in inland water ecosystem is trampled in a such restrained confinement. This condition makes the habitats as well as the biodiversity vulnerable toward environmental changes and human activities (Dudgeon et al. 2006).

Apart from being the home of incredible biodiversity, inland water ecosystems offer significant roles in human life. Human well-being is somehow extremely

M. Maghfiroh (✉) · L. Subehi · A. Satya · A. Rahmat
Research Center for Limnology and Water Resources, National Research and Innovation Agency (BRIN), Bogor, Indonesia
e-mail: mira006@brin.go.id

A. C. Santos-Borja
Southeast Asian Limnological Network (SEALNET), Bogor, Indonesia

© The Author(s), under exclusive license to Springer Nature Singapore Pte Ltd. 2025
A. C. Santos-Borja et al. (eds.), *Progress on Ecosystem Restoration of Tropical Inland Waters*, https://doi.org/10.1007/978-981-96-2284-9_2

connected to inland water. In tropical area, where some of the largest river basins such as Amazon River, Congo River, Mekong River are located, millions of people rely their lives on rivers and tributaries (Cantonati et al. 2020). Mekong River, one of the world's longest rivers, is the central resource that serves 70 million people for transportation, food supply and other daily purposes (Hudson-Rodd and Shaw 2003). More than a half of the global population lives in areas less than 3 km from surface freshwater bodies (Kummu et al. 2011) and this indicates that inland water ecosystem is the pivotal resource of water supply and other benefits. Collectively, these benefits are called ecosystem service. As mentioned in Millenium Ecosystem Assessment (2005), ecosystem service refers to any benefit that human beings obtain from ecosystems into which the services are considered as provisioning, regulating, cultural and supporting.

- Provisioning services include the production of fish, water retention for multiple uses, extraction of materials for pharmaceuticals.
- Regulating services can be reflected through flood control, storm protection, water regulation and soil retention.
- Ecosystems are often used as communal or cultural space for recreation, religious place and aesthetic landscape.
- Additionally, ecosystems also sustain nutrient cycling which delivers supporting service (MEA 2005).

Understanding the concept of ecosystem services and its valuation benefits in better identifying the issues and challenges of lake–river basin systems. An assessment such as Ecosystem Service Shared Value Assessment is adaptable to the practice of some approaches such as Integrated Lake Basin Management and Integrated Lentic Lotic Basin Management (Nakamura et al. 2022).

Despite the enormous services that inland water ecosystems provide, it is so unfortunate that these ecosystems are prone to threats and/or pressures that induce ecosystem changes. The schematic illustration in Fig. 2.1 shows interactions between inland water ecosystem with its services and humans. Generally, inland water ecosystem provides ecosystem services that are utilized by humans (green arrows). However, threats and pressures emerge (dashed red arrows) as the results of human activities, e.g., agriculture, urbanization, industry, aquaculture, tourism (Dudgeon et al. 2006; Matthews 2016). To some extent, these threats lead to ecosystem changes which eventually affect ecosystem services (Carpenter et al. 2011) as well as human well-being.

2.1 What Degrading Inland Water Ecosystems Show Us

The tropics are populous regions and even one third of the global population lives in tropical Asia (Cantonati et al. 2020). Around 1.2 billion people or 30% of the population in this region highly depend their basic needs on the nature which also involve water supply (Fedele et al. 2021). The growing number of population is an increasing

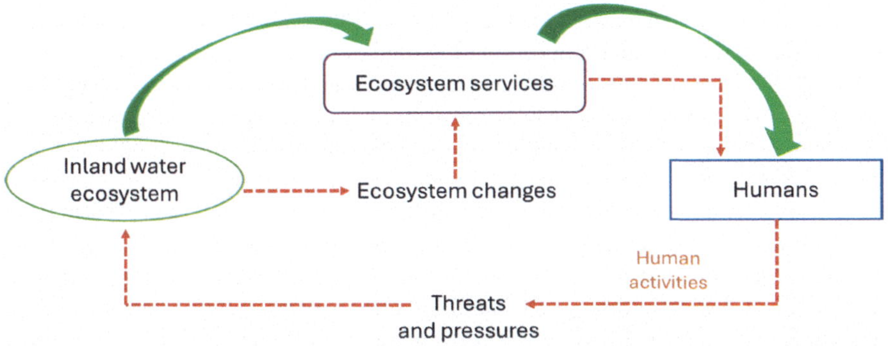

Fig. 2.1 Interactions between inland water ecosystem with its ecosystem services and humans (green arrows). The emergence of threats results in ecosystem changes that affect ecosystem services and human well-being (dashed red arrows)

pressure to the inland water ecosystem due to greater water demand to supply for the population. Furthermore, a large number of populations result in intense human activities (Ganivet 2020). Human driven pressures and threats (e.g., point source pollution, non-point (diffuse) pollution, water abstraction, habitat degradation) can come from various anthropogenic activities which affect on water quality (chemical, physical, biological) and quantity and finally bring the consequences of environmental impacts such as loss of species or habitat, drought or flood, eutrophication (Friberg 2014).

Water pollution, especially freshwater pollution, is a critical issue that impacts water security and, thus, the identification of pollution sources is helpful to establish restoration strategies and mitigate the potential threats on the aquatic environment for sustainable water security. Water pollution can be generated from many sources, however, diffuse and point source pollutions are commonly discussed to simplify the classification of water pollution sources (Holt 2000). Unlike point source pollutions that are well-defined and cause direct discharge to the watercourses (Holt 2000), diffuse pollutions encompass true non-point sources alongside the minor point sources which are independently inconsequential but collectively substantial sources in a catchment (Ferrier et al. 2005). Diffuse, or sometimes also refers to non-point pollutions can contaminate freshwater through agricultural practices, land runoff or dispersed dwellings (Holt 2000). Point source pollutions are usually derived from recognizable spots such as collective sewer system, animal factory (Singh et al. 2020) and wastewater treatment plant (Holt 2000). In Biñan River Basin, the Philippines, the discharged load from industrial point source contributed around 12%, whereas livestock contribution was 23% and it was estimated that 6388 metric tons of treated waste load was discharged into the main outlets (Amaya et al. 2012). While in Indonesia, around 61–65% of urban greywaters were discharged directly to the river or ditch without proper treatment (Wulan et al. 2022).

In Southeast Asia, it is concerning that only 46% of population used sanitation services and only 26% of domestic wastewater flows were safely treated, whereas in Africa, the percentages were even lower with 23% and 24% for population that used sanitation services and proportion of treated domestic wastewater, respectively (WHO, 2023). Poor sanitation and hygiene and untreated domestic wastewater permit higher risks of interrelated transmission of human or animal excrete, thereby the pathogens can travel from the medium via aquatic environment and finally contaminate drinking water (Prüss et al. 2002). Fecal contamination by protozoan parasite *Cryptosporidium* was detected in the surface water of Laguna Lake, Philippines which implied that the contamination from sewage or agricultural sources, e.g., swine and cattle (dela Peña et al. 2021). An investigation on bacterial contamination in Rangkui River, located in Bangka Island Indonesia, showed a high concentration of fecal coliform (> 11,000 MPN per 100 mL) in the downstream area (Fig. 2.2) which hinted the contamination was possibly influenced by the establishment of residential district and traditional market (Maghfiroh et al. 2019). In addition, potential human-associated pathogenic taxa (e.g., *Staphylococcus, Turicella, Faecalibacterium*) in urban rivers (Kisat and Auji Rivers) of Kenya were recovered from the middle and lower catchment areas which were characterized by heavily urbanized zone such as residential and industrial site (Odhiambo et al. 2023). This study also revealed that taxonomic biomarkers associated with greywater pollution (*Brevibacterium, Staphylococcus, Cutibacterium, Streptococcus, Corynebacterium, Micrococcus, Rothia*), heavy metal tolerant bacteria (*Amycolatopsis*) and denitrifying bacteria (*Bradyrhizobium*) in heavily N and P contaminated urban river were also detected in the mid and lower catchment zones (Odhiambo et al. 2023).

Fecal contamination is not the only signal that suggests the water is polluted. Taxonomic composition and community structure of macroinvertebrates can be useful to assess the water quality of an aquatic ecosystem. Due to biological and ecological characteristics, macroinvertebrates are commonly identified to examine the local conditions of a stream (Begum et al. 2023). Some of the taxa (e.g., *Chironomidae, Moinidae, Hirudinae, Oligochaeta, Simuliidae, Platyhelminthes, Syrphidae, Prosobranchia, Hydropsychidae, Hyroptilidae, Cyclopoidae*) are less sensitive to pollution

(a) (b)

Fig. 2.2 The downstream catchment areas of Rangkui River, Indonesia are characterized by **(a)** housing area and **(b)** traditional market (Maghfiroh et al. 2019). Photo credit: Heni Novianti

while other taxa (e.g., *Megaloptera, Trichoptera, Ephemeroptera, Plecoptera, Mesogastropoda, Coleoptera, Zygoptera, Isopoda, Anisoptera, Decapoda, Amphipoda, Diptera*) are more sensitive to polluted streams (Uherek and Pinto Gouveia 2014; Begum et al. 2023). In tropical lotic and lentic environments, macroinvertebrate communities are often observed for biomonitoring. Biomonitoring of Agbede wetlands in Nigeria within eighteen months demonstrated that overall water quality status was fairly good based on Pollution Tolerance Index (PTI) which was determined from the macroinvertebrate community structure such as the presence of *Ephemeropterans, Trichopterans,* and *Coleopteran* (Dirisu and El Surtasi 2021). Apart from the observation of macroinvertebrates, cyanobacteria, macrophytes and fish community structures can also relatively reflect relevant responses to ecological perturbations (Frieberg 2014). More details and discussions on other biotas are reviewed in Part 2.

Another crucial issue of inland water ecosystem degradation is regarding the significant changes of physicochemical properties of the water. The occurrence of harmful and/or toxic chemical substances is one of the major concerns. Organic pollutants consist of many substances such as Polycyclic aromatic hydrocarbons (PAHs), Polychlorinated biphenyls (PCBs), Polychlorinated dibenzo-p-dioxins and dibenzofurans (PCDD/Fs), trichloroethene, carbon tetrachloride (Perelo 2010), herbicides, pesticides, pharmaceuticals (i.e., antibiotics), phthalates, and other Persistent Organic Pollutants (POPs) (Sousa et al. 2018). Various types of organic pollutants (e.g., aliphatic hydrocarbons, ether, ketone, phenol, phthalate, fatty acid ester, pesticide) were found in lake water of Tonle Sap, Cambodia and derived from disposal of aquaculture agriculture, and other human activities (Meas et al. 2021). PCBs and DDTs (dichlorodiphenyltrichloroethane) in the sediments from Mekong River Basin were also detected as high as 310 μg/kg and 52 μg/kg, respectively, and the concentration of contaminants in the Vientiane canal was significantly higher than those in mainstream of Phong River and/or Mekong River which indicated that the sources had arisen from urbanized areas (Sudaryanto et al. 2011). A more detailed explanation of other emerging pollutants is discussed separately in Part 2. Inorganic pollutants can often be found in the form of heavy metals, inorganic nitrogen (nitrate, nitrite, ammonium), inorganic phosphate (Singh et al. 2020). Metal substances had already contaminated river water of the Tonle Sap–Bassac (Cambodia), the Saigon (Vietnam), the lower Chao Phraya (Thailand) and the Citarum (Indonesia). A study revealed that the concentrations of Pb, Zn, Al, Cd, Co, Mn and Ni were enhanced in the rivers by external sources rather than natural inputs, additionally, anthropogenic origins such as discharges of both industrial and domestic wastewater and non-point sources (i.e., street dust and road runoff) were the main donors of metal contamination into these rivers (Chanpiwat and Sthiannopkao 2014).

Non-point pollutions from agricultural activities via runoff promote the increasing concentrations of nutrients such as inorganic nitrogen and/or phosphate in the water bodies (Xia et al. 2020), thereby, eutrophication due to nutrient enrichment prevails as it is evidently seen in some lakes. Some examples of eutrophic lakes in Southeast Asia are Lake Maninjau in Indonesia (Sulastri et al. 2022), Laguna Lake in the Philippines (Herrera and Nadaoka 2021). Eutrophication is predominantly identified by rapid growth of algae and other planktons, the deterioration of water quality, the decline of dissolved oxygen (DO) and the mass death of fish (Zhang et al. 2020). Eutrophic water usually shows distinguished qualities of environmental variables such as higher concentrations of total dissolved solids, chlorophyll a, total nitrogen, and total phosphorus compared to those of oligotrophic and mesotrophic water (Callisto et al. 2014). In response to eutrophication, tropical lakes are relatively more sensitive than temperate lakes to the increase of nutrient and display the changes of water quality and biotic communities proportionately and, thus, more strict regulation of nutrient and organic loading is required (Lewis 2000). Eutrophication in lakes and the occurrence of cyanobacteria are discussed more deeply in Part 2.

2.2 The Ecosystem Restoration Process

Owing to tremendous services that inland water ecosystems provide us, ecosystem restoration has become an inevitable responsibility to assist the recovery and ensure the conservation of the environment. FAO, SER and IUCN CEM specified a standard practice to assist the ecosystem restoration (FAO et al. 2023). The guidelines incorporate five components of the restoration process (Fig. 2.3). The first component is assessment of the conditions of the degraded ecosystem. This contains constructive information on socioeconomic and cultural along with ecological attributes. The visions and targets of restoration should be clearly determined prior to further steps. In various situations, the targets or the goals of the restoration vary which can be based upon aesthetic values, fisheries, or recreational purposes (Jamwal et al. 2023). Some others may focus on flood risk control (Veról et al. 2020) or preventing biodiversity loss and sustaining ecosystem services (Piczak et al. 2023). Other common goals can be related to improving water quality, stabilizing channels, increasing biodiversity, and improving habitat (Palmer et al. 2014).

The second component refers to planning and design of the restoration process. Effective planning helps to obtain successful accomplishment of the project. The planning and design can contain the selection of approaches or technologies for implementation. Selection of appropriate approaches and technologies considers numerous aspects such as cost effectiveness, challenges and feasibility, likelihood of success, risk assessment and management, stakeholder participations, regulations (and law) and other influencing elements. It is noteworthy to remember that an individual technology or approach, depending on the circumstances, has strengths and weaknesses, thereby modifications and combinations are encouraged throughout the process. Treatments of organic polluted aquatic sediment may include the application of bioremediation (Perelo 2010) and a floating treatment wetland may be preferable for lake's water quality improvement and microhabitat provision (Henny and Kurniawan 2019). In improving an effective formulation and systematic control of peatland restoration in Southeast Asia, a community-led 5R (rewetting, reduction of fire, revegetation, revitalization of local livelihood and reporting and monitoring) approach (Terzano et al. 2022) can be worth adopting.

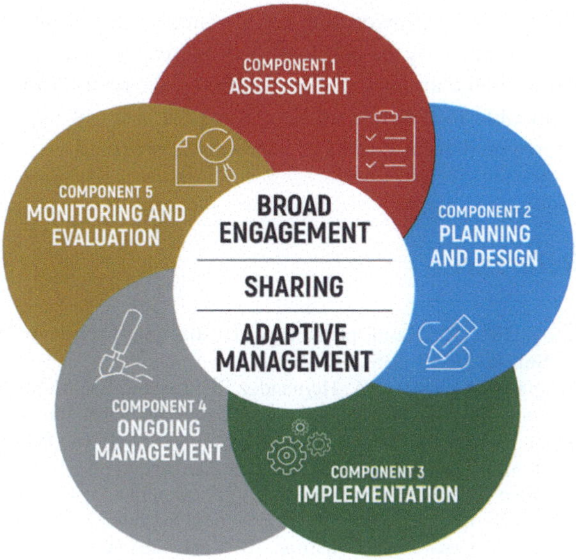

Fig. 2.3 Five components of restoration process. *Source* Food and Agriculture Organization of the United Nations. Reproduced with permission

The implementation (the third component) of the restoration process often entails adaptive practices to confront any hindrance and improve the operation performance. Adaptive restoration allows examination of pivotal variables as well as periodical review of findings which may involve the scientific standpoint for result interpretations and corrections (Zedler 2005). Involvement of researchers during the process can be advantageous in suggesting the amendment or improvement at the restoration sites. The last two components of the restoration process are the ongoing management and the monitoring and evaluation. The ongoing management requires participation of involving stakeholders. Local communities and other stakeholders such as academic and research institutions should nurture active commitment through effective communication. In addition, active commitments among the stakeholders enable long-term monitoring and proper evaluation of the restoration process. Monitoring of the restoration becomes fundamental since robust monitoring and a solid plan for adaptive management guide the successful restoration efforts or, otherwise, those lead to corrective actions (Holl 2020). Monitoring parameters are ample and selecting monitoring parameters depends on the underlying goals of restoration. Those parameters can be physical habitat and functions (topography, hydrology, water quality and others), biological factors (reproduction and mortality, undesirable species, community composition and many others) or socioeconomic (educational programs, aesthetic values and others) (Holl 2020). Microbial source tracking (MST) can also be considered for monitoring the water quality of an ecosystem such as lake. In Laguna Lake (the Philippines), DNA finger printing of *E. coli* and biomarkers based on host-associated *Bacteroides* 16S rDNA were studied to determine the sources of fecal pollution (dela Peña et al. 2021; Malajacan et al. 2023). In addition, monitoring of a tropical shallow eutrophic lake in Huizhou, China from

2004 to 2011 showed that clear-water tendency and the decrease of TP and Chl-a were obtained by the application of biomanipulation (Liu et al. 2018). This restoration effort can be an example for shallow lake restoration; however, it is not necessarily applicable to other similar ecosystems without appropriate considerations on five intertwined components of restoration process.

References

Adolph EF (1925) Some physiological distinctions between freshwater and marine organisms. Biol Bull 48(5):327–335

Amaya FL, Gonzales TA, Hernandez EC, Luzano EV, Mercado NP (2012) Estimating point and non-point sources of pollution in Biñan River Basin, the Philippines. APCBEE Proc 1:233–238

Balian EV, Segers H, Martens K, Lévéque C (2008) The freshwater animal diversity assessment: an overview of the results. Springer, Dordrecht, pp 627–637

Baseggio G (1974) The composition of sea water and its concentrates. In Fourth symposium on salt (1:351–358). Northern Ohio Geological Society

Begum W, Goswami L, Sharma BB, Kushwaha A (2023) Assessment of urban river pollution using the water quality index and macro-invertebrate community index. Environ Dev Sustain 25(8):8877–8902

Callisto M, Molozzi J, Barbosa JLE (2014) Eutrophication of lakes. Eutrophication Causes Conseq Control 2:55–71

Cantonati M, Poikane S, Pringle CM, Stevens LE, Turak E, Heino J, Richardson JS, Bolpagni R, Borrini A, Cid N, Čtvrtlíková M (2020) Characteristics, main impacts, and stewardship of natural and artificial freshwater environments: consequences for biodiversity conservation. Water 12(1):260. https://doi.org/10.3390/w12010260

Carpenter SR, Stanley EH, Vander Zanden MJ (2011) State of the world's freshwater ecosystems: physical, chemical, and biological changes. Annu Rev Environ Resour 36(1):75–99. https://doi.org/10.1146/annurev-environ-021810-094524

Chanpiwat P, Sthiannopkao S (2014) Status of metal levels and their potential sources of contamination in Southeast Asian rivers. Environ Sci Pollut Res 21:220–233

Dirisu AR, El Surtasi EI (2021) Bioindicators of lotic and lentic ecosystems in Agbede wetlands (Southern Nigeria), using macroinvertebrate tools. Scientific African 14:01000

Dudgeon D, Arthington AH, Gessner MO, Kawabata ZI, Knowler DJ, Lévêque C, Naiman RJ, Prieur-Richard AH, Soto D, Stiassny ML, Sullivan CA (2006) Freshwater biodiversity: importance, threats, status and conservation challenges. Biol Rev 81(2):163–182. https://doi.org/10.1017/S1464793105006950

dela Peña LBRO, Labrador KL, Nacario MAG, Bolo NR, and Rivera WL (2021) Microbial source tracking of fecal contamination in Laguna Lake, Philippines using the library-dependent method, rep-PCR. J Water Health 19(5):762–774

dela Peña LBRO, Vejano MRA, Rivera WL (2021) Molecular surveillance of *Cryptosporidium* spp. for microbial source tracking of fecal contamination in Laguna Lake, Philippines. J Water Health 19(3):534–544

FAO, SER, IUCN CEM (2023) Standards of practice to guide ecosystem restoration. A contribution to the United Nations Decade on Ecosystem Restoration. Summary report. Rome, FAO. https://doi.org/10.4060/cc5223en

Fedele G, Donatti CI, Bornacelly I, Hole DG (2021) Nature-dependent people: mapping human direct use of nature for basic needs across the tropics. Glob Environ Chang 71:102368

Ferrier RC, D'Arcy BJ, MacDonald J, Aitken M (2005) Diffuse pollution-what is the nature of the problem? Water Environ J 19(4):361–366

Friberg N (2014) Impacts and indicators of change in lotic ecosystems. Wiley Interdiscip Rev Water 1(6):513–531
Ganivet E (2020) Growth in human population and consumption both need to be addressed to reach an ecologically sustainable future. Environ Dev Sustain 22(6):4979–4998
Geist J, Hawkins SJ (2016) Habitat recovery and restoration in aquatic ecosystems: current progress and future challenges. Aquat Conserv Mar Freshwat Ecosyst 26(5):942–962
Hahn MW (2006) The microbial diversity of inland waters. Curr Opin Biotechnol 17(3):256–261. https://doi.org/10.1016/j.copbio.2006.05.006
Henny C, Kurniawan R (2019) Application of floating treatment wetlands in a highly eutrophic lake, Indonesia: a new tool for lake restoration and provision of microhabitat. In: IOP conference series: earth and environmental science, vol 380, no 1, IOP Publishing, p 012003
Herrera EC, Nadaoka K (2021) Temporal dynamics and drivers of lake ecosystem metabolism using high resolution observations in a shallow, tropical, eutrophic lake (Laguna Lake, Philippines). J Great Lakes Res 47(4):997–1020
Holl K (2020) Monitoring and adaptive management. In: Primer of ecological restoration. Island Press, Washington, DC, pp 34–46
Holt MS (2000) Sources of chemical contaminants and routes into the freshwater environment. Food Chem Toxicol 38:S21–S27
Hudson-Rodd N, Shaw BJ (2003) Mekong river development: whose dreams? which visions? Water Int 28(2):268–275
Jamwal P, Carvalho L, Bhattacharyya S, Muttepawar P (2023) The benefits of restoring urban lakes in the tropics. Environ Monit Assess 195(11):1294
Kummu M, De Moel H, Ward PJ, Varis O (2011) How close do we live to water? A global analysis of population distance to freshwater bodies. PLoS ONE 6(6):20578. https://doi.org/10.1371/journal.pone.0020578
Lewis WM Jr (2000) Basis for the protection and management of tropical lakes. Lakes Reserv Res Manag 5(1):35–48
Liu Z, Hu J, Zhong P, Zhang X, Ning J, Larsen SE, Chen D, Gao Y, He H, Jeppesen E (2018) Successful restoration of a tropical shallow eutrophic lake: strong bottom-up but weak top-down effects recorded. Water Res 146:88–97
Maghfiroh M, Novianti H, Lukman, Nurtjahya E (2019) Bacterial indicators reveal water quality status of Rangkui River, Bangka Island, Indonesia. In IOP conference series: earth and environmental science, vol 380, no 1. IOP Publishing, p 012008
Malajacan GT, Nacario MAG, Obusan MCM, Rivera WL (2023) Host-associated bacteroides 16S rDNA-based markers for source tracking of fecal pollution in Laguna Lake Philippines. Microorganisms 11(5):1142
Matthews N (2016) People and fresh water ecosystems: pressures, responses and resilience. Aquatic Procedia 6:99–105
Meas M, Heu R, Ma L, Eang K, Siev S (2021) Occurrence, transportation, regulation and treatment methods of organic pollutants in surface water: a review on case of Tonle Sap Lake. Cambodia. In: Proceedings of the 13th AUN/SEED-Net regional conference on chemical engineering 2020 (RCChE-2020) jointly held with the 5th international symposium on conservation and management of Tropical Lakes, pp 151–154
Millenium Ecosystem Assessment (2005) Ecosystems and human well-being: wetlands and water. World Resources Institute
Nakamura M, Olago D, Rast W, Muhandiki VS, Raini J, Chidammodzi CL, Matsumoto S. Waki H (2022) Ecosystem service shared value assessment (ESSVA) for integrated lake basin management (ILBM). 滋賀大学環境総合研究センター研究年報 19:23–50
Odhiambo KA, Ogola HJO, Onyango B, Tekere M, Ijoma GN (2023) Contribution of pollution gradient to the sediment microbiome and potential pathogens in urban streams draining into Lake Victoria (Kenya). Environ Sci Pollut Res 30(13):36450–36471
Palmer MA, Hondula KL, Koch BJ (2014) Ecological restoration of streams and rivers: shifting strategies and shifting goals. Annu Rev Ecol Evol Syst 45(1):247–269

Perelo LW (2010) In situ and bioremediation of organic pollutants in aquatic sediments. J Hazard Mater 177(1–3):81–89
Piczak ML, Perry D, Cooke SJ, Harrison I, Benitez S, Koning A, Peng L, Limbu P, Smokorowski KE, Salinas-Rodriguez S, Koehn JD (2023) Protecting and restoring habitats to benefit freshwater biodiversity. Environ Rev. https://doi.org/10.1139/er-2023-0034
Prüss A, Kay D, Fewtrell L, Bartram J (2002) Estimating the burden of disease from water, sanitation, and hygiene at a global level. Environ Health Perspect 110(5):537–542
Reid AJ, Carlson AK, Creed IF, Eliason EJ, Gell PA, Johnson PT, Kidd KA, MacCormack TJ, Olden JD, Ormerod SJ, Smol JP (2019) Emerging threats and persistent conservation challenges for freshwater biodiversity. Biol Rev 94(3):849–873. https://doi.org/10.1111/brv.12480
Saccò M, White NE, Harrod C, Salazar G, Aguilar P, Cubillos CF, Meredith K, Baxter BK, Oren A, Anufriieva E, Shadrin N (2021) Salt to conserve: a review on the ecology and preservation of hypersaline ecosystems. Biol Rev 96(6):2828–2850. https://doi.org/10.1111/brv.12780
Singh J, Yadav P, Pal AK, Mishra V (2020) Water pollutants: origin and status. In: Sensors in water pollutants monitoring: role of material, pp 5–20
Sousa JC, Ribeiro AR, Barbosa MO, Pereira MFR, Silva AM (2018) A review on environmental monitoring of water organic pollutants identified by EU guidelines. J Hazard Mater 344:146–162
Sudaryanto A, Isobe T, Takahashi S, Tanabe S (2011) Assessment of persistent organic pollutants in sediments from Lower Mekong River Basin. Chemosphere 82(5):679–686
Sulastri, Nomosatryo S, Henny C, Sulawesty F (2022) Functional groups of Phytoplankton and their relationship with environmental factors in Lake Maninjau, West Sumatra, Indonesia. In: IOP conference series: earth and environmental science, vol 1062, no 1. IOP Publishing, p 012012
Terzano D, Attorre F, Parish F, Moss P, Bresciani F, Cooke R, Dargusch P (2022) Community-led peatland restoration in Southeast Asia: 5Rs approach. Restor Ecol 30(8):e13642
Uherek CB, Pinto Gouveia FB (2014) Biological monitoring using macroinvertebrates as bioindicators of water quality of Maroaga Stream in the Maroaga Cave System, Presidente Figueiredo, Amazon Brazil. Int J Ecol 2014(1):308149
Veról AP, Bigate Lourenço I, Fraga JPR, Battemarco BP, Linares Merlo M, Canedo de Magalhães P, Miguez MG (2020) River restoration integrated with sustainable urban water management for resilient cities. Sustainability 12(11):4677
Wetzel RG (2001) Freshwater ecosystems. In: Encyclopedia of biodiversity: second edition. Elsevier Inc., pp 560–569
WHO (2023) World health statistics 2023: monitoring health for the SDGs, sustainable development goals. World Health Organization, Geneva. Licence: CC BY-NC-SA 3.0 IGO
Wulan DR, Hamidah U, Komarulzaman A, Rosmalina RT, Sintawardani N (2022) Domestic wastewater in Indonesia: generation, characteristics and treatment. Environ Sci Pollut Res Int 29(22):32397
Xia Y, Zhang M, Tsang DC, Geng N, Lu D, Zhu L, Igalavithana AD, Dissanayake PD, Rinklebe J, Yang X, Ok YS (2020) Recent advances in control technologies for non-point source pollution with nitrogen and phosphorous from agricultural runoff: current practices and future prospects. Appl Biol Chem 63:1–13
Zedler J (2005) Ecological restoration: guidance from theory. San Francisco Estuary Watershed Sci 3(2)
Zhang Y, Luo P, Zhao S, Kang S, Wang P, Zhou M, Lyu J (2020) Control and remediation methods for eutrophic lakes in the past 30 years. Water Sci Technol 81(6):1099–1113

Part II
Threats, Problems, and Impacts

Chapter 3
Lakes

Zati Sharip, Fatimah Md. Yusoff, and Salmah Zakaria

Lakes are important water resources varying in size, depth, shape, volume, and residence time. Large lakes larger than 100 km^2 account for nearly 90% of the total surface area and volume of the world's lakes (Jenny et al. 2020). There are only 1709 large lakes in the world which make up 0.2% of the total (Jenny et al. 2020). The majority of lakes are smaller in size, but they also contribute to various functions. Although all of these water bodies represent 0.5% of the freshwater system, they are of great importance to mankind due to their numerous ecosystem services. Lewis (1996) reported that tropical lakes only formed approximately 10% of the total number of world lakes as most of the lakes originated from glacial activities which were limited to temperate areas. Approximately 60% of temperate lakes are glacial lakes, while the rest were formed by permafrost and other processes. In the tropics, lakes were mainly created by river/flood plain activities (40%), coastal processes (20%), aeolian processes (20%), volcanic activities (10%), and others (10%), including man-made lakes (Lewis 1996).

Like many other freshwater systems, lakes are threatened by human pressures that focus on resources that provide services of economic value without similarly considering other services such as regulating and supporting. As a result, the conditions of the lakes continue to deteriorate, leading to changes in quantity, quality, and biological diversity (Brönmark and Hansson 2002; Beeton 2002; Mammides 2020). Lewis (2000) reported that more than half of tropical lakes are associated

Z. Sharip (✉)
National Water Research Institute of Malaysia (NAHRIM), Seri Kembangan, Malaysia
e-mail: zati@nahrim.gov.my

F. Md. Yusoff
Department of Aquaculture, Faculty of Agriculture, University Putra Malaysia, Serdang, Selangor, Malaysia

F. Md. Yusoff · S. Zakaria
Academy of Sciences of Malaysia, Kuala Lumpur, Malaysia

© The Author(s), under exclusive license to Springer Nature Singapore Pte Ltd. 2025

A. C. Santos-Borja et al. (eds.), *Progress on Ecosystem Restoration of Tropical Inland Waters*, https://doi.org/10.1007/978-981-96-2284-9_3

with rivers, and thus degradation of water quality in rivers would have direct negative impacts on lakes in the tropics. Biogeographically, anthropogenic pressure shows varying patterns with significant increases in tropical and sub-tropical regions, which is probably due to the rapid development of the areas (Mammides 2020). Threats to lake sustainability and ecosystem health, such as siltation and sedimentation, inorganic and organic pollution, habitat modification, and introduced species, are all human-induced (Nakamura 1997; Brönmark and Hansson 2002). Climate change poses significant challenges for sustainable lake management, particularly large lakes (Brönmark and Hansson 2002; Beeton 2002). Although threats, problems, and impacts are similar in most lakes of the world, the responses in tropical lakes are relatively different from those in the temperate region (Lewis 2000). Higher solar radiation, warmer waters, absence of marked seasonal variations, relatively constant temperature and light availability, and relatively high rainfall in tropical areas have important impacts on biogeochemical processes and hydrochemical characteristics of tropical lake basins, compared to temperate lakes (Nilssen 1984; May et al. 2021). With high annual temperatures and abundant sunlight in tropical areas, coupled with high nutrients from anthropogenic activities, eutrophication in tropical lakes can be accelerated due to the incessant growth of phytoplankton leading to dissolved oxygen depletion and toxic chemical accumulation (Koh et al. 2019). The persistence of eutrophication in tropical lakes can be a long-term phenomenon due to relatively stable climatic regimes compared to temperate lakes. Basic ecological processes such as nutrient cycle, energy flow, food chain/food web, energy flow, and organismic interactions (grazing, competition, predation, symbiosis, and adaptation) occur at faster and more consistent rates in the tropical lakes compared to their counterparts in temperate areas which undergo seasonal changes annually. Sim et al. (2025) reported that cyanobacteria were the dominant form of phytoplankton in freshwater tropical lakes that received high nutrient inputs. While cyanobacterial dominance in temperate lakes can be seasonal, the dominance in tropical lakes can be relatively permanent depending on the supply of nutrients. In tropical lakes, biological communities are intricately tied to nutrient concentrations and biogeochemical cycles (Sim et al. 2025).

This chapter briefly reviews threats, problems, and impacts on the lake, particularly the tropical water bodies, toward ecosystem restoration, management, and conservation progress. The main causes of threats were human-driven, either in the lake resulting from disposal and release of aquaculture effluents or fish pens, within the lake basin such as the release of sullage and sewage, improper agricultural practices or fertilizer applications and deforestation, or beyond the lake catchment through deposition of airborne particles from forest fires (Ogutu-Ohwayo et al. 1997; Cohen et al. 2019). Climate change exacerbates the threat, problems, and impacts by changing lentic water's hydrological regimes and physical characteristics. Certain threats can have cascading effects and cause other threats and problems affecting the use of the lake for human consumption and impacting ecosystem services and function.

3.1 Major Threats and Problem Facing Lakes and Their Associated Impact

Major factors contributing to lake degradation are pollution, acidification, habitat change, and loss, over-harvesting of lake resources, introduced species, climate change, and unsustainable economic activities (livestock, aquaculture, and other economic activities with organic matter discharge into lakes). Pollutants that can cause deteriorating lake environment and poor lake water quality include organic waste, heavy metals, pesticides, silt and sediments, acid rain, and emerging pollutants (endocrine disrupting chemicals, steroids, hormones, drugs, antibiotics, and microplastics). Pollutants in lakes can originate from both allochthonous and autochthonous sources. Allochthonous sources from point sources could be controlled but non-point sources confront many complexities and uncertainties related to multiple users in the catchment area such as agricultural lands, industrial areas, and commercial/urban centers. Autochthonous sources can be the primary drivers of hypoxic conditions in enriched water bodies with high densities of phytoplankton blooms during the drought when the allochthonous inputs are minimal (Lv et al. 2022). Zhu et al. (2018) reported that the accumulation of organic matter in the bottom sediments can be a major source of inorganic nutrients that can drive eutrophication in lakes, although there are no allochthonous nutrient inputs. Shallow lakes are more vulnerable to eutrophication due to readily available nutrients from the lake bottom in addition to allochthonous contributions (Zhou et al. 2022). In lakes, excessive nutrients, especially phosphorus and nitrogen, can cause eutrophication, resulting in harmful algal blooms, poor water quality, hypoxia, fish kills, a decline in biodiversity, and serious economic losses (Sayer et al. 2016). In fact, eutrophication impairs the ability of the lake ecosystem to provide services, including recreation, fishing, and water supply provision (Jenny et al. 2020).

3.2 Eutrophication and Harmful Algal Blooms

Eutrophication is a serious global issue that could affect the quality of water in lakes and their ecosystem services. Eutrophication is known as the enrichment of nutrients, either phosphorus and/or nitrogen, in waterbodies that result in water quality degradation, excessive growth of plants and/or algae, decline in fisheries, and decrease in biodiversity that could eventually affect the health and livelihood of the surrounding communities. The word 'eutrophic' originates from the Greek term meaning well-nourished. Eutrophication of lakes became a major problem starting in the 1960s due to excess nutrients from agriculture practices, poor waste management, and the usage of phosphate detergents (Beeton 2002). Accelerated eutrophication in lakes is mainly due to the synergic action of anthropogenic activities and climate change (Pereira and Mulligan 2023). According to the survey of the International Lake Environment Committee (ILEC) between 1988 and 1993, the percentage of eutrophic lakes by regions are as follows: Asia Pacific Region (54%), Europe (53%), North America (43%), South America (41%), and Africa (28%) (ILEC 2002; Nakamura 1997). In North America, eutrophication remains a severe problem and challenge in many tropical waterbodies in Florida, such as Lake Okeechobee and Lake Apopka. According to Lewis (2000), tropical lakes are more sensitive to increased nutrients than temperate lakes, as shown by higher proportionate changes in water quality and biotic communities in response to eutrophication in the former compared to the latter. Barjau-Aguilar et al. (2022) showed that nutrients in a tropical reservoir could accumulate in the sediment and become a source of nutrient supply that could maintain the eutrophic status of a lake even in the absence of external loading. Tay et al. (2022) reported that lakes with high sediment concentrations and high internal phosphorus recycling (IPR) could experience unexpected algal blooms and delays in the lake recovery process.

Eutrophication was a major issue in lakes located in developed countries, but it was projected to become less significant due to improvements in waste management and nutrient regulation (Brönmark and Hansson 2002). In Japan, eutrophication in Biwa Lake has led to the women's soap movement against eutrophication and algal bloom (Nakamura 1997). The use of ordinary detergents promoted algal blooms and red tide in the lake in the 1970s. As a result, the women's movement called for the detergent to be switched to phosphorus-free material. This has abated the eutrophication of Japanese lakes such as Lake Biwa and Lake Kasumigaura. Eutrophication will remain an increasingly significant threat in developing countries (Brönmark and Hansson 2002). This includes Lake Victoria and Tana in Africa, Lake Taihu and Dianchi in China, Tonle Sap Lake in Cambodia, and Laguna de Bay Lake in the Philippines that experience severe eutrophication (Ayele and Atlabachew 2021; Otieno et al. 2022; Ouma et al. 2016; Zhang et al. 2020).

Lake eutrophication also remains a concern in tropical countries such as Malaysia, where a desk study by the National Water Research Institute of Malaysia found that more than 60% out of 104 major lakes that various researchers have studied are categorized as eutrophic while the remaining are mesotrophic and oligotrophic (Sharip et al. 2020). Shallow urban lakes in Malaysia are vulnerable to eutrophication since these water bodies receive high inputs of nutrients from their economically active watersheds. Used mining pools are numerous in some states like Perak and Selangor, where mining was once an important economic activity. These water bodies are relatively shallow, do not have natural outflows, and with long residence time, they can rapidly become eutrophic (Yusoff et al. 1997; Zhou et al. 2022). A recent study indicated that more than 85% of Malaysia's studied ponds are experiencing eutrophication (Sharip et al. 2024). Many urban lakes in Uganda, Sri Lanka, Indonesia, and Brazil also suffer from eutrophication (Oliver et al. 2019; Nakagiri et al. 2023). Unlike rural areas, where nutrients may originate from the use of fertilizers in agricultural areas, the source of nutrients in urban water bodies mostly comes from either inefficient or inadequate wastewater treatment facilities or gray water from domestic and commercial areas (Sharip and Mohamad 2019; Oliver et al. 2019). In some urban tropical lakes in Africa, sewerage coverage is very low where most waste bathrooms, latrines, and septic tanks enter nearby drains and ultimately end in the lake (Nakagiri et al. 2023).

Eutrophic lakes are typified by a decline in water quality, high concentration of chlorophyll *a*, high levels of toxic nitrogenous compounds, low transparency, and low levels of nightly and/or hypolimnetic dissolved oxygen. Eutrophication also alters species composition and causes biodiversity loss (Rollwagen-Bollens et al. 2018). Enhanced growth of certain species due to their advantage in up-taking nutrients promotes blooms, causing a regime shift from a clear water state with high transparency to a turbid water environment (Scheffer 2007). The increase in nutrients, especially phosphorus, promotes the expansion of algal growth, particularly with cyanobacterial (blue-green algae) dominance, such as Planktothrix and Microcystis (Zhang et al. 2021; Massey et al. 2020). These cyanobacterial species have competitive advantages over the other phytoplankton groups because most of them have the floating mechanisms to escape light limitation under eutrophic conditions (Walsby 1981). Those cyanobacterial species with heterocysts can synthesize their own inorganic nitrogen to overcome inorganic nitrogen shortages (Kumar et al. 2010). Thus, cyanobacteria become dominant in eutrophic lakes and initiate the episodic harmful algal blooms.

Reservoirs, especially in tropical regions, have an excess of phosphorus, causing algal to grow linearly, with increasing total phosphorus (TP) concentrations and nitrogen becoming the limiting nutrient that promotes algal blooms. The linear relationship between Chlorophyll *a*, the indicator of algal productivity, and TP remains unclear when a weak relationship between TP and Chlorophyll *a* was found in lakes with a mean annual temperature > 25 °C or in the tropical regions (Quinlan et al. 2021). Lake characteristics, such as water chemistry, morphometry, and temperature, influence the relationship between nutrient concentrations and algal productivity (Quinlan et al. 2021). Other factors, such as sediment–water phase nutrient flux, drastic environmental fluctuations, and phytoplankton-consumer interactions, may interfere with the TP-Chlorophyll *a* relationship (Yusoff & Lock 1995). Dissolved oxygen, pH, and temperature and their fluctuations affect the sediment nutrient release, further increasing the eutrophic status in the overlying factors (Liu et al. 2022). In addition, Harvey and Menden-Deuer (2012) reported that phytoplankton has mechanisms to minimize predation and alter its community structure.

In some reservoirs with high nutrient inputs, such as Sembrong reservoir in Johor Malaysia, water hyacinth infestation spreads quickly across the lake surface, and blue-green algae forms > 90% of the phytoplankton community (Sharip et al. 2020). This is very different from oligotrophic lakes where diatom species such as *Cyclotella* spp. and nutritious golden brown and/or green algae are known to be dominant (Ardiles et al. 2012; Rollwagen-Bollens et al. 2018). Changes in phytoplankton community structure in eutrophic lakes would also affect the rest of the lake food low chain due to high abundance of inedible cyanobacterial species in the epilimnion (Yusoff et al. 1998; Rose et al. 2020). Shakri et al. (2020) showed that zooplankton *Moina micrura* fed with toxic cyanobacteria, *Microcystis aeruginosa*, and *M. viridis* had reduced fecundity, longevity, and population density compared to those fed

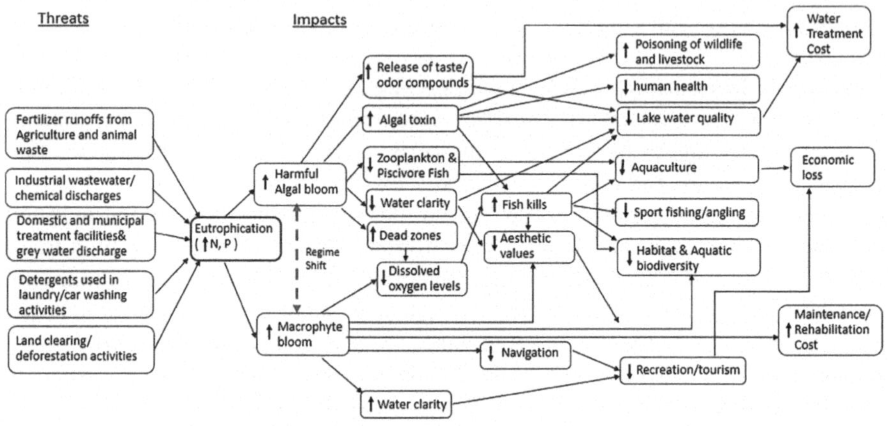

Fig. 3.1 Threats and impacts of eutrophication

with *Chlorella vulgaris*. Eutrophication also shaped the macroinvertebrate community, decreasing total richness and diversity and increasing pollution-tolerant species (Zhang et al. 2020).

Eutrophication results in the occurrence of a 'dead zone' in aquatic systems. Dead and decomposed algal bloom biomass sinks to the bottom, resulting in accumulation of dead organic matter and oxygen depletion, especially at night due to the consumption of oxygen for respiration and bacterial decomposition without replacement of dissolved oxygen from the lake surface. Low dissolved oxygen in the water column at night and hypoxic conditions at the lake bottom could result in fish kills, which is typical in shallow urban ponds (Fig. 3.1).

Dead zones have been reported in various temperate lakes, with the first occurrence reported in Chesapeake Bay, USA, in 1970 followed by the Laurentian Great Lakes, particularly the central basin of Lake Erie (Chislock et al. 2013; Dybas 2005; Bailey et al. 2020). Dead zones in Lake Erie, e.g., appear during the warm summer season in the shallow central area of the lake, caused by fertilizer runoff from large farms during low water conditions (Dybas 2005). Unlike temperate water bodies, the extent of dead zones in tropical lakes remains unknown due to limited published data. Tropical lakes experience only wet and dry seasons, and the appearance of dead zones may be more common in enriched environments due to hot temperatures all year round. LaBrie and Maranger (2024) used big data and machine learning approaches to predict the presence and development of hypoxia in the hypolimnion of lakes. Many factors control toxic microalgal blooms and their toxins. When physical parameters such as light and temperature are adequate for photosynthesis, nutrients, especially phosphorus and nitrogen, become limiting. Then, they act as the main drivers of the phytoplankton growth and formation of algal blooms. Harmful algal blooms (HABs) due to nutrient enrichment are one of the major causes of water quality deterioration, massive mortality of aquatic species, and disruption of safe resources (foods, drinking water, and recreation) for human consumption. Various species of toxic microalgae from cyanobacteria, dinoflagellates, and euglenophytes are commonly found in HAB communities. Typical phytoplankton genera that produce algal bloom are *Microcystis*, *Anabaena*, *Ocillatoria*, and *Planktothrix* (Massey et al. 2020; Mowe et al. 2015). Compared to temperate lakes, reports on algal bloom in tropical lakes are sporadic (Mowe et al. 2015). Blooming of *Microcystis*, *Cylindrospermopsis*, and *Anabaena* is widely known in African lakes (Mowe et al. 2015). González and Roldán (2020) also reported that eutrophication is characterized by the dominance of cyanobacterial genera such as *Microcystis*, *Anabaena*, *Planktothrix*, *Oscillatoria*, and *Cylindrospermopsis* due to increased nutrient concentrations especially nitrogen and phosphorus from inefficient wastewater treatment plants, agricultural lands, residential areas, industrial-commercial centers.

Microcystis blooms were also reported throughout tropical Asia, while *Cylindrospermopsis* commonly occurs in tropical lakes in Australia (Mowe et al. 2015). Arif et al. (2023) noted that various factors influence the seasonal occurrence of bloom-forming cyanobacteria; *Microcystis* was more influenced by nutrients than temperature, while *Planktothrix* was more sensitive to temperature. Another study using ribosomal RNA (rRNA) sequencing methods in Taihu Lake found a change in *Planktothrix* and *Microcystis* dominance of the 2016 cyanobacterial bloom due to temperature, dissolved forms of nitrogen, and interactions between the two algae species (Zhang et al. 2021).

Cyanobacterial blooms are commonly associated with toxic odor compounds such as geosmin and 2-MIB (2-methylisoborneol), which impart an unpleasant taste to water and fish, besides the release of toxins that are poisonous to animals and humans. Chislock et al. (2013) reported that HABs induced by cyanobacteria release several off-flavor compounds (e.g., methylisoborneal and geosmin) that affect fish culture in aquaculture ponds and taint the tastes of municipal drinking water systems, causing significant financial losses for state and regional economies. Although a variety of bacteria and fungi produce geosmin, cyanobacteria are major producers of geosmin (John et al. 2018). Massive proliferation of HABs causes lakes to lose their ecosystem functions, such as being unfit for supplying water for drinking and direct contact recreation (Mowe et al. 2015). Common toxins detected in tropical lakes are microcystin, anatoxin, and cylindrospermopsin produced by *Microcystis*, *Anaebaena*, *Cylindrospermopsis*, and *Aphanizomenon*, respectively (Massey et al. 2020; Mowe et al. 2015). Few studies in tropical waterbodies in Vietnam, Thailand, Singapore, and the Philippines reported microcystin levels associated with *Microcystis* spp. blooms above the safe WHO drinking water guideline level of 1 $\mu g\ L^{-1}$ (Mowe et al. 2015). Cylindrospermopsin was first discovered in tropical Australian waters with the poisoning of more than 140 people, resulting in health issues such as vomiting and bloody diarrhea (Massey et al. 2020). Microcystin-producing cyanobacteria have also been reported in tropical Lake Okeechobee, Florida (Massey et al. 2020). The toxin impact of MCs on human organs is more profound in the liver and, subsequently, other organs such as the kidneys, heart, nervous system, and gastrointestinal tract (Akinnawo 2023).

Algal blooms were reported when the concentration of phosphate and nitrate was in the range of 0.01–2.5 ug/L and 0.01–15.89 ug/L, respectively (Sidabutar et al.

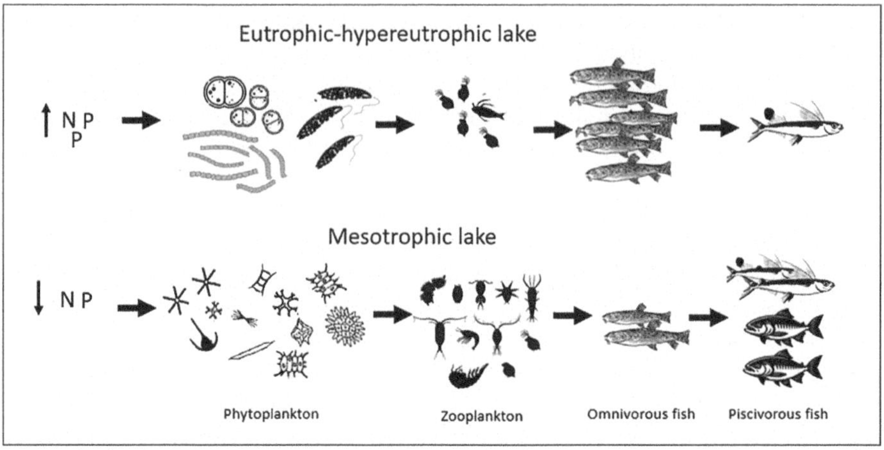

Fig. 3.2 Changes in lake food web with nutrient concentration

2020). Rollwagen-Bollens et al. (2018) reported that at the peak of cyanobacterial bloom, its carbon biomass reached levels as high as 20 μg/L, with total Chlorophyll *a* often exceeding 750 μg/L. As a result, copepod feeding rates decreased significantly, approaching zero, due to the detrimental effects of toxic cyanobacterial blooms on zooplankton. However, micro-zooplankton grazing rates increased during this period, suggesting that these small-sized zooplankton may be consuming bacterial cells from the decomposing cyanobacterial biomass (Yusoff et al. 2020). Figure 3.2 illustrates the changes in the food web caused by eutrophication.

Eutrophication can also induce macrophyte bloom in different waterbodies. High nutrient concentration in water caused excessive growth of floating macrophytes such as water hyacinth and Kariba weeds due to the uptake of nutrients in the water from their root. Their wide coverage on the lake surface blocks sunlight penetration to the lake bottom and reduces the aesthetic values. Certain submerged macrophytes grow in abundance due to the ability to uptake nutrients in the water through their stems and from sediment through their root (Schooler et al. 2009; Sousa 2011). The massive growth of macrophyte covering the lake disrupted boat navigation as the plant entangled on boat engines and can be endangering humans doing recreational swimming. Excessive growth of macrophytes also caused alteration in water quality by reducing oxygen from air dissolved in water due to cover on the lake surface. The respiration activities and decomposition of the plants also consume oxygen, further depriving oxygen contents in the water column. Removal of these macrophytes increases maintenance costs due to the need to use specialized harvesting equipment, and the plant can regrow from stem remains. For some submerged species, the stems are easily broken, and the use of equipment can further proliferate their growth, making them hard to eradicate and leading to substantial rehabilitation costs (Schooler et al. 2009).

Eutrophication caused significant financial losses of $2.2 billion per year in temperate freshwater, with the loss of lakefront property value contributing the largest economic losses at $0.3 billion–$2.8 billion per year. This is followed by a loss in recreational use of $0.37 billion–$1.16 billion per year (Dodds et al. 2009). Pretty et al. (2003) estimated the damage cost of freshwater eutrophication to be USD105–160 million/year due to additional costs of nutrients, algal toxins, and pollutant removal in drinking water, reduced aesthetic and recreational costs, net economic losses from the tourist industry. Globally, nutrient enrichment and the resulting eutrophication continue to cause severe effects on many sectors of the economy. Thus, the prevention of eutrophication, which results in water quality degradation, toxic resources and environment, loss and extinction of species, and economic loss of aquatic resources/foods, should be prioritized and accomplished. Silvino and Barbosa (2015) reported that an integrated analysis of the lake trophic state, morphometry, land occupation, and land use is important in providing information for effective management strategies to achieve successful restoration, efficient conservation, and sustainable ecosystem services. Aligned with stakeholder participation, good governance and continuous monitoring, external and internal nutrient inputs need to be effectively managed by using holistic approach to attenuate the availability of nutrients as the key to long-lasting eutrophication free lakes.

3.3 Silt and Sedimentation

Silt and sedimentation are the most common forms of reservoir pollutants due to uncontrolled land-based activities. Sedimentation occurs when eroded silt materials and particulate matter from the basin are carried by fluid flow, either by suspension or roll along the bottom of a waterway and are deposited on the bed or bottom of the water body when fluid velocity is lost. Land clearing activities and deforestation, which disturb the original trees and topsoil, are typically the causes of erosion and sedimentation. Rainfall weakens and loosens the soils in exposed land areas, making them more susceptible to erosion.

Tropical lakes are more prone to sedimentation due to extensive land-based activities in the catchment. In a tropical region with high rainfall, erosion and runoff over denuded land can deposit high silt and sediment loadings into lakes, causing severe sedimentation. The primary cause of the increase in the sedimentation rate was land use change, which was followed by the combined effects of land activities and climate change (Rodríguez-Bolaña et al. 2023).

Sedimentation is a major threat to many dams, reducing their ability to store water; with an estimated yearly average of 0.5–1% (ICOLD 2023). The global loss of reservoir storage capacity reported in Palmieri et al. (2001) is around 0.8%, resulting in a loss of revenue of $13 billion/year. A more recent survey by the International Commission on Large Dams (ICOLD) found that the global average value for sedimentation rate is around 0.96% (ICOLD 2023). According to Pereira and Mulligan (2023), sedimentation will account for 26% of global storage loss for large reservoirs by 2050. Reservoir sedimentation poses a threat to sustainable energy and food production in many regions of the world, especially in Asia, where it is predicted that in 2035, 80% of the useful storage capacity for hydropower production will be lost, while in 2025, 70% of the storage volumes used for irrigation will be lost to sedimentation. The average annual sedimentation rate of China's reservoirs is 2.3%, exceeding the capacity loss rates observed in the global reservoirs (Hu et al. 2018). In India, the Central Water Commission conducted a nationwide analysis of capacity survey data of 263 reservoirs in 2020, including tropical lakes, and found large differences in the sedimentation rate of reservoirs in the country (Central Water Commission 2020). Some reservoirs were heavily sedimented and no longer have usable capacity, such as Nizamsagar and Dhukwan reservoirs (Asthana and Khare 2022). One of the lakes with the highest sedimentation rates is South Africa's Welbedacht Reservoir, where during its first 30 years of use, an annual sedimentation rate of roughly 3% was recorded (Basson et al. 2022). Within the first nine years of operation, the reservoir lost roughly 50% of its original storage capacity, and after just 45 years, it was losing roughly 95% of its original capacity (Basson et al. 2022). Climate, physiography, and hydrometeorology are some factors that affect the rate of sedimentation of reservoirs (Central Water Commission 2020). Soil erosion and sediment transport rates increased with increased in rainfall intensity and frequency (Dash and Maity 2023).

Similarly, tropical reservoirs in Ethiopia also experienced sedimentation; the mean sediment yield in micro-reservoirs was 1900 t km^{-2} $year^{-1}$, which is significantly higher than the Global and African averages, which are between 1000 and 1500 t km^{-2} $year^{-1}$, thus reducing their life span by half (Adugna and Cherie 2021). Amasi et al. (2021) review the average sedimentation rate of East African hydropower reservoirs, which ranges between 0.52 and 327% per year. The sediment volume loss by 0.26–0.91% per year for 9 irrigation reservoirs with percentage storage capacity loss ranged from 8.54 to 34.12% in Ghana with annual sediment inflows of 1594–355,017 m^3 (Adongo et al. 2020).

There are only a few man-made lakes in Malaysia where sedimentation has been reported. The Ringlet Lake in Pahang is one of the country's lakes that have experienced severe sedimentation problems. The sedimentation rate of Ringlet Reservoir has been increasing; the rate was reported to have increased to an average of almost sixfold in 2008 (139,712 m^3 per year) compared to 1965 (25,000 m^3 per year) (Luis et al. 2016). A more recent study indicated a storage loss of 46% for Ringlet Lake (Razad et al. 2023). The loss of storage has caused water levels in the lake to rise rapidly at 0.59 m per h and together with unusual downpours, 13 times that of normal monsoon rain, resulting in mud flooding in downstream areas and four deaths (Floodlist 2013).

Other Malaysian lakes in highland areas, such as Jor and Mahang, experience high sedimentation rates that need to be regularly dredged due to erosion of agricultural areas, mining, and deforestation activities in the surrounding hill slopes (Choy and Hamzah 2001). Recent findings show much higher storage loss, about 54%, in Jor Lake compared to Ringlet Reservoir (Razad et al. 2023). Natural lakes associated with rivers, such as Lake Chini and Lake Bera in the Pahang River flood plain and oxbow lakes along Kinabatangan River, can become turbid during the rainy season due to silt and sediments from the upstreams where logging and land clearings for agriculture are common (Shuhaimi-Othman et al. 2007; Sharip et al. 2019). Olson and Morton (2018) reported that Tonle Sap Lake also has high sedimentation problems due to increased population and development rates that affect the lake depth and pose navigation problems.

Sedimentation is also a major problem for Tondano Lake, the largest lake in Sulawesi, Indonesia (Moningkey et al. 2022). Bathymetric analysis between 2015 and 2020 showed that the silting rate of Tondano Lake was 4.4 cm per year, compared to a silting rate of 13.41 cm per year from 2004 to 2010 (Moningkey et al. 2022). According to the predictions of Basson (2009), sediments will fill an average of 80% of reservoir capacity differently on several continents, with Asia being the earliest in 2035, followed by Central East and North America in 2060, Europe and Russia in 2080, and then Africa in the year 2100.

High silt and sediment load cause high water turbidity, negatively affecting primary production due to light limitation and low nutrient supply. Wondie et al. (2007) reported low water transparency and nutrients due to high silt loading in tropical Lake Tana, Ethiopia. In addition, increased silt contents and suspended solids decreased fish larvae' growth and survival rates (Suede et al. 2017).

3.4 Introduced and Invasive Species

The introduction of alien species has resulted in negative ecological impacts worldwide. Some species are invasive and threaten global biodiversity, ecosystem functioning, and economic growth, while others are benign. Introduced species have been found to have certain characteristics that make them invasive in new environments. These include rapid growth, short reproductive cycles, high environmental tolerance, low predation pressure, broad diet, and efficient dispersal mechanisms (Brönmark and Hansson 2002; Keane 2002). Invasive species are a major threat to large lakes due to their dispersal ability and environmental tolerances (Jenny et al. 2020). For example, the spread of zebra mussels by releasing ballast waters from ships not only caused the extinction of native unionid mussels but also affected intake pipes and economic losses and altered nutrient pathway biological communities (Brönmark and Hansson 2002; Jenny et al. 2020). In tropical countries such as Malaysia, there are about 24 introduced species (Chong et al. 2010), originally as aquaculture, aquarium trade, and/or recreational species. Introducing species into the Malaysian environment can negatively impact indigenous biotic communities, resulting in the decline of native species, the spread of new diseases, altering natural habitats, and changing ecosystem functions. However, the ecological impacts in lakes and rivers have not been thoroughly assessed for most species.

3.4.1 Fishes

Those involving freshwater fish are among the animal taxa with the most well-documented biological invasions. The most recent Intergovernmental Science-Policy Platform on Biodiversity and Ecosystem Services (IPBES) Invasive Alien Species Assessment Reports provide a plethora of information on the characteristics of invasive alien fishes, invaded areas, and invasion pathways. Fish biological invasions can occur via various routes, such as inter-oceanic channels, ballast waters, purposeful introductions for aquaculture or fishing, ornamental uses, and escapes (Seebens et al. 2023). However, there are certain differences in the extent of occurrence and impact of invasive species across regions; North America, Europe, and Australasia have a more established record than other regions (Seebens et al. 2023).

In Africa, many freshwater fish have been intentionally introduced throughout the continents to support fisheries production via aquaculture farming or promote recreational fishing (Seebens et al. 2023). Records on the introduction of these alien fishes were limited by research efforts and funding. Published reports concentrated on the introduction of Nile perch (*Lates niloticus*), Tanganyika sardine (*Limnothrissa miodon*), and tilapias, into many tropical lakes of the Rift Valley area, such as Lakes Victoria, Tanganyika, and Malawi, resulting in large population densities of alien fish and associated fisheries important for subsistence (Seebens et al. 2023). The introduction of a few fish species such as *L. niloticus* and *Oreochromis niloticus* in Lakes Victoria and Kyoga in the 1950s and early 1960s led to the domination of the two fish species in the 1980s and the extinction of a number of endemic species (Ogutu-Ohwayo and Hecky 1991) including 200 species of endemic cichlid fish (Bacher et al. 2023). This introduced species also represents the greatest threat to Lake Malawi's native fishes (Stauffer et al. 2022). According to Stauffer et al. (2022), the success of tilapia colonization of Lake Malawi is attributed to its rapid growth rate, large size compared to native *Oreochromis* species, and diverse repertoire of feeding options. In Malaysia, for example, Ng et al. (2017) reported 16 introduced species out of 166 fish species in Sabah, and again, there is no information on their impacts on the indigenous species. Nine of 82 fish species reported in Pahang River were introduced (Zulkafli et al. 2015).

Although this forms only 11% of the total fish species, their impacts on the local fish communities should not be ignored. Samat et al. (2016) reported on native species of South America, the sailfin catfish (*Pterygoplichthys pardalis*) that has successfully colonized local rivers and lakes in Malaysia due to its wide tolerance to habitat conditions, successful reproductive capacity, and absence of indigenous predators. In this way, this exotic species can easily outcompete the indigenous species. *Oreochromis* sp. (black tilapia), which is commonly cultured in various Malaysia lakes, canals, and rivers, can be considered invasive as they have a high tolerance to environmental changes and can rapidly multiply to displace the indigenous species (Yusoff 1990; Chong et al. 2010). Efforts have been made to minimize the impacts of tilapia in the wild, such as the production of male-only tilapia or sterile hybrids. Tilapia hybrid has become one of the major cultured species in the country, and it is one of the main commercially cultured species in the cage aquaculture industry located mainly in lakes, rivers, and estuaries.

Besides tilapia, Peacock bass is also a threat to tropical lakes, especially in developing countries where the species is continuously introduced as an ornamental attraction, flooding, and sport fishing (Sastraprawira et al. 2020; Khaleel et al. 2021). This fish species is increasingly becoming a major concern in tropical African and Asian lakes and is widespread in lakes such as Lake Chenderoh and Timah Tasoh in Malaysia (Khaleel et al. 2021). The historical presence of peacock bass in Malaysia was reported to have first appeared in Kampar lakes and has spread to 10 states in the country, with higher rates of success in the lakes (84.38%) compared to other water bodies (Khaleel et al. 2021). Other Asian countries that suffer from the invasion of peacock bass in lakes are Thailand and Indonesia. Their predatory behavior on macro-invertebrates, zooplanktons, and small fishes becomes fully piscivorous, making Peacock bass one of the top predators in most of the invaded freshwater habitats. According to a review by Sastraprawira et al. (2020), in addition to its predatory behavior, this cichlid can also spread disease-causing bacteria and parasites into new habitats, which can be dangerous for other native fish species.

3.4.2 Invertebrates

Introduced invertebrates have also had various negative impacts on lakes, such introduced invertebrates have also had various negative impacts on lakes, such as alteration in the ecological processes, threats to the survival of native species through predation, reproductive interference, or hybridization, and the spread of deadly diseases (Bacher et al. 2023; Lodge et al. 2012). Introduced crayfish can also act as vectors for parasites and pathogens such as *Cherax quadricarinatus* bacilliform virus and the nematode *Gnathostoma spinigerum*, and causing plagues that reduce the production of native species (Lodge et al. 2012), while invasive rusty crayfish (*Faxonius rusticus*) caused a reduction in water clarity (Szydlowski et al. 2023). Naqiuddin et al. (2016) noted that the Australian red claw crayfish *C. quadricarinatus* has established its population in East and West Malaysia, but their ecological impacts have not been observed. An invasive Caribbean bivalve *Mytilopsis sallei* (dreissenidae) has been reported to have colonized inland waters in Southern Johor and Singapore (Tan and Morton 2006).

Rama Rao et al. (2018) reported the negative impacts of golden apple snails, *Pomacea* spp., introduced from tropical South America, and spread rapidly into Malaysian aquatic ecosystems, including lakes. *Pomacea canaliculata* is among the world's top 100 invasive species that have caused huge economic loss and ecosystem damage (Horgan et al. 2014). This golden apple snail is also invasive in the lower Mekong countries such as Vietnam, Cambodia, Lao PDR, and Thailand. Other invasive non-native fauna in Thailand includes giant African snails (Miththapala 2007). Similar to fish, the negative impact of introduced invertebrates is the predation-induced extinction of native communities. Both lake prey communities (snails) and the associated habitat (macrophytes) showed recovery when invasions of introduced *F. rusticus* declined (Szydlowski et al. 2023).

3.4.3 Macrophytes

The most common and well-studied invasive macrophyte species impacting many lakes worldwide are two submerged plants, namely *Myriophyllum spicatum* and *Hydrilla verticillata*, an emergent plant, *Phragmites australis*, and floating macrophyte, water hyacinth (Evangelista et al. 2014). The most common family of invasive macrophytes is Hydrocharitaceae, followed by Poaceae, and the most studied invasive macrophytes were *Egeria densa* Planch and *Elodea canadensis* Michx (Bora and Padial 2023).

Among the introduced species, *H. verticillata*, a species from the group Hydrocharitaceae, is considered one of the worst aquatic invasive plants worldwide. This rooted submerged macrophyte, which is native to Asia, has invaded aquatic systems recorded outside its native range in tropical Florida lakes in the United States in 1960 and expanded its distribution to almost all continents, for example, in South America in 2005 (Zhu et al. 2015). The non-native species are also better competitors in South America, infesting various reservoirs and lakes such as Itaipu and Proto Primavera reservoirs, and are likely to put more strain on habitats than other native Hydrocharitaceae (Sousa 2011). *Myriophyllum aquaticum* from tropical and sub-tropical South America is not only the dominant invasive submerged plant in temperate lakes but has also caused severe infestation in tropical waterbodies in Africa and Asia. The plant forms floating mats, and the breakaway of the floating mats can disrupt pump operations in hydroelectric plants (Sheppard et al. 2006). Few studies found that a thick underwater mat of *M. spicatum*, also known as spiked watermilfoil, has the ability to hybridize with native North American milfoils, resulting in hybrid watermilfoil with better reproduction and has been linked to native macrophytes and invertebrates declining in North American water bodies (Glisson and Larkin 2021; Sheppard et al. 2006).

Cabomba sp., another submerged macrophyte from South America, becomes invasive in non-native environments. *Cabomba caroliniana* is becoming a weed of concern in many countries, including tropical lakes in Northern Australia (Roberts and Florentine 2021; Schooler et al. 2009). Major ecological consequences include the formation of dense mats that altered biodiversity, impaired navigation and had health consequences on recreational activities such as diving (Schooler et al. 2009). The species can grow through broken stems, making eradication difficult (Schooler et al. 2009). In Malaysia, *Cabomba furcata*, another species of the genus Cabomba, is a nuisance weed in Chini Lake (Siti-Munirah and Chew 2010; Sharip et al. 2014). Sharip et al. (2014) reported that a native South American species, *C. furcata*, displaced the native *Nelumbo nucifera* in Chini Lake in 2010. *C. furcata* has become a pest in numerous lakes, such as Bukit Merah, Beris, and many ponds in Malaysia. Due to invasive species' high potential negative impacts on native aquatic species and ecosystems, their population characteristics and colonization should be studied and managed.

In terms of floating plants, *Eichhornia crassipes*, or the water hyacinth, is considered the other world's most prevalent invasive macrophyte species that have been introduced globally and choking many tropical and sub-tropical freshwater systems, including lakes and reservoirs in 62 countries (Seebens et al. 2023). Water hyacinths originated in South America and this plant has become invasive in Victoria Lake, Tana Lake, and other Rift Valley lakes in Africa, requiring comprehensive management measures (Williams et al. 2005; Tegene and Ayele 2014; Asmare et al. 2020). Water hyacinth is also considered the most invasive species in Malaysia (Faridah-Hanum and Latiff 2021). Swe et al. (2021) reported that water hyacinth is common free-floating macrophyte in shallow tropical lakes.

Many lakes in Malaysia experience severe infestation of water hyacinths, including Ringlet Lake, and urban ponds such as Gopeng and Bercham, Perak (Sharip et al. 2024). Water hyacinth also causes a severe infestation in many other tropical lakes, such as Rawapening Lake and Lake Toba in Indonesia, and Mekong Delta and Nong Han wetland in Thailand (Maulidyna et al. 2021; Nasution et al. 2016; Soeprobowati 2015; Rayan et al. 2021). Otieno et al. (2022) showed that water hyacinth impacted biotic and abiotic factors, requiring multiple measures for eradication. The negative impacts caused by water hyacinth in Rawapening Lake include hindering navigation by fisherman and their fishing gear due to the overgrowth of water hyacinth, resulting in a reduction in fish catches (Maulidyna et al. 2021). Like many other macrophytes, water hyacinth creates dense mats of floating aquatic plants when the environment is suitable. As a result, these dense mats can spread to cover the larger or complete surface area of water bodies. This reduces the light penetration needed for phytoplankton and submerged plants, thus depleting dissolved oxygen and altering the composition of invertebrate communities (Hill et al. 2020).

Other floating macrophytes that can also form dense mats, block out light penetration, and produce anaerobic conditions that are detrimental to the existing biodiversity, interfering with recreational fishing and altering physiochemical characteristics of invaded water bodies and subsequently causing local extinction of native species are *Salvinia molesta*, *Azolla* sp. and *Pistia stratiotes* (Sheppard et al. 2006). *S. molesta* covered both lakes in the Mekong Delta and oxbow lakes along the Kinabatangan River, such as Danau Tungog, affecting tourism and biodiversity (Miththapala 2007; EPPO 2017).

Most introduced macrophytes were first introduced as aquarium refugees and spread to other lakes, where they were sold as "oxygen plants" (Seebens et al. 2023). These introduced plants became widespread through aquarium escapees, and their diverse environmental tolerances contributed to their invasive success in various ecosystems (Bora and Padial 2023). The negative impacts of introduced and invasive macrophytes include the shift of native biota, hybridization, ecological disturbance, disruption of navigation, water quality changes, and financial loss (Peh 2010). Table 3.1 highlights some of the threats and impacts of introduced species in different lakes worldwide.

Table 3.1 Threats and impacts of introduced species in various tropical lakes

Types	Threats	Known infested areas	Impact
Fish	Nile perch	Lake Victoria (Africa)	Predation-induced extinction
	L. niloticus, *O. niloticus*	Lake Victoria	Extinction of native fish
	Oreochromis sp.	Malaysian lakes	Outcompete native fish
Macrophytes	*Hydrilla verticillata*	Itaipu Reservoir and other lakes in Parana Watershed (Brazil)	Physico-chemical and biotic effects
	Water hyacinth	Lake Victoria (Africa), Rawapening and Toba lakes (Indonesia), Lake Ringlet and Lake Sembrong (Malaysia), Tana Lake (Ethiopia)	Depleting dissolved oxygen, affect dam operation
	Salvinia molesta	Lake Victoria (Africa), Danau Tungog (Malaysia)	Altered navigation/ reduced tourism, deplete biodiversity
	C. furcata	Lake Chini and Lake Bukit Merah (Malaysia)	Displacement of Lotus, altered navigation
	C. *caroliniana*	Lake Macdonald (*Australia*)	Disrupt navigation

3.5 Climate Change

Climate change has been identified as one of the major threats to lakes (Jenny et al. 2020; Woolway et al. 2020; Woolway and Merchant 2022). Climate change affects the physical characteristics of the lakes, including the mixing processes, surface temperature, stratification onset, stability, and evaporation rates (Jenny et al. 2020). Climate change was reported as one of the two main threats impacting large lakes (Jenny et al. 2020). Warmer climates and long droughts increase the desiccation rate and hasten the shrinkage and disappearance of lakes in arid regions (Liu et al. 2013).

The pervasive impacts of climate change include the loss of ice cover, particularly in deeper lakes with large fetch (Woolway et al. 2020). The deep lake requires a longer period of time to freeze due to wind action (Woolway et al. 2020). A shorter duration of ice was also reported as a result of climate change impacts; northern hemisphere lakes, e.g., recorded 28 days shorter on average over 150 years. Higher air temperature and increase in ice cover caused less frequency of mixing regime but also alters evaporation rates that reduce water levels and surface area and affect storage capacities (Wang et al. 2018). According to a study by Wang and colleagues, by the end of the century, the global-mean annual evaporation rates are expected to rise to 16% (Wang et al. 2018). Because lakes have a large surface area, they are highly exposed to atmospheric changes. Climate change is affecting the overall surface energy budgets of lakes. Global warming increases the lake's surface temperature, increases short and long-wave radiation, and reduces near-surface wind speeds, all of which will alter turbulent heat fluxes, vertical mixing, and heat storage (Woolway et al. 2020).

Stratification is predicted to become more intense in warm tropical lakes where there are no season-induced temperature differences (Cohen et al. 2016). Based on temperature records, the tropical lake in Ecuador has shown prolonged periods of thermal stratification, particularly when air temperatures were highest and wind speeds were lowest (Labaj et al. 2018). Similar increases in stratification impacts of global warming have been documented in other tropical lakes, such as Lake Tanganyika and Lake Kariba in Africa (Marshall 2017; Verburg and Hecky 2009). Lake Kariba experienced a temperature increase since the 1990s, with lower vertical temperature fluctuations and reduced dissolved oxygen in the upper 20 m of the lake (Marshall 2017). In Lake Tanganyika, both the surface and 1000 m depth temperatures have increased by 1.3 °C and 0.2 °C, respectively, and this differential temperature heating has increased the density gradient in the water column, reducing the potential for vertical mixing and more stable stratification, which has an impact on water quality (Verburg and Hecky 2009). The surface water temperature of Lake Tanganyika is predicted to increase by 3 °C based on the future climate change projection scenario under the high emission representative concentration pathways (RCPs) 8.5 (Sterckx et al. 2023). The three-dimensional hydrodynamic model of this scenario revealed that stratification is predicted to increase in the upper 150 m of the water column, decreasing mixing and preventing the formation of the thermocline, ultimately leading to the closure of deep-water ventilation of the south basin (Sterckx et al. 2023).

Temperature, net evaporation, and precipitation are the three climatic factors most influencing salt lakes (Williams 2002). Solute concentration increases with high net evaporation and lower precipitation. A study by Shinhu et al. (2023) reported that the increase in evapotranspiration in Lake Victoria induced by rising temperature will lead to a decline in river discharge. However, the combined effect of climate change and land use alteration in this lake will cause an increase in extreme discharge events (Shinhu et al. 2023). Since the greatest increases in temperature projected by the climate change model will occur in semi-arid regions, the effects on salt lakes will be highest. This includes permanent salt lakes shrinking in size and becoming more salinized while temporary salt lakes continue to be drier for a longer period (Williams 2002). Such changes will affect the diversity of biota, including the loss of less tolerant salinity species.

There are increasing publications showing the impact of climate change on lake ecosystems (Kerman et al. 2010; Kosten et al. 2012). This includes changes in water quality, increase in phytoplankton biomass, and alteration in community composition (Vilhena et al. 2010). A review by Wiley and MacPherson (2023) examines how climate change contributes to trends in bloom frequency or severity. Early onset of phytoplankton bloom was predicted for Upper Lake Constance, Europe, based on model simulations induced by high air temperatures and low wind speeds

(Peeters et al. 2007). Simulations of three flood underflow events in Lake Burragorang, Australia, showed that a large algal bloom in the lake is triggered when a high volume of inflow enters the lake at low water level, caused by the underflow constituents mixed into the surface layer (Vilhena et al. 2010). Climate change may exacerbate the biological invasion such as by enabling invasive plants to proliferate and widen their ranges (Lozano 2021). The overall impact of global warming was illustrated in a graphic by Moss et al. (2011), which includes changes in ecosystem processes such as an increase in respiration rates leading to a reduction in dissolved oxygen, an increase in mineralization and sedimentation resulting in the release of methane and other greenhouse gases.

3.6 Pollutants and Micropollutants Contamination

Pollutants and micropollutant contaminants have been a common problem in tropical lakes. Water pollution can originate from discharges from industry, mining, and agriculture activities, as well as atmospheric transport. Industrial discharges, which include waste or byproducts, usually contain heavy metals that are not only toxic to humans and wildlife but also not suitable for agriculture development such as irrigation and fish farming (Ho et al. 2006; Sial et al. 2006; Saravanakumar et al. 2022). Contaminants that various types of industry have discharged in different countries include mercury (Hg), arsenic (As), cadmium (Cd), copper (Cu), chromium (Cr), and lead (Pb) (Ho et al. 2012). As, Cd, Hg, and Pb are listed in the top 10 hazardous substances by the World Health Organization and the Agency for Toxic Substance and Disease Registry (ATSDR) (Reddy 2017). Due to their high toxicity, a review on global metal contamination in waterbody that results in public health impact includes arsenic, causing cancer and kidney injuries, and Cu, Cd, Cr, Ni, Pb, and Zn, causing Asthma and skin disease (Reddy 2017).

Many industrial discharges contain high heavy metals exceeding the national water quality standards, such as iron contents in all the effluents (Sial et al. 2006). Certain industries discharge different heavy metals that are of concern, such as copper in effluents of textile and sewage and manganese in ghee industry effluents (Sial et al. 2006). The source of cadmium detected in the urban lake, Hussain Sagar Lake in India, was due to the discharge of effluents from batteries and ferrous and non-ferrous metal manufacturing industries (Ayyanar and Thatikondar 2020). In Puteri Lake, Malaysia, the high iron content is due to natural factors, such as the lake, which was created from iron mining activities, still contains high iron ore reserves (Sulaiman et al. 2018).

A review on pollution in tropical African lakes indicated heavy metals residues such as Cadmium in fishes and birds in Lake Nakuru, Lake Tanganyika, Lake Edward, and Lake George (Ogutu-Ohwayo et al. 1997). High metal residues in fish are a great concern due to their large human consumption. Some heavy metals are trapped or adsorbed in the sediment, causing catastrophic effects on benthic biota and habitat

(Redwan and El-Haddad 2022). Heavy metal concentrations in sediments have been reported to be in varying degrees in various other tropical water bodies.

Some heavy metals are a great concern of contaminants due to their transport mechanism through the atmosphere. Mercury, e.g., is not only persistent in the environment but also can experience biomagnification in the aquatic food web. Like many other heavy metals, Mercury does not degrade in an environment (Pacyna 2020). The release of mercury is increasing in Asia due to industrial applications, coal burning, cement production, and mining practices (Pacyna 2020). Atmospheric deposition, volcanic influences, and climate change are some of the sources of Mercury in the tropical lakes in South America (Yang et al. 2023). Such long-term contamination has induced biomagnification in the food web, while continued discharge led to atmospheric contamination due to the ability of gaseous mercury emitted to retain longer within the atmosphere (Roberts et al. 2020). Mercury biomagnification was reported in Amazonian floodplain lakes where concentration was highest in predatory pirarucu and piranha (Nyholt et al. 2022).

3.6.1 Pesticides

Micropollutants are becoming a threat due to human concerns for their health. Micropollutants are generally pollutants that occur at trace levels. These chemicals could influence the endocrine system. Agrochemicals or pesticides have been reported in many reservoirs with extensive farming and plantation within the catchment. Organochlorine pesticides (OCP), chlorinated hydrocarbon compounds, were the most widely applied pesticides worldwide. It generally can be divided into three main groups, namely (i) dichlorodiphenylethanes such as DDT and its metabolites, dicofol and metolachlor, (ii) chlorinated cyclodienes such as aldrin, dieldrin, endrin, chlordane, endosulfan, heptachlor, and (iii) chlorinated benzenes and cyclohexanes such as hexachlorocyclohexane (HCH), hexachlorobenzene (HCB), and toxaphene (Horak et al. 2021). Many of these chemicals, which were widely applied in the 1950s, are now being banned in many countries, yet their presence in the environment, especially in stagnant water and sediment, remains due to their persistent breaking down.

Organochlorine pollution (OCP) such as DDT and Chlordane was reported in large tropical Lake Malawi affecting fish and animals (Beeton 2002). Restriction in the use of OCPs worldwide led to reduced inputs of the substance. Despite being banned in many countries, DDT was observed to occur in tropical Sembrong Lake, Malaysia (Sharip et al. 2017) and Laguna del Cisne, Uruguay (Rodríguez-Bolaña et al. 2023). Various pesticides and emerging contaminants were detected in the Guarapiranga reservoir, Brazil (López-Doval et al. 2017). In Volta Lake Ghana, DDT was not detected in water and sediment samples but was present in fish samples such as Nile tilapia and African catfishes' (Gbeddy et al. 2015). Similarly, pesticide residues, including OCPs, were widespread in fishes of a sub-tropical lake, Laguna del Cisne, Uruguay, requiring continuous monitoring to prevent human and aquatic health impacts (Rodríguez-Bolaña et al. 2023).

A review by Horak et al. (2021) reported the detection of high DDT residue concentrations in sediment samples of Lake Sibaya, Southern Africa's largest natural freshwater lake and biota, including crocodiles and fish communities in iSimangaliso Wetland Park, South Africa. Similarly, HCHs, DDTs, and heptachlor were also detected in Dongting Lake, together with sixteen other types of OCPs (Cao et al. 2021). However, their concentration was still safe for aquatic organisms and human health (Cao et al. 2021). Organophosphorus pesticides (OPs) such as chlorpyrifos, which has a long half-life in soil, have also been detected in the surface waters of Rietvlei Nature Reserve and Albasini Dam (Horak et al. 2021). The impact of EDCs, such as OCPs and OPs, include hormonal changes and intersex in fish.

3.6.2 Hormones and Hormone-Like Compounds

The compounds such as estrone, 17-alpha-ethinylestradiol (17α-EE2), 17-Beta-20 estradiol, testosterone, and progesterone, e.g., pharmaceuticals are some of the emerging contaminants that are of concern which can cause endocrine disruption with potential environmental impacts. Most hormones found in lake water originate from wastewater and sewage treatment plant effluents, agricultural runoff, and discharge waters from urban areas (Sacdal et al. 2020). These trace chemicals were detected in Laguna Lake, Philippines, and many lakes in China, such as in Wuhan and along the Yangtze River (Ana and Espino 2020; Sacdal et al. 2020; Yang et al. 2015; Zhang et al. 2022). Hormones such as 17α-ethinylestradiol (EE2) were found to be in high concentration in Honghu Lake and Wuhan Lakes in China (Yang et al. 2015; Zhang et al. 2022). Generally, 17α-ethinylestradiol (EE2) is a commonly used hormone for human medications, livestock, and aquaculture. The impact of 17α-ethinylestradiol (EE2) contamination includes delayed sexual maturity and altered sex determination (Aris et al. 2014). Since there are no local guidelines on these emerging contaminants in the Philippines, their detection of these contaminants was reckoned as parameters of contamination for monitoring in the lake water (Ana and Espino 2020). Antibacterial agents such as triclosan and triclocarban, which are commonly used in toothpaste and soaps, can be toxic to aquatic life (Olaniyan et al. 2016).

Other chemicals that are very harmful and categorized as EDCs include Bisphenol A and phthalates, a category of EDCs that have been widely reported. Bisphenol A is commonly used to produce industrial polycarbonate plastics and epoxy resins. Phthalates are chemicals used as plasticizers, lubricants, and solvents in various industries, including the manufacturing of bottles and containers (Gonsioroski et al. 2020). According to Gonsioroski et al. (2020), both types of EDCs have caused various health impacts. The occurrence of Bisphenol A in a lake was documented in the Guarapiranga reservoir, Brazil (López-Doval et al. 2017). There is still limited information on the impact of Bisphenol A contamination in lakes.

Other contaminants of emerging concern are fluorinated substances such as per and poly-fluoroalkyl substances (PFAS), which are listed by the United States Environmental Protection Agency (USEPA) due to their environmental impact and adverse impacts on human health, animals and biota. These trace chemicals occur in the lake environment, exceeding the advisory level for drinking purposes (Podder et al. 2021). Few studies documented that short-chain PFCAs were dominant in water and long-chain PFSAs in sediments (Goodrow et al. 2020; Chen et al. 2023). Additionally, PFAS were found in fish tissue at 11 targeted temperate waterbodies (Goodrow et al. 2020; Miranda et al. 2023). A study focusing on different types of fluorinated substances in biota, all perfluorooctanoic sulfonic acid (PFOS), PFDA, PFUnA, and PFDoA compounds were identified as the most prevalent in fish tissue, with PFOS having the highest concentrations, with the average level of PFOS in various fish species was high enough to warrant advisories against eating fish. Different PFAS types were detected in high concentrations with fish types; Perfluoroalkyl carboxylates (PFCA) concentrations were higher in prey fish than in predators (Miranda et al. 2023). The highest levels of PFAS in fish were found downstream of a military installation, indicating the potential release of the chemicals (Goodrow et al. 2020).

Studies of PFAS contamination in tropical lakes are limited. Direct discharge and atmospheric deposition were identified as primary sources of PFAS in surface water (Podder et al. 2021; Chen et al. 2023). The sources of PFAS in lakes in the southern Tibetan Plateau largely come from long-range atmospheric transport from South Asia, while lakes in northern areas were formed by local emissions and tourism activities (Chen et al. 2023). A study on per- and poly-fluoroalkyl substances (PFAS) in deep lakes found a general correlation between the level of lake catchment urbanization and PFAS contamination in lake fish (Valsecchi et al. 2021). Engineered nanoparticles with 1–100 nm dimensions, such as titanium oxides, ferum oxides, and zinc oxides, are also endocrine disruptors of concern and have been detected in various waterbodies (Reddy 2017).

3.6.3 Microplastics

Microplastics are becoming a major concern due to the large amount of plastic waste in many temperate and tropical lakes. Despite these concerns, published records on microplastic pollution for tropical water bodies are still sparse, while data increasing records in the marine environment. Malaysia, e.g., has had a booming plastics industry over the past 50 years and the waste from this product usually ends up in rivers and lakes (Sulaiman et al. 2023). Evidence shows that microplastics have been found in Cempaka Lake and Setiu Wetland in Malaysia (Suardy et al. 2020). In Indonesia, increasing publication on microplastic in lakes has been documented. Microplastic was found in Lake Maninjau and Lake Singkarak in West Sumatera, Lake Towuti in Sulawesi, and Lake Beratan in Bali (Yusuf et al. 2022; Henny et al. 2022; Deswati et al. 2023; Watiniasih et al. 2023). In Lake Beratan, microplastics were found in water, sediment, and biota, such as fish and shrimps (Watiniasih et al. 2023). Microplastics are small plastic particles with a size of < 5 mm that can accumulate in the environment through primary sources such as manufacturing discharges and secondary sources such as the degradation of large plastic products (Su et al. 2016). In the latter case, the microplastic pollution may originate from municipal wastewater discharges, combined storm sewer overflows and aquaculture activities (Mai et al. 2023). Few studies have shown the high microplastic levels in stagnant waterbodies, with higher concentrations occurring at the lake outlet (Malla-Pradhan et al. 2022) or in heavily contaminated areas (Su et al. 2016). In another study in Lake Singkarak, Indonesia, microplastic was more abundant at the lake inlet (Henny et al. 2022).

Fibers were the most common type of microplastic found in Phewa Lake, Nepal (Malla-Pradhan et al. 2022), and Lake Naivasha, Kenya (Migwi et al. 2020). Fibers could originate from domestic effluents such as washing clothes in households or laundry activities (Eerkes-Medrano et al. 2015). Films and fragments were another type of microplastic found in the two lakes (Migwi et al. 2020). Fragments usually come from the weathering of plastic-based products such as packaging or plastic bags and drinking bottles (Deswati et al. 2023). Further microscopic analysis of these fibers, films, and fragments revealed that they were composed of polypropylene, polyethylene, and polyester (Migwi et al. 2020). Similarly, fibers were the most abundant microplastics in three tropical urban lakes in Hanoi, Vietnam, followed by fragments whose main chemical composition is rayon and polyethylene terephthalate (Mai et al. 2023).

Fibers were the main type of microplastic found in biota, such as fish and shrimps in Lake Belatan, contributing more than 64% (Watiniasih et al. 2023). Similarly, in Lake Titicaca, an ancient tropical lake in South America, fibers were also more abundant in fish stomachs, while fragments were more abundant in water (Loayza et al. 2022). Mallik et al. (2021) reported that microplastics can cause health problems in aquatic animals, such as gill malfunction, reduced reproductive capacity, and increased vulnerability to diseases. Toxic compounds in plastic, or known as additives, such as plasticizers are endocrine disruptors that affect human health (Gonsioroski et al. 2020).

3.6.4 Microbial

Microbial contaminant remains an important problem in most developing countries in the trophic due to sanitary and domestic waste discharges and inadequate treatment facilities. In Africa, microbial contamination has shown a great fluctuation in the numbers of bioindicators along the African lakes in the past years (Tahoun 2023). Significant amounts of sewage, along with other industrial, agricultural, municipal and domestic wastes, are still discharged directly, untreated, into most lakes in Africa, endangering human health and polluting the environment (Tahoun 2023; Nakagiri et al. 2023). The untreated human sewage or runoff from animal husbandry facilities into the lake's stream are African lakes' main microbial pollution sources (Tahoun 2023). Similarly, higher levels of economic development activities, including livestock and agricultural intensity, sewerage and toilets, population, river proximity, and urban dominance, lead to higher levels of bacterial coliforms (Moshi et al. 2022). In Malaysia, microbial waste such as coliform, *Escherichia coli*, and waterborne parasites dominate many urban lakes, exceeding the national water quality standards (Sharip and Mohamad 2019; Onichandran et al. 2013). However, contamination of these fecal indicator bacteria varied temporally along urbanization gradients (Sharip and Mohamad 2019).

Compared to developing countries, tropical lakes in developed countries have fewer microbial contamination problems due to good wastewater treatment facilities. In a study by Hoyer et al. (2006) on 99 tropical lakes in Florida, USA, fecal contamination in the Florida lakes was not widespread. Bacterial contamination of reservoirs and lakes in developed countries such as the USA and Germany was generally low and below the existing water quality standard (Hoyer et al. 2006; Reitter et al. 2021). The study, however, found that the content of microbes such as *E. coli* and coliform in waste with septic tanks was no different from those with central wastewater treatment systems (Hoyer et al. 2006). Their study also found that total coliform and *E. coli* counts were much lower in open-water samples than littoral samples, and bacterial counts were positively correlated to lake trophic status and aquatic bird abundance (Hoyer et al. 2006). Seasonal and precipitation effects on bacterial contamination varied between sites. Few studies reported seasonal microbial pollutants where higher contamination occurred during the wet season, which is associated with storm water run-offs from multiple sources (Ouma et al. 2016; Asfi et al. 2023).

Microbial pollution is a concern as it causes diseases that affect human health. Pathogenic bacteria include *Salmonella* sp. and *Vibrio* species; *Vibrio* species which can cause cholera and bacillary dysentery or shigellosis, respectively, that results in salmonellosis and typhoid fever. *E. coli* can cause acute diarrhea and gastroenteritis disease. Fecal indicator bacteria (FIB) monitoring in Lake Bunyonyi, Africa, revealed high *E. coli* and enterococci counts in the samples, rendering them bacteriologically unsuitable for drinking purposes (Saturday et al. 2021). However, the water was safe for recreational purposes, with 97.2% of the water samples having *E. coli* and enterococci below recreational standards (USEPA). FIB monitoring along nearshore in Lake Malawi found that fifty-three percent of water samples exceeded the USEPA Recreational Water Quality Criteria, and 90% exceeded the WHO drinking water standard (Tyner et al. 2018). Table 3.2 gives the various threats, source, and risks of different contaminants.

Table 3.2 Threats, sources, and health risks of contaminants

Pollutants	Sources	Health risk
Heavy metals		
1. Mercury	Industrial application, coal burning, cement production and mining practices, atmospheric deposition	Cancer
2. Arsenic	Industrial application, mining practices	Cancer, kidney injuries
3. Cadmium	Industrial application, mining practices	Asthma and skin disease
Pesticides		
1. Organochlorine (OCPs)	Agricultural activities	Hormonal changes and intersex in fish
2. Organophosphorus pesticides	Agricultural activities	Hormonal changes and intersex in fish
Hormones	Wastewater and sewage treatment plants effluents, hospital wastes, agricultural runoff, discharge waters from urban areas	Delayed sexual maturity and altered sex determination
PFAS	Direct discharges of wastewater and sewage treatment plants effluent and urban areas, atmospheric deposition	Hormonal changes and intersex in fish
Microplastics	Households or laundry activities	Gill malfunction, reduced reproductive capacity, increase vulnerability to diseases
Microbial		
1. Pathogenic bacteria (*E. coli*, *Salmonella* sp., *Vibrio* species)	Sewage, wastewater, septic tanks, sanitary and domestic waste discharges, runoff from animal husbandry facilities	Cholera, diarrheas and gastroenteritis disease
2. Pathogenic parasites	Human and/or animal excreta	Giardiasis, cryptosporidiosis and amoebiasis
3. Virus	Runoff from animal husbandry facilities	Diarrheas and gastroenteritis

3.7 Water Abstraction

Water abstraction upstream for irrigation threatens a few arid lakes in Australia, such as Coongie Lakes on Cooper Creek and the Paroo Lakes in the northwestern Murray Darling catchment. The water diversion has caused many lakes to shrink, such as the Aral, Ebinur, Chad, and Orumieh lakes. The Aral Lake declined tremendously due to water diversion for cotton production in the 1960s, while the diversion in Chad Lake between 1980 and 1990 led to only 1/3 of the lake remaining with water (Pham-Duc et al. 2020). Many tropical lakes also suffer from water abstraction due to increasing water demands for agricultural purposes. Lakes Ziway and Abyata in Ethiopia experience significant direct water withdrawals by farmers for irrigation, which impacts water levels. If left unmanaged, the lake will become terminal lakes and eventually become saline (Jansen et al. 2007; Goshime et al. 2019).

Diversion of water for irrigation from the Perkerra, Molo, and Ol Arabel rivers, the main tributaries to Lake Baringo in Africa, caused the lake level to fall to 1.7 m in 2003, compared to an average depth of 8 m in the early 1970s (Odada et al. 2002). The lake size also shrank by 54% to 108 km^2 in 2001 compared to 1976 records (Odada et al. 2002). The combined effects of agriculture and climate contribute to half of the arid lakes in Gobi, China, having fully become desiccated by 2020s (Liu et al. 2013).

3.8 Shoreline Modification

Shoreline modification poses a threat to many lakes, including in developing countries. Urban lakes, in particular, are experiencing changes to their shorelines, with natural banks being replaced by retaining walls to accommodate human settlements, development, and tourism facilities (Lorenz et al. 2017). Yet, the impacts of these changes on littoral habitats of tropical lakes have not been widely studied. Lake shoreline development will remove littoral habitats, subsequently reducing littoral community diversities and impacting the food web.

Alteration of a lake shoreline structure in a Japanese lake significantly impacts fish behavior and affects their migration from lakes to nearby channels (Mitsuo et al. 2023). Retaining walls have been found to reduce the total densities and number of macroinvertebrate taxa, likely due to physical factors such as an increase in bottom current velocity and transport mechanism (Ostendorp et al. 2020). Shoreline modification involving structure can also affect the physical influence of waves and littoral slopes and diminish the natural wetland habitats (Jenny et al. 2020). Our reviews found limited published scientific studies assessing shoreline modification on tropical lakes and their ecosystem.

3.9 Acidification

Acidification is a lake threat that originates outside the lake catchment where it occurs due to the meteoric deposition of strong acids such as sulfuric and nitric acid originating from the burning of high-sulfur fossil fuels and automobile exhausts (Nakamura 1997; Johansson et al. 2023). A decreased water pH resulting from inputs of acid anions and/or dissolution of atmospheric carbon dioxide led to increased acidification. Large lakes have a higher potential to experience acidification from local atmospheric deposition due to the high ratio of catchment to surface area (Jenny et al. 2020). It is anticipated that the lakes will also result in a higher concentration of Al complexed with naturally occurring dissolved organic ligands such as fulvic acid (Johansson et al. 2023).

Although many acidified lakes are concentrated in the Scandinavian and Northern continents, acidification occurs in tropical lakes, particularly those exposed to acid rain in industrialized and mining regions. For example, Dunas Lake, a tropical pond in Northeast Brazil, experiences acidification resulting from deposition from industries and domestic waste activities in its lake basin (Araújo et al. 2013), requiring rehabilitation. Acidification was also found in high-altitude lakes in Taiwan, where metals such as iron and manganese were released due to redox reactions and subsequent low metal/Al ratios occurred in surface sediments (Arthur Chen et al. 2004). Many foreign sources, including dust from the arid regions of China, have been blamed for the acid rain statistical records in the early 2000s, which may have led to the acidification of Taiwan's lakes (Arthur Chen et al. 2004). Lake acidification before the 1950s was most likely caused by long-distance movements of acidic material from Japan via the northeast monsoon (Arthur Chen et al. 2004).

Acidification of lakes and streams can lead to adverse effects on aquatic biota, particularly those sensitive to alkaline water (Tranvik 2021). Some studies have recorded a decline in fish population including mortality of salmonid fish populations such as in Scandinavian waterbodies. Future acidification trends, especially in the tropical region, still need to be studied further despite a continued increase of atmospheric carbon dioxide (CO_2) due to the effects of algal bloom in decreasing CO_2 solubility (Jenny et al. 2020). Findings from such studies will support informed management and restoration efforts.

3.10 Remarks

Threats to lakes are driven by environmental changes within or beyond the lake catchment, as well as climate change. Major global lake issues, including in the tropics, are eutrophication, sedimentation, invasive species, and pollution. Certain lakes also experience shoreline modification and acidification. For tropical lakes, effective management should be focused on management of nutrients, organic matter and other pollutant inflows into lakes, prevention of invasive species, and intervention of negative anthropogenic impacts in lake basins. The impacts of these threats and issues include affecting human and environmental health. The problem, if not addressed, will cause lake ecosystem to deteriorate affecting its ability to provide numerous ecosystem services.

References

Adugna TM, Cherie DA (2021) A review on reservoirs sedimentation problems in Ethiopia. Asian J Adv Res Rep 15(3):1–8. https://doi.org/10.9734/ajarr/2021/v15i330372

Adongo TA, Kyei-Baffour N, Abagale FK, Agyare WA (2020) Assessment of reservoir sedimentation of irrigation dams in Northern Ghana. Lake Reservoir Manag 36(1):87–105. https://doi.org/10.1080/10402381.2019.1659461

Akinnawo SO (2023) Eutrophication: causes, consequences, physical, chemical and biological techniques for mitigation strategies. Environ Challenges 12:100733. https://doi.org/10.1016/j.envc.2023.100733

Amasi A, Wynants M, Blake W, Mtei K (2021) Drivers, impacts and mitigation of increased sedimentation in the hydropower reservoirs of East Africa. Land 10(6). https://doi.org/10.3390/land10060638

Ana KMS, Espino MP (2020) Occurrence and distribution of hormones and bisphenol A in Laguna Lake, Philippines. Chemosphere 256:127122. https://doi.org/10.1016/j.chemosphere.2020.127122

Araújo CVM, Cohin-de-Pinho SJ, Chastinet CBA, Machado SL, da Silva EM (2013) The natural rehabilitation of an anthropogenically acidified tropical lake: two decades of monitoring. Chemosphere 90(3):887–894. https://doi.org/10.1016/j.chemosphere.2012.06.029

Ardiles V, Alcocer J, Vilaclara G, Oseguera LA, Velasco L (2012) Diatom fluxes in a tropical, oligotrophic lake dominated by large-sized phytoplankton. Hydrobiologia 679:77–90. https://doi.org/10.1007/s10750-011-0853-7

Arif S, Djebbari N, Belhaoues S, Touati H, Bensouilah M (2023) Spatio-temporal dynamics of cyanobacterial abundance and toxicity in a Mediterranean hypereutrophic lake. J Oceanol Limnol 41(5):1834–1848. https://doi.org/10.1007/s00343-022-1313-0

Aris AZ, Shamsuddin AS, Praveena SM (2014) Occurrence of 17α-ethynylestradiol (EE2) in the environment and effect on exposed biota: a review. Environ Int 69:104–119. https://doi.org/10.1016/j.envint.2014.04.011

Arthur Chen CT, Wu JT, Wang BJ, Huang KM (2004) Acidification and trace metals of lakes in Taiwan. Aquat Geochem 10:33–57

Asfi MF, Razak SBA, Zulkifli MH, Sharip Z (2023) Spatio-temporal factors influencing coliform and *Escherichia coli* contamination in gross pollutant trap and wetland inlet, Putrajaya, Malaysia. Int J Energy Water Resour 7:487–498. https://doi.org/10.1007/s42108-022-00233-w

Asmare T, Demissie B, Nigusse AG, GebreKidan A (2020) Detecting spatiotemporal expansion of water hyacinth (*Eichhornia crassipes*) in Lake Tana, Northern Ethiopia. J Indian Soc Remote Sens 48:751–764

Asthana BN, Khare D (2022) Reservoir sedimentation. In: Asthana BN, Khare D (eds) Recent advances in dam engineering. Springer International Publishing, Berlin, pp 265–288. https://doi.org/10.1007/978-3-030-32278-6_12

Ayele HS, Atlabachew M (2021) Review of characterization, factors, impacts, and solutions of lake eutrophication: lesson for lake Tana, Ethiopia. Environ Sci Pollut Res 28(12):14233–14252. https://doi.org/10.1007/s11356-020-12081-4

Ayyanar A, Thatikonda S (2020) Distribution and ecological risks of heavy metals in lake Hussain Sagar, India. Acta Geochim 39(2):255–270

Bacher S, Galil B, Nunez M, Ansong M, Cassey P, Dehnen-Schmutz K, Fayvush G, Hiremath A, Ikegami M, Martinou A, McDermott SM, Preda C, Vilà M, Weyl OLF (2023) Impacts of invasive alien species on nature, nature's contributions to people, and good quality of life. In: Roy HE, Pauchard A, Stoett P, Renard Truong T (eds) Thematic assessment report on invasive alien species and their control of the intergovernmental science-policy platform on biodiversity and ecosystem services, chap 4. Bonn, Germany, pp 1–222

Bailey A, Meyer L, Pettingell N, Macie M, Korstad J (2020) Agricultural practices contributing to aquatic dead zones. Ecol Pract Appl Sustain Agric 373–393. https://doi.org/10.1007/978-981-15-3372-3_17

Barjau-Aguilar M, Merino-Ibarra M, Ramírez-Zierold JA, Castillo-Sandoval SF, Vilaclara-Fatjó G, Guzmán-Arias AP, Macek M, Alcántara-Hernández RJ, Sánchez-Carrillo S, Valdespino-Castillo PM, Sacristán-Ramírez A, Quintanilla-Terminel JG, Monroy-Ríos E, Díaz-Valenzuela J, Lestayo-González JA, Oscar A, Gerardo-Nieto OA, De Zayas RG (2022) Nitrogen and phosphorous retention in tropical eutrophic reservoirs with water level fluctuations: a case study using mass balances on a long-term series. Water 14:2144. https://doi.org/10.3390/w14142144

Basson G (2009) Management of siltation in existing and new reservoirs. ICOLD, Paris

Basson GR, Efthymiou N, Annandale GW (2022) Management of extreme sedimentation in a reservoir in a semi-arid region. In: Proceedings of the 39th IAHR world congress

Beeton AM (2002) Large freshwater lakes: present state, trends, and future. Environ Conserv 29(1):21–38. https://doi.org/10.1017/S0376892902000036

Bora LS, Padial AA (2023) A global review on invasive traits of macrophytes and their link to invasion success. Acta Limnol Bras 35:e20

Brönmark C, Hansson LA (2002) Environmental issues in lakes and ponds: current state and perspectives. Environ Conserv 29(3):290–307. https://doi.org/10.1017/S0376892902000218

Cao F, Li Z, He Q, Lu S, Qin P, Li L (2021) Occurrence, spatial distribution, source, and ecological risk assessment of organochlorine pesticides in Dongting Lake China. Environ Sci Pollut Res 28(24):30841–30857. https://doi.org/10.1007/s11356-021-12743-x

Central Water Commission (2020) Compendium on sedimentation of reservoirs in India

Chen Y, Wei L, Luo W, Jiang N, Shi Y, Zhao P, Ga B, Pei Z, Li Y, Yang R (2023) Occurrence, spatial distribution, and sources of PFASs in the water and sediment from lakes in the Tibetan Plateau. J Hazard Mater 443:130170

Chislock MF, Doster E, Zitomer RA, Wilson AE (2013) Eutrophication: causes, consequences, and controls in aquatic ecosystems. Nat Educ Knowl 4(4):10

Chong VC, Lee PKY, Lay CM (2010) Diversity, extinction risk and conservation of Malaysia fishes. J Fish Biol 79(9):2009–2066

Choy FK, Hamzah FB (2001) Cameron highlands hydroelectric scheme: landuse change—impacts and issues. In: Honningsvag B, Midttomme GH, Repp K, Vaskinn KA, Westeren T (eds) Hydropower in the new millennium. AaBalkema Publishers, Leiden, pp 215–221

Cohen AS, Gergurich EL, Kraemer BM, McGlue MM, McIntyre PB, Russell JM, Simmons JD, Swarzenski PW (2016) Climate warming reduces fish production and benthic habitat in Lake Tanganyika, one of the most biodiverse freshwater ecosystems. Proc Natl Acad Sci 113(34):9563–9568

Cohen AS, Kaufman L, Ogutu-Ohwayo R (2019) Anthropogenic threats, impacts and conservation strategies in the African great lakes: a review. In: Limnology, climatology and paleoclimatology of the east African lakes, pp 575–624

Dash SS, Maity R (2023) Effect of climate change on soil erosion indicates a dominance of rainfall over LULC changes. J Hydrol Reg Stud 47:101373. https://doi.org/10.1016/j.ejrh.2023.101373

Deswati D, Tetra ON, Hayati M, Putra A, Fitri WE, Suparno S, Pardi H (2023) Preliminary detection of microplastics in surface water of Maninjau Lake in Agam, Indonesia. Aquac Aquarium Conserv Legislation 16(5):2601–2614

Dodds WK, Bouska WW, Eitzmann JL, Pilger TJ, Pitts KL, Riley AJ, Schloesser JT, Thornbrugh DJ (2009) Eutrophication of U.S. freshwaters: analysis of potential economic damages. Environ Sci Technol 43(1):12–19. https://doi.org/10.1021/es801217q

Dybas CL (2005) Dead zones spreading in world oceans. Bioscience 55(7):552–557. https://doi.org/10.1641/0006-3568(2005)055[0552:DZSIWO]2.0.CO;2

Eerkes-Medrano D, Thompson RC, Aldridge DC (2015) Microplastics in freshwater systems: a review of the emerging threats, identification of knowledge gaps and prioritisation of research needs. Water Res 75:63–82. https://doi.org/10.1016/j.watres.2015.02.012

EPPO (2017) Pest risk analysis for *Salvinia molesta*. EPPO, Paris

Evangelista H, Thomaz SM, Umetsu CA (2014) An analysis of publications on invasive macrophytes in aquatic ecosystems. Aquat Invasions 9(4). https://doi.org/10.3391/ai.2014.9.4.10

Faridah-Hanum I, Latiff A (2021) Invasive alien species in Malaysia. Invasive alien species: observations and issues from around the world, vol 2, pp 151–167
Floodlist (2013) What went wrong at Lake Ringlet? https://floodlist.com/asia/floods-lake-ringlet-malaysia. Accessed on 10 October 2023
Gbeddy G, Glover E, Doyi I, Frimpong S, Doamekpor L (2015) Assessment of organochlorine pesticides in water, sediment, African cat fish and *Nile tilapia*, consumer exposure and human health implications, Volta Lake, Ghana. J Environ Anal Toxicol 5(4):297
Glisson WJ, Larkin DJ (2021) Hybrid watermilfoil (*Myriophyllum spicatum* × *Myriophyllum sibiricum*) exhibits traits associated with greater invasiveness than its introduced and native parental taxa. Biol Invasions 23(8):2417–2433
Goodrow SM, Ruppel B, Lippincott RL, Post GB, Procopio NA (2020) Investigation of levels of perfluoroalkyl substances in surface water, sediment and fish tissue in New Jersey, USA. Sci Total Environ 729:138839
Gonsioroski A, Mourikes VE, Flaws JA (2020) Endocrine disruptors in water and their effects on the reproductive system. Int J Mol Sci 21(6):1929. https://doi.org/10.3390/ijms21061929
González EJ, Roldán G (2020) Eutrophication and phytoplankton: some generalities from lakes and reservoirs of the Americas. In: Microalgae—from physiology to application. IntechOpen. https://doi.org/10.5772/intechopen.89010
Goshime DW, Absi R, Ledésert B, Dufour F, Haile AT (2019) Impact of water abstraction on the water level of Lake Ziway, Ethiopia. WIT Trans Ecol Environ 239:67–78
Harvey EL, Menden-Deuer S (2012) Predator-induced fleeing behaviors in phytoplankton: a new mechanism for harmful algal bloom formation? PLoS ONE 7(9):e46438. https://doi.org/10.1371/journal.pone.0046438
Henny C, Rohaningsih D, Suryono T, Santoso AB, Waluyo A (2022) Microplastic pollution in the surface water of Lake Singkarak, Indonesia. IOP Conf Ser Earth Environ Sci 1118(1):012050. https://doi.org/10.1088/1755-1315/1118/1/012050
Hill MP, Coetzee JA, Martin GD, Smith R, Strange EF (2020) Invasive alien aquatic plants in South African freshwater ecosystems. In: van Wilgen BW, Measey J, Richardson DM, Wilson JR, Zengeya TA (eds) Biological invasions in South Africa. Springer International Publishing, Cham, pp 97–114
Ho YC, Show KY, Guo XX, Norli I, Abbas FMA, Morad N (2012) Industrial discharge and their effect to the environment. In: Kuan-Yeow S, Xinxin G. Rijeka (eds) Industrial waste, chap 1. IntechOpen
Horak I, Horn S, Pieters R (2021) Agrochemicals in freshwater systems and their potential as endocrine disrupting chemicals: a South African context. Environ Pollut 268:115718. https://doi.org/10.1016/j.envpol.2020.115718
Horgan FG, Stuart AM, Kudavidanage EP (2014) Impact of invasive apple snails on the functioning and services of natural and managed wetlands. Acta Oecol 54:90–100
Hoyer MV, Donze JL, Schulz EJ, Willis DJ, Canfield DE Jr (2006) Total coliform and *Escherichia coli* counts in 99 Florida lakes with relations to some common limnological factors. Lake Reservoir Manag 22(2):141–150. https://doi.org/10.1080/07438140609353891
Hu HH, Deng AJ, Chen JG, Shi HL, Ye M (2018) A research overview of the siltation loss controls and capacity recovery processes in China's reservoirs. MATEC Web Conf 246:01086
ICOLD (2023) Sedimentation and sustainable use of reservoirs and river systems. CRC Press
International Lake Environment Committee (ILEC) (2002) Lake Biwa Research Institute. 1988–1993 Survey of the state of the World's lakes; International Lake Environment Committee, Otsu and United Nations Environment Programme, Nairobi, Kenya
Jansen H, Hengsdijk H, Legesse D, Ayenew T, Hellegers P, Spliethoff P (2007) Land and water resources assessment in the Ethiopian Central Rift Valley: project: ecosystems for water, food and economic development in the Ethiopian Central Rift Valley, Alterra
Jenny J-P, Anneville O, Arnaud F, Baulaz Y, Bouffard D, Domaizon I, Bocaniov SA, Chèvre N, Dittrich M, Dorioz J-M, Dunlop ES, Dur G, Guillard J, Guinaldo T, Jacquet S, Jamoneau A, Jawed Z, Jeppesen E, Krantzberg G, Lenters J, Leoni B, Meybeck M, Nava V, Nõges T, Nõges P,

Patelli M, Pebbles V, Perga M-E, Rasconi S, Ruetz CR, Rudstam L, Salmaso N, Sapna S, Straile D, Tammeorg O, Twiss MR, Uzarski DG, Ventelä A-M, Vincent WF, Wilhelm SW, Wängberg S-Å, Weyhenmeyer GA (2020) Scientists' warning to humanity: rapid degradation of the world's large lakes. J Great Lakes Res 46(4):686–702

John N, Koehler AV, Ansell BR, Baker L, Crosbie ND, Jex AR (2018) An improved method for PCR-based detection and routine monitoring of geosmin-producing cyanobacterial blooms. Water Res 136:34–40. https://doi.org/10.1016/j.watres.2018.02.041

Keane R (2002) Exotic plant invasions and the enemy release hypothesis. Trends Ecol Evol 17(4):164–170

Kerman M, Battarbee RW, Moss B (2010) Climate change impacts on freshwater ecosystems. Blackwell Publishing Ltd., West Sussex

Khaleel AG, Ismail N, Ahmad-Syazni K (2021) Introduction of invasive peacock bass (*Cichla* spp.), its rapid distribution and future impact on freshwater ecosystem in Malaysia. Croatian J Fish 79(1):33–46

Koh HL, Tan WK, Teh SY and Tay CJ (2019) Water quality simulation for rehabilitation of a eutrophic lake in Selangor, Malaysia. IOP Conf Ser Earth Environ Sci 380:012006

Kosten S, Huszar VLM, Bécares E, Costa LS, van Donk E, Hansson L-A, Jeppesen E, Kruk C, Lacerot G, Mazzeo N, De Meester L, Moss B, Lürling M, Nõges T, Romo S, Scheffer M (2012) Warmer climates boost cyanobacterial dominance in shallow lakes. Glob Change Biol 18(1):118–126

Kumar K, Mella-Herrera RA, Golden JW (2010) Cyanobacterial heterocysts. Cold Spring Harb Perspect Biol 2(4). https://doi.org/10.1101/cshperspect.a000315

Labaj AL, Michelutti N, Smol JP (2018) Annual stratification patterns in tropical mountain lakes reflect altered thermal regimes in response to climate change. Fundam Appl Limnol 191(4):267–275

LaBrie R, Maranger R (2024) Predicting the presence of hypoxic hypolimnia in lakes at large spatial scales. Limnol Oceanogr 69(2):355–366. https://doi.org/10.1002/lno.12488

Lewis WM (1996) Tropical lakes: how latitude makes a difference In: Schiemer F, Boland KT (eds) Perspectives in tropical limnology. SPB Academic Publishing B.V., Amsterdam, pp 43–64

Lewis WM (2000) Basis for the protection and management of tropical lakes. Lakes Reservoir Res Manag 5(1):35–48. https://doi.org/10.1046/j.1440-1770.2000.00091.x

Liu H, Yin Y, Piao S, Zhao F, Engels M, Ciais P (2013) Disappearing lakes in semiarid northern China: drivers and environmental impact. Environ Sci Technol 47(21):12107–12114

Liu B, Zhang X, Tong Y, Ao W, Wang Z, Zhu S, Wang Y (2022) Quantification of nutrient fluxes from sediments of Lake Hulun, China: implications for plateau lake management. Sustainability 15(11):8680. https://doi.org/10.3390/su15118680

Loayza E, Barrientos ACT, Janssens GP (2022) Evidence of microplastics in water and commercial fish from a high-altitude mountain lake (Lake Titicaca). PeerJ 10:e14112. https://doi.org/10.7717/peerj.14112

Lodge DM, Deines A, Gherardi F, Yeo DC, Arcella T, Baldridge AK, Barnes MA, Chadderton WL, Feder JL, Gantz CA, Howard GW (2012) Global introductions of crayfishes: evaluating the impact of species invasions on ecosystem services. Annu Rev Ecol Evol Syst 43:449–472

López-Doval JC, Montagner CC, de Alburquerque AF, Moschini-Carlos V, Umbuzeiro G, Pompêo M (2017) Nutrients, emerging pollutants and pesticides in a tropical urban reservoir: spatial distributions and risk assessment. Sci Total Environ 575:1307–1324. https://doi.org/10.1016/j.scitotenv.2016.09.210

Lorenz S, Pusch MT, Miler O, Blaschke U (2017) How much ecological integrity does a lake need? Managing the shores of a peri-urban lake. Landsc Urban Plan 164:91–98. https://doi.org/10.1016/j.landurbplan.2017.04.007

Lozano V (2021) Distribution of five aquatic plants native to South America and invasive elsewhere under current climate. Ecologies 2:27–42. https://doi.org/10.3390/ecologies2010003

Luis J, Sidek LM, Jajarmizadeh M (2016) Impact of sedimentation hazard at Jor Reservoir, Batang Padang hydroelectric scheme in Malaysia. IOP Conf Ser Earth Environ Sci (IOP Publishing)

Lv S, Li X, Wang R, Wang Y, Dong Z, Zhou T, Liu Y, Lin K, Liu L (2022) Autochthonous sources and drought conditions drive anomalous oxygen-consuming pollution increase in a sluice-controlled reservoir in eastern China. Sci Total Environ 841:156739. https://doi.org/10.1016/j.scitotenv.2022.156739

Mai H, Hoi Bui V, Duong Dao T, Duong Nguyen T, Thien Nguyen D, Khanh Vu T, Hoang ACN, Quan Tran M, Stragy E (2023) Distribution of microplastics in surface water of tropical urban lakes: a case study in Ha Noi, Vietnam. Vietnam J Earth Sci 45(3):388–404. https://vjs.ac.vn/index.php/jse/article/view/18595

Malla-Pradhan R, Suwunwong T, Phoungthong K, Joshi TP, PradhaN BL (2022) Microplastic pollution in urban Lake Phewa, Nepal: the first report on abundance and composition in surface water of lake in different seasons. Environ Sci Pollut Res 29(26):39928–39936

Mallik A, Xavier KM, Naidu BC, Nayak BB (2021) Ecotoxicological and physiological risks of microplastics on fish and their possible mitigation measures. Sci Total Environ 779:146433. https://doi.org/10.1016/j.scitotenv.2021.146433

Mammides C (2020) A global assessment of the human pressure on the world's lakes. Glob Environ Change 63:102084

Marshall BE (2017) An assessment of climate change and stratification in Lake Kariba (Zambia–Zimbabwe). Lakes Reserv Res Manag 22(3):229–240

Massey IY, Al Osman M, Yang F (2020) An overview on cyanobacterial blooms and toxins production: their occurrence and influencing factors. Toxin Rev 1–21. https://doi.org/10.1080/15569543.2020.1843060

Maulidyna A, Alicia F, Agustin HN, Dewi IR, Nurhidayah I, Dewangga A, Kusumaningrum L, Nugroho GD, Jumari J, Setyawan AD (2021) Economic impacts of the invasive species water hyacinth (*Eichhornia crassipes*): case study of Rawa Pening Lake, Central Java, Indonesia. Int J Bonorowo Wetlands 11(1)

May L, Aura CM, Becker V, Briddon CL, Carvalho1, Dobel AJ, Jamwal P, Kamphuis B, Marinho MM, McGowan S, Nandini S, Nyamweya C, Ongore C, Sarma SSS, Wishart MJ (2021) Getting into hot water: water quality in tropical lakes in relation to their utilisation. Int Conf Ocean Earth Sci IOP Conf Ser Earth Environ Sci 789:012021. https://doi.org/10.1088/1755-1315/789/1/012021

Migwi FK, Ogunah JA, Kiratu JM (2020) Occurrence and spatial distribution of microplastics in the surface waters of lake Naivasha, Kenya. Environ Toxicol Chem 39(4):765–774. https://doi.org/10.1002/etc.4677

Miranda DA, Zachritz AM, Whitehead HD, Cressman SR, Peaslee GF, Lamberti GA (2023) Occurrence and biomagnification of perfluoroalkyl substances (PFAS) in lake Michigan fishes. Sci Total Environ 895:164903

Miththapala S (2007) A strategy for addressing issues of aquatic invasive alien species in the Lower Mekong Basin. IUCN: Mekong Wetland Biodiversity Programme and Regional Species Conservation Programme, The World Conservation Union (IUCN), Colombo, Asia

Mitsuo Y, Ohira M, Tsunoda H, Yuma M (2023) Effects of anthropogenic shoreline alteration on fish emigration from small lakes. Limnology 24(3):217–225. https://doi.org/10.1007/s10201-023-00720-x

Moningkey AT, Rampengan MMF, Tumengkol AA, Kumaat JC (2022) Study of bathymetry and sedimentation in Tondano Lake. IOP Conf Ser Earth Environ Sci 986(1):012038

Moshi HA, Shilla DA, Kimirei IA, O'Reilly C, Clymans W, Bishop I, Loiselle SA (2022) Community monitoring of coliform pollution in Lake Tanganyika. PLoS ONE 17(1):e0262881

Moss B, Kosten S, Meerhoff M, Battarbee RW, Jeppesen E, Mazzeo N, Havens K, Lacerot G, Liu Z, Meester LD, Paerl H, Scheffer M (2011) Allied attack: climate change and eutrophication. Inland Waters 1:101–105

Mowe MA, Mitrovic SM, Lim RP, Furey A, Yeo DC (2015) Tropical cyanobacterial blooms: a review of prevalence, problem taxa, toxins and influencing environmental factors. J Limnol 24:205–224. https://doi.org/10.4081/jlimnol.2014.1005

Nakagiri A, Mutyaba A, Onyutha C, Okurut K (2023) Abating eutrophication on urban lakes: a case study of Kabaka's Lake, Uganda. Water Pract Technol 18(8):1970–1979. https://doi.org/10.2166/wpt.2023.116

Nakamura M (1997) Preserving the health of the world lakes. Environment 39(5):16–40

Naqiuddin AS, Rahim KAA, Long SM, Nicholas FFF (2016) The spread of the Australian redclaw crayfish (Cherax quadricarinatus von Martens, 1868) in Malaysia. J Sustain Sci Manag 11(2):31–38

Nasution MIT, Awal SMS, Permana DM (2016) The methods of preventing water hyacinth as aquatic pollution in Lake Toba caused by agricultural waste. Int J Environ Sci Dev 7:630–633

Ng CKC, Abdullah F, Biun H, Ibrahim MK, Mustapha S, Sade A (2017) A working checklist of the freshwater fish diversity for habitat management and conservation work in Sabah, Malaysia, North Borneo. Biodivers J Biol Divers 18(2):560–574. https://doi.org/10.13057/biodiv/d180217

Nilssen JP (1984) Tropical lakes—functional ecology and future development: the need for a process-orientated approach. Hydrobiologia 113:231–242. https://doi.org/10.1007/BF00026611

Nyholt K, Jardine TD, Villamarín F, Jacobi CM, Hawes JE, Campos-Silva JV, Srayko S, Magnusson WE (2022) High rates of mercury biomagnification in fish from Amazonian floodplain-lake food webs. Sci Total Environ 833:155161

Odada EO, Onyando J, Obudho PA (2002) Lake Baringo. ILEC, Kusatsu

Olaniyan LW, Mkwetshana N, Okoh AI (2016) Triclosan in water, implications for human and environmental health. SpringerPlus 5(1):1639. https://doi.org/10.1186/s40064-016-3287-x

Oliver S, Corburn J, Ribeiro H (2019) Challenges regarding water quality of eutrophic reservoirs in urban landscapes: a mapping literature review. Int J Environ Res Pub Health 16(1):40. https://doi.org/10.3390/ijerph16010040

Olson KR, Morton LW (2018) Tonle Sap Lake and River and confluence with the Mekong River in Cambodia. J Soil Water Conserv 73(3):60A

Ogutu-Ohwayo R, Hecky RE (1991) Fish introductions in Africa and some of their implications. Can J Fish Aquat Sci 48(S1):8–12. https://doi.org/10.1139/f91-299

Ogutu-Ohwayo R, Hecky RE, Cohen AS, Kaufman L (1997) Human impacts on the African great lakes. Environ Biol Fishes 50:117–131. https://doi.org/10.1023/A:1007320932349

Onichandran S, Kumar T, Lim YAL, Sawangjaroen N, Andiappan H, Salibay CC, Chye TT, Ithoi I, Dungca JZ, Sulaiman WYW, Ling LY, Nissapatorn V (2013) Waterborne parasites and physico-chemical assessment of selected lakes in Malaysia. Parasitol Res 112:4185–4191

Ostendorp W, Hofmann H, Teufel L, Miler O (2020) Effects of a retaining wall and an artificial embankment on nearshore littoral habitats and biota in a large Alpine Lake. Hydrobiologia 847(2):365–389. https://doi.org/10.1007/s10750-019-04099-8

Otieno D, Nyaboke H, Nyamweya CS, Odoli CO, Aura CM, Outa NO (2022) Water hyacinth (*Eichhornia crassipes*) infestation cycle and interactions with nutrients and aquatic biota in Winam Gulf (Kenya), Lake Victoria. Lakes Reservoirs Sci Policy Manag Sustain Use 27(1):e12391. https://doi.org/10.1111/lre.12391

Ouma S, Ngeranwa J, Juma K, Mburu D (2016) Seasonal variation of the physicochemical and bacteriological quality of water from five rural catchment areas of Lake Victoria basin in Kenya. J Environ Anal Chem 3(170):2

Pacyna JM (2020) Recent advances in Mercury research. Sci Total Environ 738:139955. https://doi.org/10.1016/j.scitotenv.2020.139955

Palmieri A, Shah F, Dinar A (2001) Economics of reservoir sedimentation and sustainable management of dams. J Environ Manag 61(2):149–163. https://doi.org/10.1006/jema.2000.0392

Peh KS-H (2010) Invasive species in Southeast Asia: the knowledge so far. Biodivers Conserv 19:1083–1099

Peeters F, Straile D, Lorke A, Livingstone DM (2007) Earlier onset of the spring phytoplankton bloom in lakes of the temperate zone in a warmer climate. Glob Change Biol 13(9):1898–1909

Pereira AC, Mulligan CN (2023) Practices for eutrophic shallow lake water remediation and restoration: a critical literature review. Water 15:2270. https://doi.org/10.3390/w15122270

Pham-Duc B, Sylvestre F, Papa F, Frappart F, Bouchez C, Crétaux J-F (2020) The Lake Chad hydrology under current climate change. Sci Rep 10(1):5498

Podder A, Sadmani AHMA, Reinhart D, Chang N-B, Goel R (2021) Per and poly-fluoroalkyl substances (PFAS) as a contaminant of emerging concern in surface water: a transboundary review of their occurrences and toxicity effects. J Hazard Mater 419:126361

Pretty JN, Mason CF, Nedwell DB, Bragg RE, Leaf S (2003) Environmental costs of freshwater eutrophication in England and Wales. Environ Sci Technol 37(2):201–208

Quinlan R, Filazzola A, Mahdiyan O, Shuvo A, Blagrave K, Ewins C, Moslenko L, Gray DK, O'Reilly CM, Sharma S (2021) Relationships of total phosphorus and chlorophyll in lakes worldwide. Limnol Oceanogr 66(2):392–404

Rao RS, Liew TS, Yow YY, Ratnayeke S (2018) Cryptic diversity: two morphologically similar species of invasive apple snail in Peninsular Malaysia. PLoS One 13(5):e0196582. https://doi.org/10.1371/journal.pone.0196582

Rayan S, Kaewdonree S, Rangsiwiwat A, Chartchumni B (2021) Distribution of aquatic plants in Nong Han wetland, Thailand. Songklanakarin J Sci Technol 43(1)

Razad AZA, Ahmad SA, Abdullah WHW, Selamat N, Samsudin SH (2023) Reservoir sedimentinflow estimation: case study for Ringletand Jor Reservoir. In: Proceedings of the 2nd international conference on dam safety management and engineering, Singapore. Springer Nature, Singapore

Reddy DHK (2017) Water pollution control technologies. In: Abraham MA (ed) Encyclopedia of sustainable technologies. Elsevier, Oxford, pp 3–22

Redwan M, El-Haddad E (2022) Heavy metal pollution in Manzala Lake sediments, Egypt: sources, variability, and assessment. Environ Monit Assess 194(6):436. https://doi.org/10.1007/s10661-022-10081-0

Reitter C, Petzoldt H, Korth A, Schwab F, Stange C, Hambsch B, Tiehm A, Lagkouvardos I, Gescher J, Hügler M (2021) Seasonal dynamics in the number and composition of coliform bacteria in drinking water reservoirs. Sci Total Environ 787:147539. https://doi.org/10.1101/2021.02.16.428560

Roberts J, Florentine S (2021) A global review of the invasive aquatic weed *Cabomba caroliniana* [A. Gray] (Carolina fanwort): current and future management challenges, and research gaps. Weed Res 62(1):75–84. https://doi.org/10.1111/wre.12518

Roberts S, Adams JK, Mackay AW, Swann GEA, McGowan S, Rose NL, Panizzo V, Yang H, Vologina E, Sturm M (2020) Mercury loading within the Selenga River Basin and Lake Baikal, Siberia. Environ Pollut 259:113814

Rodríguez-Bolaña C, Pérez-Parada A, Tesitore G, Goyenola G, Kröger A, Pacheco M, Gérez N, Berton A, Zinola G, Gil G, Mangarelli A (2023) Multicompartmental monitoring of legacy and currently used pesticides in a subtropical lake used as a drinking water source (Laguna del Cisne, Uruguay). Sci Total Environ 874:162310. https://doi.org/10.1007/s10530-021-02514-7

Rollwagen-Bollens G, Lee T, Rose V, Bollens SM (2018) Beyond eutrophication: Vancouver Lake, WA, USA as a model for assessing multiple, interacting biotic and abiotic drivers of harmful cyanobacterial blooms. Water 10(757). https://doi.org/10.3390/w10060757

Rose V, Bollens SM, Zimmerman J (2020) Effects of grazing and nutrients on phytoplankton blooms and microplankton assemblage structure in four temperate lakes spanning a eutrophication gradient. Water 13(8):1085. https://doi.org/10.3390/w13081085

Sacdal R, Madriaga J, Espino MP (2020) Overview of the analysis, occurrence and ecological effects of hormones in lake waters in Asia. Environ Res 182:109091

Samat AF, Yusoff FM, Arshad A, Ghaffar MA, Nor SM, Magalhaes ALB, Das SK (2016) Reproductive biology of the introduced sailfin catfish *Pterygoplichthys pardalis* (Pisces: Loricariidae) in Peninsular Malaysia. Indian J Fish 63(1):35–41

Saravanakumar K, De Silva S, Santosh SS, Sathiyaseelan A, Ganeshalingam A, Jamla M, Sankaranarayanan A, Veeraraghavan VP, MubarakAli D, Lee J, Thiripuranathar G, Wang MH (2022) Impact of industrial effluents on the environment and human health and their remediation using MOFs-based hybrid membrane filtration techniques. Chemosphere 307(Pt 1):135593. https://doi.org/10.1016/j.chemosphere.2022.135593

Sastraprawira SM, Razak IHA, Shahimi S, Pati S, Edinur HA, John AB, Ahmad A, Kumaran JV, Martin MB, Chong JL, Chowdhury AJK (2020) A review on introduced *Cichla* spp. and emerging concerns. Heliyon 6(11). https://doi.org/10.1016/j.heliyon.2020.e05370

Saturday A, Lyimo TJ, Machiwa J, Pamba S (2021) Spatial and temporal variations of faecal indicator bacteria in Lake Bunyonyi, South-Western Uganda. SN Appl Sci 3(7):697

Sayer CD, Davidson TA, Rawcliffe R, Langdon PG, Leavitt PR, Cockerton G, Rose NL, Croft T (2016) Consequences of fish kills for long-term trophic structure in shallow lakes: implications for theory and restoration. Ecosystems 19:1289–1309. https://doi.org/10.1007/s10021-016-0005-z

Scheffer M (2004) Ecology of shallow lakes. Dordrecht, The Netherlands, Kluwer Academic Publishers

Schooler S, Cabrera-Walsh W, Julien M (2009) *Cabomba caroliniana* gray (Cabombaceae). In: Muniappan R, Reddy GVP, Raman A (eds) Biological control of tropical weeds using arthropods. Cambridge University Press, Cambridge, pp 88–107

Seebens H, Meyerson L, Rahlao S, Lenzner B, Tricarico E, Aleksanyan A, Courchamp F, Keskin E, Saeedi H, Tawake A (2023) Trends and status of alien and invasive alien species. In: Roy HE, Pauchard A, Stoett P, Renard Truong T (eds) Thematic assessment report on invasive alien species and their control of the intergovernmental science-policy platform on biodiversity and ecosystem services. IPBES Secretariat, Bonn, Germany

Shakri A, Yusoff FM, Ismail IS, Toda T (2020) Reduced reproductive capacity in *Moina micrura* Kurz, 1875 exposed to toxic *Microcystis* spp. Asian Fish Sci 33(1):42–49

Sharip Z, Hashim N, Suratman S (2017) Occurrence of organochlorine pesticides in a tropical lake basin. Environ Monit Assess 189(11):560. https://doi.org/10.1007/s10661-017-6274-y

Sharip Z, Mohamad MF (2019) Microbial contamination in urban tropical lentic water bodies and ponds along urbanisation gradient. J Trop Agric Sci 42(1):165–184

Sharip Z, Zaki ATA, Salmah Z (2014) Flooding effects on the population dynamics of *Cabomba furcata* and *Nelumbo nucifera* in a shallow floodplain wetland. Wetlands 34(4):713–723. https://doi.org/10.1007/s13157-014-0536-z

Sharip Z, Yusoff FM, Jamin A (2019) Seasonal water quality and trophic status of shallow lentic waters and their association with water levels. Int J Environ Sci Technol 16(8):4851–4862

Sharip Z, Yusoff FM, Johar Z, Amin MZM, Noh MNM (2020) Inventory and classification of lentic ecosystems in Malaysia. Seri Kembangan, NAHRIM

Sharip Z, Zulkifli MH, Abd Wahab MNF, Johar Z, Amin MZM (2024) Assessing trophic state and water quality of small lakes and ponds in Perak. J Teknologi 86(2):51–59. https://doi.org/10.11113/jurnalteknologi.v86.20566

Sheppard A, Shaw R, Sforza R (2006) Top 20 environmental weeds for classical biological control in Europe: a review of opportunities, regulations and other barriers to adoption. Weed Res 46(2):93–117. https://doi.org/10.1111/j.1365-3180.2006.00497.x

Shinhu RJ, Amasi AI, Wynants M, Nobert J, Mtei KM, Njau KN (2023) Assessing the impacts of land use and climate changes on river discharge towards Lake Victoria. Earth 4(2):365–383

Shuhaimi-Othman M, Lim EC, Mushrifah I (2007) Water quality changes in Chini Lake, Pahang, West Malaysia. Environ Monit Assess 131(1–3):279–292

Sial RA, Chaudhary MF, Abbas ST, Latif MI, Khan AG (2006) Quality of effluents from Hattar Industrial Estate. J Zhejiang Univ Sci B 7(12):974–980. https://doi.org/10.1631/jzus.2006.B0974

Sidabutar T, Srimariana ES, Wouthuyzen S (2020) The potential role of eutrophication, tidal and climatic on the rise of algal bloom phenomenon in Jakarta Bay. IOP Conf Ser Earth Environ Sci 429(1):012021. https://doi.org/10.1088/1755-1315/429/1/012021

Silvino RF, Barbosa FAR (2015) Eutrophication potential of lakes: an integrated analysis of trophic state, morphometry, land occupation, and land use. Braz J Biol 75(3):607–615

Sim ZY, Goh KC, Nur Hanisah S, Mao F, He Y, Gin KYH (2025) Influence of phytoplankton, bacteria and viruses on nutrient supply in tropical waters. J Environ Sci 151:174–186. https://doi.org/10.1016/j.jes.2024.02.032

Siti-Munirah MY, Chew MY (2010) Cabombaceae, a new family record for Peninsular Malaysia. Malayan Nature Journal 62(3): 241-248

Soeprobowati TR (2015) Integrated lake basin management for save Indonesian lake movement. Procedia Environ Sci 23368–374. https://doi.org/10.1016/j.proenv.2015.01.053

Sousa W (2011) *Hydrilla verticillata* (Hydrocharitaceae), a recent invader threatening Brazil's freshwater environments: a review of the extent of the problem. Hydrobiologia 669:1–20. https://doi.org/10.1007/s10750-011-0696-2

Stauffer JR, Chirwa ER, Jere W, Konings AF, Tweddle D, Weyl O (2022) Nile Tilapia, Oreochromis niloticus (Teleostei: Cichlidae): a threat to native fishes of Lake Malawi? Biol Invasions 24:1585–1597. https://doi.org/10.1007/s10530-022-02756-z

Sterckx K, Delandmeter P, Lambrechts J, Deleersnijder E, Verburg P, Thiery W (2023) The impact of seasonal variability and climate change on lake Tanganyika's hydrodynamics. Environ Fluid Mech 23(1):103–123. https://doi.org/10.1007/s10652-022-09908-8

Su L, Xue Y, Li L, Yang D, Kolandhasamy P, Li D, Shi H (2016) Microplastics in Taihu Lake, China. Environ Pollut 216:711–719

Suardy NH, Tahrim NA, Ramli S (2020) Analysis and characterization of microplastic from personal care products and surface water in Bangi, Selangor. Sains Malaysiana 49(9):2237–2249

Suede BC, Wilkens JL, Kennedy AJ (2017) Effects of suspended sediment on early life stages of smallmouth bass (*Micropterus dolomieu*). Arch Environ Contam Toxicol 27(1):119–131

Sulaiman NH, Khalit SI, Sharip Z, Samsudin MS, Azid A (2018) Seasonal variations of water quality and heavy metals in two ex-mining lakes using Chemometric Assessment Approach. Mal J Fund Appl Sci 14(1):67–72

Sulaiman RNR, Bakar AA, Ngadi N, Kahar INS, Nordin AH, Ikram M, Nabgan W (2023) Microplastics in Malaysia's aquatic environment: current overview and future perspectives. Glob Challenges 7(8):2300047. https://doi.org/10.1002/gch2.202300047

Swe T, Lombardo P, Ballot A, Thrane JE, Sample J, Eriksen TE, Mjelde M (2021) The importance of aquatic macrophytes in a eutrophic tropical shallow lake. Limnologica 90:125910. https://doi.org/10.1016/j.limno.2021.125910

Szydlowski DK, Elgin AK, Lodge DM, Tiemann JS, Larson ER (2023) Long-term macrophyte and snail community responses to population declines of invasive rusty crayfish (*Faxonius rusticus*). Ecol Appl 33(3):e2818. https://doi.org/10.1002/eap.2818

Tahoun U (2023) Microbial pollution in African lakes. In: El-Sheekh M, Elsaied HE (eds) Lakes of Africa, chap 14. Elsevier, Amsterdam, pp 403–427

Tan KS, Morton B (2006) The invasive Caribbean bivalve *Mytilopsis sallei* (Dreissenidae) introduced to Singapore and Johor Bahru, Malaysia. Raffles Bull Zool 54(2):429–434

Tay CJ, Koh HL, Mohd Hafiz M, Teh SY (2022) Assessing the role of internal phosphorus recycling on eutrophication in four lakes in China and Malaysia. Eco Inform 72:10183

Tegene S, Ayele N (2014) Prevalence and intensity of water hyacinth infestation in the water bodies of Rift Valley, Ethiopia. J Agric Nat Resour Sci 1(2):118–126

Tranvik LJ (2021) Acidification of inland waters: this article belongs to Ambio's 50th anniversary collection. Theme: acidification. Ambio 50(2):261–265. https://doi.org/10.1007/s13280-020-01441-6

Tyner EH, Jensen P, Ngochera M, Bootsma HA (2018) *Escherichia coli* (*E. coli*) distribution in the Lake Malawi nearshore zone. J Great Lakes Res 44(6):1281–1288. https://doi.org/10.1016/j.jglr.2018.10.011

Valsecchi S, Babut M, Mazzoni M, Pascariello S, Ferrario C, De Felice B (2021) Per- and polyfluoroalkyl substances (PFAS) in fish from European lakes: current contamination status, sources, and perspectives for monitoring. Environ Toxicol Chem 40(3):658–676. https://doi.org/10.1002/etc.4815

Verburg P, Hecky RE (2009) The physics of the warming of Lake Tanganyika by climate change. Limnol Oceanogr 54(6, Part 2):2418–2430. https://doi.org/10.4319/lo.2009.54.6_part_2.2418

Vilhena LC, Hillmer I, Imberger J (2010) The role of climate change in the occurrence of algal blooms: Lake Burragorang, Australia. Limnol Oceanogr 55(3):1188–1200. https://doi.org/10.4319/lo.2010.55.3.1188

Walsby AE (1981) Cyanobacteria: planktonic gas-vacuolate forms. In: Starr MP, Stolp H, Trüper HG, Balows A, Schlegel HG (eds) The prokaryotes. Springer, Berlin. https://doi.org/10.1007/978-3-662-13187-9_10

Wang W, Lee X, Xiao W, Liu S, Schultz N, Wang Y, Zhang M, Zhao L (2018) Global lake evaporation accelerated by changes in surface energy allocation in a warmer climate. Nat Geosci 11(6):410–414. https://doi.org/10.1038/s41561-018-0114-8

Watiniasih NL, Hendrawan IG, Nuarsa IW, Wiradana PA (2023) Investigation of microplastic contamination in sediments, water and aquatic biota in lake beratan, tabanan regency, bali province – Indonesia. J Ecol Eng 24(3):323–332

Wiley DY, McPherson RA (2023) The role of climate change in the proliferation of freshwater harmful algal blooms in inland waterbodies of the United States. Earth Interact

Williams WD (2002) Environmental threats to salt lakes and the likely status of inland saline ecosystems in 2025. Environ Conserv 29(2):154–167. https://doi.org/10.1017/S0376892902000103

Williams AE, Duthie HC, Hecky RE (2005) Water hyacinth in Lake Victoria: why did it vanish so quickly and will it return? Aquat Bot 81(4):300–314. https://doi.org/10.1016/j.aquabot.2005.01.003

Wondie A, Mengistu S, Vijverberg J, Dejen E (2007) Seasonal variation in primary production of a large high altitude tropical lake (Lake Tana Ethiopia): effects of nutrient availability and water transparency. Aquat Ecol 41(2):195–207. https://doi.org/10.1007/s10452-007-9080-8

Woolway RI, Merchant CJ (2022) Lakes in hot water: the impacts of a changing climate on aquatic ecosystems. BioScience 72(11):1050–1061. https://doi.org/10.1038/s41561-019-0322-x

Woolway RI, Kraemer BM, Lenters JD, Merchant CJ, O'Reilly CM, Sharma S (2020) Global lake responses to climate change. Nat Rev Earth Environ 1(8):388–403. https://doi.org/10.1038/s43017-020-0067-5

Yang Y, Cao X, Zhang M, Wang J (2015) Occurrence and distribution of endocrine-disrupting compounds in the Honghu Lake and East Dongting Lake along the Central Yangtze River, China. Environ Sci Pollut Res 22(22):17644–17652. https://doi.org/10.1007/s11356-015-4980-y

Yang H, Macario-González L, Cohuo S, Whitmore TJ, Salgado J, Peréz L, Schwalb A, Rose NL, Holmes J, Riedinger-Whitmore MA, Hoelzmann P, O'Dea A (2023) Mercury pollution history in tropical and subtropical American lakes: multiple impacts and the possible relationship with climate change. Environ Sci Technol 57(9):3680–3690

Yusoff FM (1990) Observation on the effects of wild tilapia population on the growth of cultured fish. J Biosci 1:101–105

Yusoff FM, Ismail MS, Law AT, Anton A (1997) Nutrient availability in tropical lakes of varying trophy. In: Williams WD, Sladeckova A (eds) International association of theoretical and applied limnology, Vol 26, Pt 2. Stuttgart, E Schweizerbart'sche Verlagsbuchhandlung. pp 420–426

Yusoff FM, Lock MA (1995) Thermal stratification and its role in controlling eutrophication in tropical a reservoir, Malaysia. In: Timotius KH, Goltenboth F (eds) Tropical limnology, vol II. Tropical lakes and reservoirs, Salatiga, Indonesia, pp 277–285

Yusoff FM, Happey-Wood CM, Anton A (1998) Vertical and seasonal distribution of phytoplankton in a tropical reservoir, Malaysia. Int Rev Hydrobiol 83:121–134

Yusoff FM, Sharip Z, Zakaria S, Noh MNM (2020) Major issues, problems and solutions for lentic water bodies inventory and classification of lentic ecosystems in Malaysia. In: Sharip Z, Yusoff FM, Johar Z, Amin MZM, Noh N (eds) Seri Kembangan, NAHRIM, pp 73–93

Yusuf MA, Yaqin K, Wicaksono EA, Tahir A (2022) Abundance and characteristics of microplastics in Lake Towuti, East Luwu, South Sulawesi. AACL Bioflux 15(2):621–631

Zhang J, Wang C, Jiang X, Song Z, Xie Z (2020) Effects of human-induced eutrophication on macroinvertebrate spatiotemporal dynamics in Lake Dianchi, a large shallow plateau lake in China. Environ Sci Pollut Res 27(12):13066–13080

Zhang Z, Fan X, Peijnenburg WJGM, Zhang M, Sun L, Zhai Y, Yu Q, Wu J, Lu T, Qian H (2021) Alteration of dominant cyanobacteria in different bloom periods caused by abiotic factors and species interactions. J Environ Sci 99:1–9

Zhang Y, Cao J, Ke T, Tao Y, Wu W, Wang P (2022) Pollution characteristics and risk prediction of endocrine disruptors in lakes of Wuhan. Toxics 10(2):93

Zhou J, Leavitt PR, Zhang Y, Qin B (2022) Anthropogenic eutrophication of shallow lakes: is it occasional? Water Res 221:118728. https://doi.org/10.1016/j.watres.2022.118728

Zhu J, Yu D, Xu X (2015) The phylogeographic structure of *Hydrilla verticillata* (Hydrocharitaceae) in China and its implications for the biogeographic history of this worldwide-distributed submerged macrophyte. BMC Evol Biol 15:1–11. https://doi.org/10.1186/s12862-015-0381-6

Zhu Y, Feng W, Liu S, He Z, Zhao X, Liu Y, Guo J, Giesy JP, Wu F (2018) Bioavailability and preservation of organic phosphorus in lake sediments: insights from enzymatic hydrolysis and 31P nuclear magnetic resonance. Chemosphere 211:50–61. https://doi.org/10.1016/j.chemosphere.2018.07.134

Zulkafli AR, Mustafa A, Mohammad Noor AA (2015) Fish diversity of Tembeling and Pahang rivers Pahang, Malaysia. Check List 11(5):1–6

Chapter 4
Groundwater Interaction with Lakes

Rachmat Fajar Lubis, Priyo Hartanto, and Nico Anatoly

Understanding the geomorphology of lake basins is essential for various scientific disciplines, including limnology (the study of inland waters), geology, and environmental science. It provides insights into the history, processes, and environmental conditions that shape lakes and contributes to our understanding of the dynamic interactions between geology, climate, and hydrology in lake ecosystems. The interaction between lakes and groundwater can pose several challenges, often intertwining hydrological, ecological, and socioeconomic issues. Base on many reference, there are common problems associated with lake–groundwater interaction such as:

1. *Water Level Fluctuations*: Changes in groundwater levels can influence the water levels of lakes. If groundwater recharge to the lake decreases, it can lead to a decline in lake levels, affecting water availability for various uses such as drinking water supply, irrigation, and recreation. Conversely, excessive groundwater discharge into the lake can cause flooding in surrounding areas.
2. *Ecological Impacts*: Alterations in groundwater flow patterns can disrupt the ecological balance of lakes and their surrounding ecosystems. For example, changes in nutrient input from groundwater can lead to eutrophication, excessive algae growth, and depletion of oxygen levels, harming aquatic life and biodiversity.
3. *Water Quality Degradation*: Groundwater contamination can adversely affect lake water quality. Pollutants such as agricultural runoff, industrial discharges, and septic system leakage can infiltrate into groundwater and subsequently enter lakes, posing risks to human health, aquatic organisms, and recreational activities. These also can affect opposingly when the lake is contaminated and influence the groundwater.

R. F. Lubis (✉) · P. Hartanto · N. Anatoly
Research Center for Limnology and Water Resources BRIN, Jakarta, Indonesia
e-mail: rach007@brin.go.id

N. Anatoly
Master Student of Groundwater Engineering ITB, Bandung, Indonesia

© The Author(s), under exclusive license to Springer Nature Singapore Pte Ltd. 2025

A. C. Santos-Borja et al. (eds.), *Progress on Ecosystem Restoration of Tropical Inland Waters*, https://doi.org/10.1007/978-981-96-2284-9_4

4. *Subsidence and Land Instability*: Excessive groundwater extraction, particularly in areas where lakes are closely connected to aquifers, can lead to land subsidence and instability. This phenomenon, known as land subsidence, can result in infrastructure damage, increased flood vulnerability, and habitat loss.
5. *Saltwater Intrusion*: In coastal areas where freshwater lakes interact with saline groundwater, over-pumping of groundwater or sea-level rise can induce saltwater intrusion into lakes. This can compromise freshwater resources, affect agricultural productivity, and threaten ecosystems adapted to freshwater conditions.
6. *Climate Change Effects*: Climate change exacerbates the complexities of lake–groundwater interactions. Altered precipitation patterns, temperature variations, and shifts in hydrological cycles can influence groundwater recharge rates, lake levels, and water quality dynamics, posing additional challenges for water resource management and adaptation.

The geomorphology of lake basins refers to the study of the physical features and landforms that characterize lakes' shape, structure, and origin. Lake basins exhibit diverse geomorphic characteristics, and a combination of geological, climatic, and tectonic processes influences their formation. In Indonesia, there are six types of the geomorphology of lake basins following previous classification (Verstappen 1983; Wetzel 2001).

1. *Tectonic activities*, such as Rift Lakes and Tectonic Basins, involve the formation of Rift Lakes in tectonic rift zones where the Earth's crust is pulling apart, creating elongated and often deep basins. Tectonic Basins result from subsidence or uplift associated with tectonic plate movements.
2. *Volcanic activities*, such as Volcanic Calderas and Crater Lakes. Volcanic Calderas are circular basins formed by the collapse of a volcanic summit, creating depressions filled with water. Volcanic Crater Lakes are lakes within volcanic craters, often resulting from explosive volcanic activity.
3. *Karst landscapes*, such as Karst Basins and Karst Spring Lakes. Karst Basins form in limestone or other soluble rocks, characterized by features like sinkholes, disappearing streams, and underground drainage. Karstic Springs are lakes fed by karstic springs emerging from underground cave systems.
4. *Riverine processes*, such as Oxbow Lakes and River-Dominated Basin Lakes. Oxbow Lakes form when a meandering river changes its course, leaving behind a curved, disconnected basin. River-dominated basins are lakes influenced by sedimentation and erosional processes associated with rivers.
5. *Sedimentary basin lakes* are bodies of water that accumulate sediments over time, leading to changes in bathymetry and lake morphology.
6. *Artificial basins*, such as reservoirs and fishponds. A reservoir is a human-made basin created by damming rivers for various purposes such as water supply, hydropower, or flood control. Fishponds are artificial basins constructed for aquaculture and fish farming.

4.1 Common Methods for Studying Groundwater-Lake Interactions

Investigating groundwater–lake exchange is crucial for understanding water flow dynamics and quality in aquatic systems. There are some methods commonly used for studying groundwater–lake exchange:

4.1.1 *Hydrogeological Surveys*

Conducting a hydrogeological survey is an essential step in investigating groundwater–lake exchange. A hydrogeological survey involves gathering information about an area's geological, hydrological, and hydrochemical characteristics to understand the groundwater flow dynamics and its interaction with a lake. Here are key components and methods involved in a hydrogeological survey for investigating groundwater–lake exchange:

- Geological Mapping: Objective: Identify and map geological formations in the study area. Methods: Conduct geological field surveys, analyze existing geological maps, and collect rock and sediment samples for analysis.
- Geophysical Mapping is crucial in investigating groundwater–lake exchange by providing information about subsurface properties without requiring invasive techniques. These geophysical methods can be used individually or in combination with a comprehensive understanding of groundwater–lake exchange. Integrating geophysical data with other hydrogeological and geochemical information enhances the accuracy and reliability of the investigation. Each method has its strengths and limitations, so selecting the appropriate combination depends on the specific characteristics of the study area.
- Borehole and Well Installation: Objective: Obtain subsurface information and install monitoring points for groundwater observation. Methods: Drill boreholes and install wells at strategic locations around the lake to measure water levels, collect water samples, and assess subsurface geology.
- Aquifer Testing: Objective: Determine hydraulic properties of the aquifer, such as transmissivity and hydraulic conductivity. Methods: Perform aquifer tests, including pumping and slug tests, to measure how the aquifer responds to water levels and pressure changes.
- Water Level Monitoring: Objective: Monitor changes in groundwater levels over time. Methods: Install piezometers and observation wells to continuously measure groundwater levels, allowing for the identification of groundwater flow patterns and fluctuations.
- Mapping Hydrogeological Units: Objective: Classify hydrogeological units based on geological and hydrological characteristics. Methods: Integrate geological, hydrogeological, and hydrochemical data to create hydrogeological maps, highlighting different subsurface units and their properties.

By integrating these methods, a hydrogeological survey provides a comprehensive understanding of the groundwater–lake interaction, including the sources, flow pathways, and hydrochemical characteristics of groundwater discharging into the lake. This information is crucial for effective water resource management and the preservation of lake ecosystems.

4.1.2 *Water Chemistry Analysis*

Water chemistry analysis is a valuable method for investigating groundwater–lake exchange, providing insights into lake water and groundwater composition. The analysis involves measuring various chemical parameters that reveal the sources, pathways, and interactions between groundwater and lake water. Here are key components of water chemistry analysis for investigating groundwater–lake exchange:

- Major ions include ions such as chloride (Cl^-), sulfate ($SO_4{}^{2-}$), bicarbonate ($HCO_3{}^-$), calcium (Ca^{2+}), magnesium (Mg^{2+}), sodium (Na^+), and potassium (K^+). Differences in major ion concentrations between lake water and groundwater can indicate potential contributions from different sources.
- Stable Isotopes of oxygen ($\delta^{18}O$) and hydrogen (δ^2H) are commonly analyzed. Isotopic signatures help trace the origin and movement of water, differentiating between lake water and various groundwater sources.
- Nutrients such as nitrate ($NO_3{}^-$), phosphate ($PO_4{}^{3-}$), and ammonia ($NH_4{}^+$). Elevated nutrient levels may suggest inputs from agricultural runoff or anthropogenic sources, influencing groundwater–lake exchange.
- pH and Alkalinity: Variations in pH and alkalinity can indicate the presence of dissolved substances and potential reactions with groundwater minerals.
- Trace Elements: Trace elements like iron (Fe), manganese (Mn), arsenic (As), and others. Interpretation: Elevated levels of certain trace elements may signify geological influences or contamination sources affecting groundwater–lake interactions.
- Redox Potential: Parameters: Redox potential (Eh). Redox conditions provide information about the oxidative or reducing environment, influencing the mobility of certain elements and redox-sensitive species.
- Total Dissolved Solids (TDS) and Conductivity: TDS and electrical conductivity. Changes in TDS and conductivity values can indicate variations in water salinity and dissolved ion concentrations.
- Isotopic Tracers (Environmental Tracers): Tracers like chlorofluorocarbons (CFCs), tritium (^3H), and carbon-14 (^{14}C). Isotopic tracers help estimate groundwater age, identify sources, and determine the relative contributions of different water masses.
- Microbial and Biological Indicators: Presence of microbial indicators like fecal coliforms or biological indicators like phytoplankton. Microbial contamination indicators can help assess groundwater's impact on lake water quality.

- Dissolved Gases by measurement of all dissolved gases, including oxygen (O_2), carbon dioxide (CO_2), and methane (CH_4). Variations in dissolved gas concentrations provide information on biological activity, redox conditions, and potential contributions from deeper groundwater.

By systematically analyzing these water chemistry parameters, we will gain a comprehensive understanding of the hydrogeochemical processes and interactions between groundwater and lake water. The results help identify groundwater discharge areas, assess water quality, and contribute to the aquatic ecosystem's overall management and protection.

4.1.3 Tracer Studies

Tracing methods are valuable tools for investigating groundwater–lake exchange as they involve the introduction of tracers into the subsurface to track the movement of water. These methods can help identify flow pathways, quantify flow rates, and determine the extent of groundwater–lake interaction. Here are several tracing methods commonly used in such investigations:

- Dye Tracing as non-toxic fluorescent dyes are injected into selected wells or boreholes. Monitor the appearance of the dye in the lake water to trace the movement of groundwater. It will provide a visual representation of flow pathways and help quantify the timing of groundwater discharge.
- Isotopic Tracers are naturally occurring isotopes or artificially introduced isotopes that are used to trace the movement of water. Examples: Oxygen-18, deuterium, and environmental tracers like chlorofluorocarbons (CFCs) or tritium. Analyzing the isotopic composition of lake water and groundwater to identify the source and pathways of water movement will offer insights into the origin and age of groundwater contributing to lake exchange.
- Tracer Gases, such as helium or sulfur hexafluoride, are injected into the subsurface. Measure the concentration of the tracer gases in the lake water to determine groundwater movement can be used to estimate groundwater flow rates and travel times.
- Temperature Tracing, such as monitoring changes in lake water temperature to trace the movement of the injected water. This will provide information on flow pathways and help estimate flow rates based on thermal gradients.
- Particle Tracing, such as fluorescent or inert tracers, into the subsurface. Track the movement of particles in the lake water to understand groundwater flow pathways and offer insights into groundwater–lake exchange's transport characteristics.
- Radioactive Tracers such as monitoring the presence and concentration of the radioactive tracer (Radon) in the lake water to trace groundwater flow can provide quantitative data on flow rates and direction.

When using tracing methods, it is essential to consider the environmental impact and potential risks associated with the tracers used. Additionally, combining different tracing techniques with other hydrogeological and geophysical methods can enhance the accuracy and reliability of the investigation. The choice of method depends on the study's specific objectives and the hydrogeological system's characteristics.

4.1.4 Modeling

The steps are to develop numerical models to simulate groundwater flow in the study area, use groundwater flow modeling software to simulate different scenarios, validate model results against field data, and predict the potential impact of groundwater on lake levels.

Above all, it is important to note that combining all these methods is often more effective in providing a comprehensive understanding of groundwater–lake exchange. Additionally, the choice of methods may depend on the specific characteristics of the study area, such as geological conditions, hydrological patterns, and the scale of investigation. Integrating field measurements with modeling approaches can further enhance the accuracy of the findings.

4.1.5 Groundwater and Lake Exchange Conceptual Model

Lakes are an example of surface water bodies that are unmoving or moving slowly (Gupta and Kumar Sharma 2023). They can receive subsurface inflow, provide outflow to the aquifers across their entire bed, or represent both of these characteristics at various locations (Winter et al. 1998). Based on Winter et al. (1998) and Woessner (2020), groundwater and lake interaction were divided into five cross-sectional conceptual models (Fig. 4.1).

4.2 Applications

The relationship between groundwater and lakes has been investigated across the countries. In tropical areas, the interaction between lakes and groundwater can present unique challenges due to the specific climatic, hydrological, and ecological conditions found in these regions. Based on many case studies (Table 4.1), studies regarding the interaction between groundwater and lakes in tropical regions are concentrated on a limited number of aspects for some problems such as:

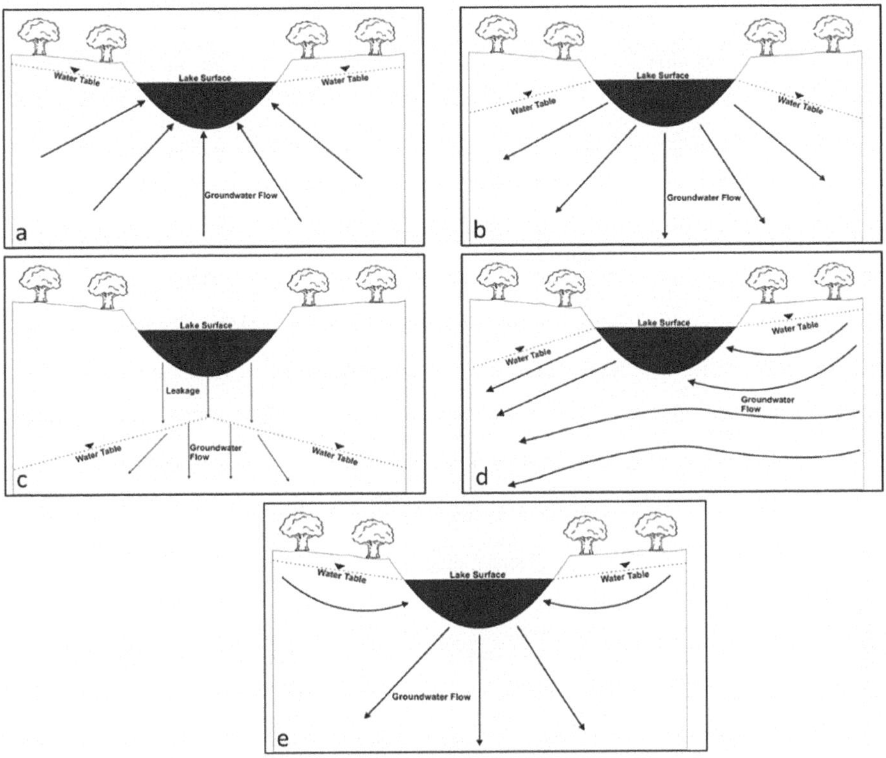

Fig. 4.1 Groundwater–lake interaction conceptual models. **a** Effluent or gaining lake. **b** Influent or losing lake. **c** Influent or losing lake perched above the water table. **d** Flow-through lake. **e** Mixed exchange lake (Winter et al. 1998; Woessner 2020)

1. *Water Quantity Issues*: Tropical regions often experience pronounced wet and dry seasons, leading to seasonal fluctuations in groundwater levels and lake water levels. Changes in precipitation patterns and intensity can affect groundwater recharge rates, leading to water scarcity or excess during different times of the year.
2. *Drought Vulnerability*: Tropical lakes and their associated groundwater systems are susceptible to droughts, which can be exacerbated by climate change. Extended dry periods can lead to reduced groundwater recharge and diminished lake levels, impacting water availability for drinking, agriculture, and ecosystems and exacerbating conflicts over water resources.
3. *Land Use Changes and Deforestation*: Deforestation and land use changes, common in tropical regions due to agriculture, urbanization, and infrastructure development, can alter hydrological processes and increase surface runoff. This can affect groundwater recharge rates, sedimentation in lakes, and nutrient loading, leading to water quality degradation and loss of biodiversity.

Table 4.1 Groundwater–lake interaction studies in Indonesia

References	Study area	Aim of study	Methods	Groundwater–lake interaction type
Marsiano et al. (2018)	ISTN and Babakan Lake	To describe the interrelationship of shallow groundwater and the lake water from ISTN Lake and Babakan Lake	Stable isotopes of oxygen ($\delta^{18}O$) and hydrogen ($\delta^{2}H$)	Influent or losing the lake
Syafalni et al. (2019)	ISTN Lake	To evaluate shallow groundwater on the bank of the ISTN Lake through lake bank filtration	Water chemistry analysis (pH, TDS) and microbiological analysis	Influent or losing the lake
Amri (2021)	Lake Sunter	To determine groundwater and lake interaction between 3 lakes (West Sunter Lake, East Sunter Lake, and North Sunter Lake)	Hydrogeological survey	Effluent or gaining lake
Apip et al. (2021)	Lake Maninjau	To quantify groundwater influx into the lake	Hydrogeological surveys and water balance model	Effluent or gaining lake
Koyari et al. (2022)	Lake Sentani	To develop an interaction model for rainfall, lake water, and groundwater in Lake Sentani, Papua, Indonesia	Hydrogeological surveys and modeling	Effluent or gaining lake
Hartanto et al. (2022)	Rawadanau Basin	To determine the source of hot springs in the Rawadanau Basin	Hydrogeological surveys, hydrochemical, and stable isotopes of oxygen ($\delta^{18}O$) and hydrogen ($\delta^{2}H$)	Mixed exchange lake
Lubis et al. (2021)	Lake Kelimutu	To determine the causes of the decrease in Lake Kelimutu's water level and investigate the correlation between groundwater and lakes	Satellite image data, hydrogeological surveys, and water chemistry analysis	Mixed exchange lake

(continued)

Table 4.1 (continued)

References	Study area	Aim of study	Methods	Groundwater–lake interaction type
Pratikno and Sidauruk (2015)	Lake Toba	To determine the interaction between Lake Toba and the spring around the lake	Stable isotopes of oxygen ($\delta^{18}O$) and hydrogen ($\delta^{2}H$)	No relation between groundwater and lake (isolated lake)
Sidauruk et al. (2018)	Lake Toba	To characterize all hydrological units in Lake Toba, including precipitation, inflows, outflows, and lake water	Stable isotopes of oxygen ($\delta^{18}O$) and hydrogen ($\delta^{2}H$)	No relation between groundwater and lake (isolated lake)

There is still a need for further development in researching the anticipation of the effects resulting from change of environmental services impact due to shifts in the equilibrium of this relationship based on climate change or anthropogenic activity.

Isolated lake conditions have been added to the groundwater–lake interaction type in Indonesia, especially in Lake Toba. Lake Toba is Indonesia's largest lake and the world's largest volcanic lake. The lake is approximately 100 km long and 30 km wide (Chesner 2012). For more than a century, numerous researchers have been fascinated by the geological history and formation of the Toba Caldera. Lake Toba began filling with water shortly after the eruption, as the caldera fill tuffs cooled. Based on current precipitation rates (2100 mm/year) and evaporation rates (1350 mm/year), the lake may have filled at least 750 mm/year (Chesner 2012). The study found that the springs of Panahatan, Manigom, Balata, Mualgoit, and Bahtio, which are located around Toba Lake, have no relationship with lake water, and they do not originate from the water of Toba Lake (Pratikno and Sidauruk 2015). This indication can be seen from the analysis of the isotope ^{18}O and ^{2}H each spring, which are more negative or depleted than Toba Lake's water. The conceptual model for the isolated lake is visualized below (Fig. 4.2).

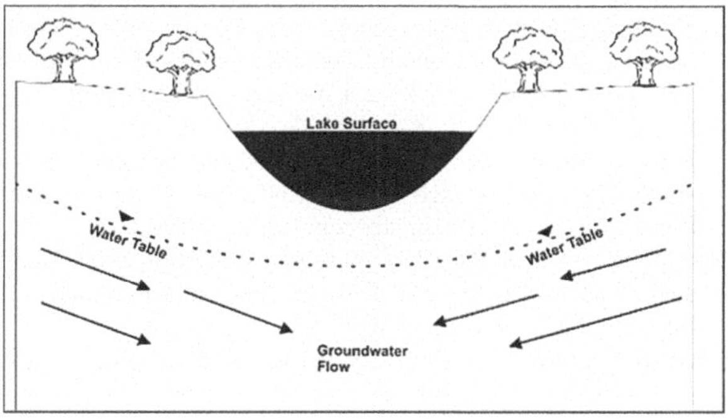

Fig. 4.2 Groundwater–lake interaction for isolated lake conceptual models

4.3 Remarks

Research on groundwater and lake exchange in tropical areas is an important and multidisciplinary field that involves hydrology, geology, ecology, and environmental science. Based on current research findings in Indonesia, four interaction models are identified, with one of the hypothetical models being the isolated model associated with its geological conditions. The development of research on the interaction between lakes and groundwater in tropical regions still requires further research collaboration between scientists, researchers, and local communities is crucial in addressing the complexities of groundwater–lake interactions in Indonesia as a tropical area. The findings from this research contribute not only to scientific knowledge but also inform conservation and management strategies for these vital water resources in Indonesia as a tropical region.

References

Amri MA (2021) The influence of surface water on groundwater based on shallow groundwater level data in the Sunter Lake area and its surroundings (in Indonesia Language). PETRO 10(3):176–179. https://doi.org/10.25105/petro.v10i3.10828

Apip, Handoko U, Ridwansyah I, Santoso A, Subehi L (2021) Water balance model aided in estimating net groundwater inflow at Lake Maninjau, West Sumatra—Indonesia. IOP Conf Ser Earth Environ Sci 789(1):012029. https://doi.org/10.1088/1755-1315/789/1/012029

Chesner CA (2012) The Toba Caldera complex. Quatern Int 258:5–18. https://doi.org/10.1016/j.quaint.2011.09.025

Gupta R, Kumar Sharma P (2023) A review of groundwater-surface water interaction studies in India. J Hydrol 621:129592. https://doi.org/10.1016/j.jhydrol.2023.129592

Hartanto P, Alam BYCSSS, Lubis RF, Ismawan I, Iskandarsyah TYWM, Sendjaja YA, Hendarmawan H (2022) The application of hydrogeochemical and stable isotope data to decipher the origin and evolution of hot springs in the Rawadanau Basin, Indonesia. Geothermics 105:102506. https://doi.org/10.1016/j.geothermics.2022.102506

Koyari E, Bisri M, Sisinggih D, Asmaranto R (2022) Interaction model of rainfall, lake, and ground waters. J Hunan Univ Nat Sci 49(9):50–57. https://doi.org/10.55463/issn.1674-2974.49.9.7

Lubis RF, Fajar O, Nugraha GU, Purwoko W, Sukmayadi D, Riyadi A, Bakti H, Purnomo BJ, Yudhanagara D, Ridwansyah I (2021) Decline of Kelimutu lake water levels: the relationship between groundwater and the lake (in Indonesia Language). Proceeding of 5rd Annual Meeting Indonesia Groundwater Expert Association (PAAI) Indonesia

Marsiano W, Syafalni S, Kuswaya W, Djamhuri MF, Pratikno B (2018) The interrelationship between ISTN lake, Babakan lake and the surrounding wells of shallow groundwater using stable isotopes δ^2H and δ^{18}O for the lake bank filtration potential. MATEC Web Conf 195:05005. https://doi.org/10.1051/matecconf/201819505005

Pratikno B, Sidauruk P (2015) Studying the relationship between Lake Toba water and water from several springs near Lake Toba (in Indonesia Language). J Appli Sci Isot Radiat 11(2):153. https://doi.org/10.17146/jair.2015.11.2.2798

Sidauruk P, Pratikno B, Pujiindiyati ER (2018) Isotopic characterization of precipitation, inflow, and outflow of Lake Toba as a first assessment of lake water balance study. Atom Indo 44(1):1. https://doi.org/10.17146/aij.2018.547

Syafalni S, Kuswaya W, Marsiano W, Zulkafli D (2019) Assessment of shallow groundwater on the bank of the ISTN lake through lake bank filtration based on aquifer properties, pH, total

dissolved solids (TDS), and microbiological analysis. MATEC Web Conf 276:04011. https://doi.org/10.1051/matecconf/201927604011
Verstappen HT (1983) Applied geomorphology: geomorphological surveys for environmental development. Amsterdam et New York, Elsevier
Wetzel RG (2001) Limnology: lake and river ecosystems, 3rd edn. Academic Press, San Diego
Winter TC, Harvey JW, Franke OL, Alley WM (1998) Ground water and surface water: a single resource
Woessner WW (2020) Groundwater-surface water exchange. The Groundwater Project, Waterloo, Ontario

Chapter 5
Climate, Lakes, and Rivers—Correlative Interactions

F. Heru Widodo, Samsul Bahri, Mahally Kudsy, Sunu Tikno, and Tukiyat

The escalating climate change phenomenon has rendered the twenty-first century a global challenge for present-day civilization (Singh et al. 2021). Climate warming directly impacts the physicochemical processes in lakes, which are well-researched. In an era marked by global climate change, it is essential to understand the complex interactions between climate, lakes, and rivers and their impacts on tropical aquatic ecosystems. The climate change problem is becoming more acute (Bazarzhapov et al. 2023). Climate, lakes, and rivers are interconnected components of Earth's ecosystems, and their interactions can have significant implications for natural processes and human activities. The interaction of correlative relationships between climate, lakes, and rivers covers:

1. Water temperature and quality
2. River flow, as well as
3. Extreme weather events.

Warmer temperatures can affect the thermal stratification of lakes, thus impacting the distribution of oxygen and nutrients. In contrast, climate change can alter rainfall patterns, affecting river flow and water quality. Likewise, rising temperatures contribute to melting glaciers, thus affecting river flow and water availability downstream. However, increasing frequency and intensity of storms due to climate change can cause flooding, impacting lakes and rivers. Human elements, geomorphology, and climate change around watersheds (which include basins downstream, middle, and upstream) impact the sedimentation process that occurs (Wahab et al. 2023).

Important implications of the interaction and correlation between climate, rivers, and lakes in the context of the environment and human life include:

1. Rainfall and Flood Patterns

F. H. Widodo (✉) · S. Bahri · M. Kudsy · S. Tikno · Tukiyat
Research Center for Limnology and Water Resources, National Research and Innovation Agency (BRIN), Jakarta, Indonesia
e-mail: flor003@brin.go.id

© The Author(s), under exclusive license to Springer Nature Singapore Pte Ltd. 2025
A. C. Santos-Borja et al. (eds.), *Progress on Ecosystem Restoration of Tropical Inland Waters*, https://doi.org/10.1007/978-981-96-2284-9_5

Climate affects rainfall patterns, affecting the volume of river and lake water. Climate change can increase the frequency and intensity of floods, hurting surrounding infrastructure and human life.

2. Clean Water Availability

   Rivers and lakes are essential clean water sources for human life and ecosystems. Climate change can affect the availability of water in rivers and lakes, which can impact the supply of clean water for residents.

3. Freshwater Ecosystems

   Rivers and lakes are habitats for various types of freshwater flora and fauna. Climate change can affect these ecosystems, including water temperature, oxygen availability, and migration patterns of aquatic animals.

4. Human Activities

   The interaction between climate, rivers, and lakes can also affect human activities such as agriculture, fishing, and transportation. Climate change can affect agricultural and fisheries productivity and the accessibility of rivers and lakes as transportation routes.

5. Public Health

   Water quality from rivers and lakes can affect the health of people who rely on these water sources for daily needs. Climate change can also affect the spread of water-related diseases.

6. Environmental Sustainability

   The harmony between climate, rivers, and lakes is essential to maintain environmental sustainability. Uncontrolled climate change can threaten the sustainability of freshwater ecosystems and hurt human life around them.

Understanding the interactions and correlations between climate, rivers, and lakes is essential to managing these natural resources sustainably and protecting the environment for generations to come. The interaction of climate change and increased anthropogenic water uptake is expected to alter surface water availability and transport of carbon (C), nitrogen (N), and phosphorus (P) in river networks (Atkinson et al. 2023). Surface water dynamics are sensitive to climate change and anthropogenic activity, providing essential feedback to both processes (Chen et al. 2023). The decline in the consistent expanse of many rivers and lakes often corresponds with increased seasonal fluctuations within these water bodies (Sayre et al. 2023).

Threats and problems of climate change can result in (1) changes in rainfall patterns, temperature changes, and extreme weather events that can pose a threat to the stability of lakes and rivers and can disrupt ecosystems, affect flora and fauna that depend on certain temperatures and rainfall conditions, (2) water scarcity due to changes in rainfall and increased evaporation due to higher temperatures can cause water scarcity in lakes and rivers, thus impacting ecosystems and water supplies for humans, (3) pollution due to climate change that can exacerbate pollution problems in lakes and rivers by affecting transportation and the fate of pollutants. Warmer temperatures, for example, can promote the growth of harmful algae; (4) habitat loss due to changes in water levels and flow patterns; and temperature changes can lead to the loss of habitat of aquatic species in lakes and rivers. For example, the Dongting Lake Basin, located in the middle of the Yangtze River region, has long been under threat from climate change, with significant increases in temperature, especially in spring. At the same time, summer, winter, and annual rainfall have also shown significant increases. However, rainfall in spring and autumn shows an insignificant decrease, and there is a clear trend of decreasing yearly river flow (Li et al. 2023).

Interaction and climate change problems, lakes, and rivers can also have an impact on:

1. Loss of biodiversity due to changes in temperature, water flow, and habitat availability that can lead to a decrease or loss of species in lakes and rivers.
2. Economic impact, due to changes in river flow patterns that impact agriculture, fisheries, and other industries that depend on water resources and the occurrence of extreme weather events can cause adverse damage, impact infrastructure and communities.
3. Human health, such as changes in water quality due to climate change that can pose risks to human health, affect drinking water sources and recreational activities, and can even have an impact on.
4. Social disruption due to resource conflicts, especially in areas already vulnerable to water scarcity. This paper will explore the interaction of correlative relationships, threats, problems, and associated impacts among these three elements, namely climate, lakes, and rivers, and highlight their important implications for efforts to restore terrestrial and aquatic ecosystems.

5.1 Effects of Climate Change on Lakes and Rivers

Climate change is a multiplying threat factor that profoundly affects communities worldwide. As an archipelagic nation stretching from Sabang to Merauke, Indonesia has a variety of lakes and rivers involving complex interactions between biotic and abiotic elements. Lake and river ecosystems significantly impact Indonesian communities' daily lives. People depend on these water resources for agriculture, fisheries, transportation, and drinking water supply. However, increasing human activities also put pressure on the sustainability of the ecosystems. This condition can pose threats and challenges, with water pollution, climate change, and deforestation being significant challenges that lake and river ecosystems face. Industrial, domestic, and agricultural waste can degrade water quality, while climate change can affect temperature and rainfall patterns, influencing the water flow of lakes and rivers. The influence of climate change on the oxygen saturation levels of surface waters globally is a crucial concern for future water quality in a warming climate. Although recent studies have extensively examined the rising river water temperatures associated with climate change, the specific effects of climate change on dissolved oxygen saturation levels have not received thorough investigation (Benateau et al. 2019).

Lakes and rivers are closely related to other ecosystems, such as forests, rice fields, and the sea. This interconnection highlights the importance of a holistic approach to preserving environmental sustainability in Indonesia. Lakes also play a role in maintaining balance in anticipating climate change, as reflected in various physical, chemical, and biological responses to climate variations and climate-induced modifications within lake basins (Williamson et al. 2009). Furthermore, according to (Woolway et al. 2019), lake variables are considered the essential climate variables. With all their uniqueness and complexity, lake and river ecosystems in Indonesia require continuous attention and support to ensure that freshwater resources remain sustainable for future generations.

Climate change significantly threatens lakes and rivers, affecting hydrological cycles and temperature regimes. Changing rainfall patterns and increasing temperatures can cause changes in water availability, affecting river flows and lake water levels. One of the main impacts of climate change is changes in rainfall and temperature patterns, which directly affect the volume and temperature of water in lakes and rivers. Decreased rainfall and increased temperatures can result in lower water levels, leading to droughts and threatening the survival of aquatic organisms. Increased water temperatures can accelerate the metabolism rate of aquatic organisms, while decreased water levels can reduce dissolved oxygen. These conditions can lead to changes in species composition and threaten the sustainability of water ecosystems. Climate change can also trigger extreme weather events such as floods, landslides, or forest fires, which can cause water pollution and damage natural habitats around lakes and rivers.

These disturbances can affect aquatic ecosystems, fish migration patterns, and the overall balance of both water bodies. The interconnected relationship between climate, lakes, and rivers faces various threats and challenges, resulting in significant and widespread impacts on ecosystems, biodiversity, and human communities. Climate change also directly impacts communities' well-being around lakes and rivers. Extreme fluctuations in weather and climate can disrupt the livelihoods of fishermen or farmers who depend on these water resources. Moreover, droughts and floods can harm the agricultural sector and result in significant economic losses. Therefore, sustainable climate change management and protecting lakes and rivers are crucial to supporting the sustainability of freshwater ecosystems and the well-being of surrounding communities.

The shrinking and drying up of the world's largest lakes and reservoirs is a severe manifestation of the adverse effects of climate change on the planet's ecosystems and human societies. The loss of these critical water resources, from Lake Chad in Africa to the Aral Sea in Central Asia and Lake Urmia in the Middle East, poses a significant threat to the livelihoods and well-being of millions of people worldwide. A study published in 2021 revealed that over half of the world's largest lakes and reservoirs had experienced a significant decline in water levels since the early 1990s, with climate change being the leading cause. The study, conducted by an international team of researchers, highlights that the Caspian Sea in Europe and Asia and Lake Titicaca in South America have been losing water at a cumulative rate of about 22 giga tons per year for almost three decades, which is equivalent to 17 times the volume of Lake Mead, the largest reservoir in the United States. A surface hydrologist at the University of Virginia who led the study, climate warming, and rising human consumption are the primary drivers of the decline in 56% of natural lakes. While climate scientists have traditionally predicted that arid regions worldwide will become increasingly dry due to climate change, the study has revealed a significant loss of water volume in moist areas. Yao et al. (2023) noted that this matter must be considered, and addressing this crisis demands a coordinated effort at both local and global levels to mitigate climate change, protect freshwater resources, and ensure the sustainability of ecosystems for future generations. The study relied on satellite measurements combined with climate and hydrological models to monitor nearly 2000 large lakes worldwide. The research found that unsustainable human consumption practices are one of the leading causes of the reduced volume of lakes, with some water bodies, such as the Aral Sea in Central Asia and the Dead Sea in the Middle East, having already fallen victim to such practices. In addition, lakes in Afghanistan, Egypt, and Mongolia are experiencing the adverse effects of rising temperatures, which increase the evaporation of water into the atmosphere. The drying up of these lakes and reservoirs ultimately affects the nearly two billion people who live nearby and depend on these water sources.

Climate is the average weather state over a large area over a long time, up to decades. Muralikrishna and Manickam (2009) stated that climate is the average weather condition in an area for several years, which describes the area's situation, whether cold, warm, dry, or wet. The climate is influenced by many factors, including the sun's altitude, latitude, length of daylight and night, solstice, equinox, distribution of land and water, elevation, and ocean current (Muralikrishna and Manickam 2009). Several factors, including temperature changes, hydrological cycles, snow melt, sea level rise, land cover change, deforestation, and extreme weather, can cause climate change in an area. Climate change can be studied from climate models. Referring to Hewitt et al. (2021), decision-makers and policymakers have widely used climate models to support climate change adaptation policies. Research and innovation in climate modeling have significantly increased public knowledge about climate and the capacity of sustainable, resilient communities (Hewitt et al. 2021).

A river is generally defined as a body of flowing water. Czuba and Allen (2023) defined rivers and streams with a specific definition. A stream is a flowing body of water with a channel width of less than 15 m, while a river is a flowing body of water with a channel width equal to or greater than 15 m. The river is more commonly used in Geosciences and Environmental Engineering, while the flow is more widely used in Ecology (Czuba and Allen 2023). Referring to Sohns (2023), rivers have three perspectives; as a boundary, as a path, and as a resource. The river is a boundary between two countries or cultures and a demarcation line between known and unknown landscapes. The river as a path means that its function is as a highway; rivers as resources provide an understanding that their function is to support development (Sohns 2023).

Climate change significantly impacts rivers, lakes, and tropical aquatic ecosystems. The impacts of climate change are specific, and their complexity varies depending on geographic location and the characteristics of lakes, rivers, or tropical aquatic ecosystems. Climate change causes changes in the water cycle and affects the distribution and availability of water in rivers and lakes and the sustainability of aquatic ecosystems. Climate change changes rainfall patterns that cause droughts and floods, affecting rivers and lakes. Drought causes a decrease in the volume of water in waters and lakes, thus changing aquatic ecosystems, while floods can damage aquatic habitat areas. Climate change impacts nutrient cycling and pollution, affecting the water quality of lakes and rivers, the health of aquatic ecosystems, and the availability of clean water for humans. Rising global temperatures result in increased water temperatures in rivers and lakes that alter thermal balance and impact aquatic organisms. Rising global temperatures are causing melting polar ice caps, which add to sea levels and affect coastal ecosystems and seawater intrusion.

In the following section, some research summaries are presented, including climate change research in several global parts.

5.1.1 Batanghari River Basin

Climate change affects the hydrological cycle and impacts water availability. Yamamoto et al. (2021) investigated that climate change in the tropics has caused environmental problems. It significantly affects the water cycle in watersheds, especially on the coast of Sumatra, which primarily consists of peatlands. Yamamoto et al. (2021) conducted a study on the projected level of flood inundation in the Batanghari River Basin (42,960 km^2) at the end of the twenty-first century related to climate change that occurs. Studies showed that flood inundation for a 20-year cycle is projected to increase the inundation volume by 3.3 times. Climate change has changed the frequency and magnitude of flooding in low-lying areas, disrupting peatland environments and damaging carbon dynamics in plantations and agricultural regions. This projected flood inundation level is essential for area managers to anticipate watershed management measures, such as land use regulations for plantations and wetland conservation, by considering the projected increase and extent of flooding in the area (Yamamoto et al. 2021).

5.1.2 Lake Atitlán, Guatemala

In the country of Guatemala in Central America there is a lake called Lake Atitlán, this lake is a beautiful volcanic lake located at an altitude of 1500 m above sea level and has a water depth of 341 m, this makes it the deepest lake in Central America. Lake Atitlan is very beautiful when viewed with the naked eye as a calm and unaffected water surface, however, this lake faces a serious threat due to the impact of climate change (Vincent 2009).

The results of the latest study have explained the worrying consequences of climate change on the fragile ecosystem. Researchers have noted that the region experienced rising temperatures, irregular rainfall patterns, and increased evaporation rates, all contributing to the gradual drying of the lake (Perreault 2020; Zubieta et al. 2021).

The occurring climate change has some impacts on Lake Atitlán which include the incidence of temperature rise, shifting rainfall patterns, and increased evaporation rates all take its toll, leading to a decrease in water levels and dramatic changes in the fragile lake ecosystem (Vincent 2009; Lalinde and Maierhofer 2020). One of the most direct impacts that occur in this lake is the disturbance of the lake's water balance. Data showed that the region had experienced a significant decline in rainfall, leading to a steady decline in lake water levels (Zubieta et al. 2021; Perreault 2020). This, in turn, has far-reaching consequences, including the loss of critical habitats for aquatic wildlife, deterioration of water quality, and exacerbations of water scarcity for communities that depend on lakes for their livelihoods (Vincent 2009). The delicate balance of the lake biological system will be disturbed by the warming of lake water caused by climate change (Vincent 2009). This warming could create conditions that favor the proliferation of invasive species, which could outperform and replace the native flora and fauna that have long flourished in lake ecosystems. Climate change that occurs in Lake Atitlán has an impact not only on its physical and ecological aspects. Reduced water volumes and declining water levels and disruption of lake ecosystems have had a major impact on communities that have long relied on lakes for their livelihoods and livelihoods (Lalinde and Maierhofer 2020).

The occurrence of warmer temperatures and increased sunlight creates ideal conditions for cyanobacteria which can cover large areas of lakes and severely disrupt aquatic ecosystems. This will have an effect on water quality and prevent oxygen and light from reaching underwater flora and fauna, leading to further ecological degradation. The potential toxicity of the cyanobacterial blooms also poses a serious health risk to communities that depend on the lake for drinking water and recreational activities (Paerl et al. 2016). In addition, changes in rainfall patterns due to climate change can lead to increased nutrient runoff from surrounding farmland, exacerbating eutrophication problems and further triggering the growth of harmful algae (May et al. 2020).

The changes in the ecology of lakes caused by climate change have dire consequences for the local communities that depend on them for their livelihoods. As the lake fish population declines, local fisheries and aquaculture operations struggle to survive. In addition, the degradation of the lake's water quality has reduced its tourism appeal, depriving many families of an important source of income.

The complex conditions of climate change, eutrophication and intersecting ecosystem changes in Lake Atitlán provide an example of the broader challenges facing Latin America and the Caribbean region. As slow events such as rising temperatures and erratic rainfall patterns continue, the region must face the daunting task of conserving its valuable natural resources and ensuring food security for its vulnerable population. Non-seasonal high rainfall conditions and increased pollution are consequences of climate change exacerbating pollution problems. During the rainy season, runoff carries significant sewage and debris into the lake from the surrounding area, especially illegal landfills near rivers. It is estimated that around 3000 tons of garbage entered the lake during this period, exacerbating the existing contamination problem.

Agricultural practices are severely affected by climate change which cause an increase in fertilizer and pesticide runoff into lakes. This runoff contributes to excess nutrients that trigger algae blooms and pose a risk to local drinking water supplies and aquatic life (Zalewski et al. 2010; Fuente et al. 2022). Disturbances in the lake's water balance also concentrate pollutants, which further degrades water quality. Overall, the interaction between climate change and local environmental management issues creates a complex situation for Lake Atitlán. To protect this vital resource, an integrated approach must consider ecological and social dimensions.

5.1.3 Lake Victoria and Chad, Africa

The impacts of climate change are causing a decline in water quality and availability, impacting food security, conflict, and health problems in African region. Although Sub-Saharan Africa has many natural resources, such as minerals, forests, wildlife, and biodiversity, these resources have not been fully utilized and do not reflect the well-being of its population. The region also has some of the world's largest tropical rainforests and the highest equatorial mountains. However, these natural resources are not evenly distributed. For example, more than 20% of tropical forests are in the Democratic Republic of Congo. Some principal rivers, such as the Congo River, the Niger River, the Nile River, the Zambezi River, as well as Lake Victoria and Lake Chad, contain more than half of the freshwater sources in Sub-Saharan Africa. Human activities, such as deforestation and agriculture, as well as greenhouse gas emissions, have disrupted the natural water cycle and affected water bodies, lakes, and reservoirs in the region.

This section discusses the impact of climate change on water resources in Sub-Saharan Africa, particularly on two large lakes important for agriculture and socio-economy, namely Lake Victoria and Lake Chad. Lake Victoria is the second-largest freshwater lake in the world and the largest in Africa, which is divided between three countries: Tanzania (49%), Uganda (45%), and Kenya (6%). Its surface area is about 68,800 km^2, with a catchment area of 193,000 km^2. The lake has an average depth of about 40 m and a maximum depth of 92 m, with a length of 344 km and a width of 240 km. Tanzania's main rivers, such as the Mara, Kagera, Mirongo, Grumeti, Mbalageti, Simiyu, and Mori, flow into the lake. The annual rainfall around the lake ranges from 1200 to 1600 mm, about the same as the average evaporation of the lake's water. Most of the area around the lake is classified as arid or semi-arid (Yunana et al. 2017).

In the 1960s, the Lake Chad Basin was the largest endorheic basin in the world, with a surface area of about 25,000 km^2, covering about 8% of Africa's total land surface. The basin has three climate zones: Sudan, the Sahara, and the Sahel, with annual rainfall decreasing from south to north. Rainfall in the Sudanese zone ranges from 500 to 1000 mm, in the Sahel zone between 100 and 150 mm, and in the

Sahara zone less than 100 mm. The depth of Lake Chad varies from 4 to 8 m in the northern part and 2–4 m in the southern part. The Lagone-Chari River accounts for 96% of the water flow to the lake. The region has a hot and dry climate, with evaporation reaching 2300 mm per year, greater than rainfall. Lake Chad has experienced significant shrinkage since the 1960s, with its surface area shrinking from about 25,000 km^2 to just 2000–15,000 km^2, depending on the season. This shrinkage is caused by drought and the use of water for irrigation. The lake's water volume has also decreased drastically, leading to reduced fish production and threatening the traditional livelihoods of the region's residents. Lake Victoria and Lake Chat have undergone significant changes due to climate change and human activities (Yunana et al. 2017).

The case study of Lake Chad and Lake Victoria explains that climate change devastates water resources. Climate change is causing considerable shrinkage and constant seasonal fluctuations in both lakes. The lack of adaptation measures due to limited technology and knowledge about the climate has left millions of people vulnerable. Therefore, it is essential to develop effective water management strategies so communities can adapt to the negative impacts of climate change, especially on water resources.

5.1.4 Lake Chini, Malaysia

Lake Chini, one of Malaysia's largest lakes, has been affected by climate change, affecting the balance of its ecosystem. This is reflected in the decline in water quality and changes in biodiversity. Soil erosion due to water is an ecological problem afflicting Lake Chini. Although some researchers have previously investigated soil erosion analysis in the Lake Chini Basin, no studies have applied such analysis using the Climate Model Intercomparison Project-6 (CMIP6) scenario (Rendana et al 2023).

The latest study by Rendana et al. (2023) is the first attempt to estimate the potential for soil erosion in Malaysia's Chini Lake under current and future conditions using the CMIP6 scenario. The results showed that the predicted average soil erosion value for the baseline scenario (2019–2021) was about 50.42 t/ha years. Mining areas in the southeast recorded the highest soil erosion values. The highest future soil erosion value (36.15 t/ha year) was obtained for SSP4.5 during 2060–2080, while the lowest value (33.30 t/ha year) was obtained for SSP2.6 during 2040–2060. According to CMIP6, the potential for future soil erosion in the study area will decrease by about 33.9% compared to the baseline (2019–2021). The erosion factor of rainfall influences the potential for soil erosion in the study area. The findings of this study are expected to contribute to the achievement of the UN 2030 Agenda for Sustainable Development (Rendana et al. 2023).

5.1.5 Lake Lanao and Angat Reservoir, Philippines

Amid the influence of large-scale atmospheric phenomena, the Philippines finds itself as one of the countries hardest hit by the effects of climate change. An in-depth study by Jose and Cruz (1999) highlighted the seriousness of this challenge. From Lake Lanao to Angat Reservoir, the country's primary water resources are threatened by extreme climate waves (Jose and Cruz 1999). Utilizing global circulation models and hydrological simulations, the study reveals a dark portrait: changing rainfall and rising temperatures will directly impact the Angat reservoir and Lake Lanao inflows. Angat Reservoir, located on the island of Luzon, is the water source for Metro Manila.

The reservoir also supplies irrigation water, serves as flood control for the downstream areas, and contributes to power generation to meet the electricity needs of the island of Luzon. Lake Lanao is the second-largest lake in the country. The lake is the leading hydroelectric power generation source on the Mindanao Island. It is essential to the country's economy in general and the Mindanao region. In Angat Reservoir, the surface flow coefficient is likely to decrease in the future, which may not be sufficient to meet water needs in the Philippines. Lake Lanao is also expected to experience a decrease in the surface flow coefficient (Jose and Cruz 1999). The consequences are clear: the country must urgently take proactive measures to address threats to its water supply. The Philippines needs innovative and collaborative solutions to secure its water future. Given the vulnerability of water resources to global warming in the Philippines, measures to address the problems facing water resources in the country need to be identified (Jose and Cruz 1999).

5.1.6 Lake Sentarum, Indonesia

Deforestation due to forest exploitation has threatened the ecosystem of Lake Sentarum in West Kalimantan, a lake officially protected as a national park area. Roslinda (2019) stated that Lake Sentarum experienced a decline in its ecosystem function and that affected various economic activities in the region. Roslinda (2019) conducted a study to determine the economic value of Lake Sentarum by calculating the total economic value (TEV) of TNDS. The study's results can be used as reference figures in financing the conservation and sustainable management of TNDS (Roslinda 2019).

Hidayat et al. (2020) stated that Lake Sentarum is a complex floodplain lake where deforestation, forest fires, monoculture agroindustry, and pollution threaten the catchment area. Hidayat et al. (2020) researched the hydrological characteristics of Lake Sentarum to reveal the dynamics of aquatic habitats due to changes in water levels. The study's results showed that the Lake Sentarum water level dynamics caused changes in water quality, nutrient availability, vegetation cover, and diversity of fish found in the area. Water quality observations showed that the aquatic environment of the Sentarum Lake area was relatively good for aquatic biota, and the diversity of Lake Sentarum fish was relatively higher during periods of high water (Hidayat et al. 2021).

Ridwansyah et al. (2021) studied that rainfall characteristics and patterns were closely related to inundation in the TNDS area. Ridwansyah et al. (2021) researched and developed a pattern of rainfall relationship with inundation in the TNDS area and found that annual rainfall in this region experienced an increasing trend with the highest peaks in December and April. This study explained that the TNDS area had a complex spatial inundation pattern and was closely related to regional rainfall. The results of this study provided critical information for area managers in making policies to protect biota and habitat in the Danau Sentarum National Park area (Ridwansyah et al. 2021).

5.1.7 Mlech Reservoir in Cambodia

Climate change is closely related to changes in the hydrological cycle. Lun et al. (2022) specified that climate change has become an increasingly important topic in managing water resources in watersheds affected by changes in the hydrological cycle. Lun et al. (2022) conducted a water balance study of the Mlech reservoir to increase water treatment capacity at the Water Treatment Plant in Kampot Province. Raw water access for water treatment plants in Kampot Province comes from Mlech Reservoir. The study calculated the reservoir's water balance by considering climate change factors, observed monthly and annual flow averages, and predicted monthly and annual flow averages with several climate change scenarios. This study is the basis for consideration of accessing raw water from the Mlech Reservoir to increase the capacity of the Water Treatment Plant in Kampot Province, Cambodia (Lun et al. 2022).

5.2 Effects of Climate Change on Aquatic Ecosystems

Climate change has significantly impacted macrophyte crops in rivers and lakes. Abd Hamid et al. (2021) studied that macrophytes are plants that quickly adapt to wet environments and one of the essential components of rivers, lakes, and other wetland ecosystems. In addition, environmental conditions significantly affect macrophyte plants, and each macrophyte species responds to the environment in different ways in different environments. Macrophytes have been widely used as biological indicators of the health of an ecosystem (Abd Hamid et al. 2021).

Temperature changes affect the growth and distribution of macrophytes, where some species may thrive at high temperatures while others may not be able to succeed. Due to global climate change, glaciers are melting, sea levels are rising, precipitation patterns are changing, and extreme weather events are more frequent and intense (Özbayram et al. 2022). Lind et al. (2022) stated that climate change has directly affected aquatic macrophytes in lakes and indirectly affected invasive species and nutrient dynamics. Climate change refers to changes in carbon dioxide, temperature, and rainfall patterns. Changes in rainfall patterns affect water levels in rivers and lakes, impacting macrophyte crops. Climate change will likely increase floating species' distribution and decrease submerged macrophytes' distribution. Climate change threatens macrophyte communities, affecting the lake ecosystem (Lind et al. 2022).

Climate change has an impact on the rate of evaporation. According to Obianyo (2019), changes in the evaporation rate from water bodies are closely related to climate change. Obianyo (2019) conducted a study on the evaporation of salinity from water bodies to determine the rate of evaporation that impacts the water cycle. The study used five salt solutions of different concentrations and the results showed that evaporation influences the water cycle, where low evaporation provides low outflow parameters. Low evaporation rates reduce cloud growth and inhibit the process of clouds becoming rain. Reduced rainfall due to cloud formation is inhibited, ultimately causing climate change (Obianyo 2019).

Embke et al. (2023) investigated that climate change due to factors such as increased rainfall, increased temperatures, snowmelt, and changes in flow has had an impact on sturgeon habitat in the Great Lakes region, which is historically already vulnerable due to overfishing in the area. The study results show that the life stage of the sturgeon lake habitat, habitat availability, and its interaction with other stressors are greatly affected by climate change. This study has provided an understanding of climate change's complex impacts on lake sturgeon fish's habitat (Embke et al. 2023).

5.3 Climate Change and Weather Modification Technology

Weather modification (WM) research activities began in the 1960s in the USA and continue to the present decade in various countries. One form of WM activity is rain enhancement. WM activities that increase rainfall can increase rainfall by 10–20%. The scope of WM research activities is not only on rain enhancement but also on reducing hail suppression.

Several countries that have implemented weather modification (MC) activities, such as the USA, Russia, Thailand, Indonesia, and other subtropical countries, have implemented WM with various methods to anticipate the decreasing availability of water sources on the surface. This kind of activity in Indonesia is known as weather modification technology or cloud seeding. Due to the impact of climate change or anomalous climatic conditions that lead to dry conditions, some of their countries carry out cloud-seeding activities to accelerate and increase rainfall. With the help of this weather modification technology, it is hoped that the inflow flowing into the river and into the lake will still be maintained and used for various purposes. For example, in Indonesia, with indications of climate change that causes reduced rainfall in the Toba watershed, the Water Resources Manager in the Toba watershed carried out weather modification technology activities in April 2021 for 30 days (Fig. 5.1). By using flare technology, which has proven to be quite effective, implementing weather modification technology provides additional rainfall of 36.3%, which also impacts the inflow to Lake Toba (Fig. 5.2) (Indonesia WMT 2021). Maintaining the flow in the river and inflow into the lake helps to sustain the wetland ecosystems.

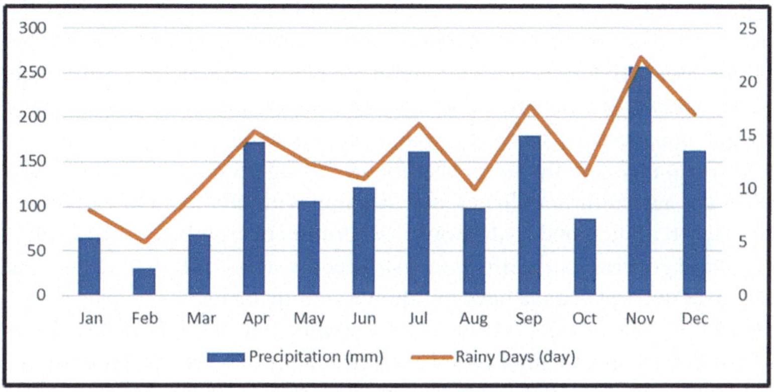

Fig. 5.1 Average area rainfall of Toba Lake and rainy days 2020

Fig. 5.2 Lake Toba (Private collection. Photo credit: BRIN/F. Heru Widodo)

5.4 Ecosystem Restoration

A law by the Water Framework Directive (WFD) has been published to protect and enhance aquatic ecosystems and promote sustainable water use in Europe. The law is intended for all aquatic ecosystems in EU countries to achieve better status by 2027. Carvalho et al. (2019) have conducted studies and provided recommendations for the monitoring and assessment system of the WFD program. Improving the WFD monitoring and assessment system can be done through earth observation networks and satellite data and combining them with statistical approaches. This system can diagnose the causes of watershed damage and provide solutions to decision-makers in watershed management. In setting watershed management measures and policies, it must deliver many benefits to all sectors, including reducing flooding, increasing benefits for the agricultural industry, and aligning with urban planning policies such as wastewater treatment and hydropower generation energy (Carvalho et al. 2019).

The Chinese government's efforts to restore ecosystems by reducing lake pollution nationally over the past decade have not shown significant results. Huang et al. (2019) conducted a study and developed a new water quality index (WQI-DET) to investigate lake pollution patterns in China based on space and time. The analysis showed that China's lake water quality has improved, but eutrophication has not been adequately addressed, and heavy metal pollution is on an upward trend. The results of the study showed that to control heavy metal pollution in China's lakes, mining activities should be reduced, and soil in polluted areas should be remediated (Huang et al. 2019).

Ecosystem restoration is a center for biodiversity conservation as part of efforts to maintain climate stability. Targets for ecosystem restoration have been set, but the priority areas considering spatial benefits and costs have not been well-identified and developed an optimum multi-criteria approach to identify the priority areas. A study showed a multiplier increase in the benefits and cost-effectiveness obtained through an optimal spatial planning approach. The results suggested that ecosystem restoration carried out in synergy significantly contributed to overcoming global challenges (Strassburg et al. 2020).

In recognition of the critical role of restoration in ecosystem health, the United Nations (UN), on March 1, 2019, declared 2021–2030 as the "UN Decade of Ecosystem Restoration". The call aims to accelerate the restoration of degraded global ecosystems and to address the climate change crisis. In addition, the call also aims at improving food security, providing clean water, and protecting biodiversity around the world. Some restoration activities include restoring 350 million km^2 of degraded terrestrial ecosystems by 2030 and increasing the global area of mangroves by 20% by 2030. The UN Decade for Ecosystem Restoration Declaration also coincides with the UN Decade of Marine Science for Sustainable Development, which aims to restore declining ocean health. These two declarations are expected to be implemented holistically and coordinated by signatory countries to accelerate the restoration of the global ecosystem called for by the United Nations (Waltham et al. 2020).

5.5 Remarks

Research on the interaction of climate change with rivers, lakes, and tropical aquatic ecosystems has been widely conducted, but much is still unknown and partially understood. The long-term impacts of climate change on lakes, rivers, and tropical aquatic ecosystems are still not fully elucidated. The effects of climate change on microorganisms, bacteria, and other microscopic organisms in tropical aquatic ecosystems are still not widely investigated. Research needs to improve to provide a specific understanding of how climate change interacts with rivers, lakes, and tropical aquatic ecosystems. In addition to improved research, conservation and adaptation efforts are needed to manage the impacts of climate change on rivers, lakes, and tropical aquatic ecosystems. Continuous monitoring of rivers, lakes, and tropical aquatic ecosystems is required. Implementing sustainable water management and increasing efforts to protect natural habitats are inevitable.

The interaction between climate, lakes, and rivers is complex and dynamic. Climate, rivers, lakes, and their correlated interactions are essential in shaping a region's physical and ecological characteristics. Climate change directly impacts rivers and lakes, affecting the health and function of river and lake ecosystems. Understanding the climate-river–lake interconnectedness is critical to develop effective strategies and adapt to the impacts of climate change. A deep understanding of the correlative interactions between climates, lakes, and rivers is essential to designing ecosystem restoration strategies to be successful in sustaining aquatic ecosystems and supporting the goals of the UN Decade of Ecosystems.

References

Abd Hamid M, Ismail SN, Mansor M (2021) Overview of macrophytes in the tropical Wetland ecosystem. Indonesian J Limnol 2(1):25–34. https://doi.org/10.51264/inajl.v2i1.12

Atkinson CL, Shogren AJ, Smith CR, Golladay SW (2023) Water availability and seasonality shape elemental stoichiometry across space and time. Ecol Appl 33(4):e2842. https://doi.org/10.1002/eap.2842

Bazarzhapov TZ, Shiretorova VG, Radnaeva LD, Nikitina EP, Sodnomov BV, Tsydypov BZ, Batomunkuev VS, Taraskin VV, Dong S, Li Z, Wang P (2023) Trend analysis of precipitation, runoff and major ions for the Russian part of the Selenga River Basin. Water 15(1):197. https://doi.org/10.3390/w15010197

Benateau S, Gaudard A, Stamm C, Altermatt F (2019) Climate change and freshwater ecosystems. Impacts Water Qual Ecol Status. https://doi.org/10.5167/uzh-169641

Carvalho L, Mackay EB, Cardoso AC, Baattrup-Pedersen A, Birk S, Blackstock KL, Borics G, Borja A, Feld CK, Ferreira MT, Globevnik L (2019) Protecting and restoring Europe's waters: an analysis of the future development needs of the water framework directive. Sci Total Environ 658:1228–1238. https://doi.org/10.1016/j.scitotenv.2018.12.255

Chen X, Zhao X, Zhao Y, Wang R, Lu J, Zhuang H, Bai L (2023) Interaction of climate change and anthropogenic activity on the spatiotemporal changes of surface water area in Horqin Sandy Land China. Remote Sens 15(7):1918. https://doi.org/10.3390/rs15071918

Czuba JA, Allen GH (2023) When does a stream become a river? River Res Appl 39(9):1925–1929. https://doi.org/10.1002/rra.4185

Embke HS, Nikiel CA, Lyons MP (2023) Potential effects of climate change on *Acipenser fulvescens* (Lake Sturgeon). US Geological Survey Open-File Report, pp viii–41. https://doi.org/10.3133/ofr20211104E

Fuente MJDL, Gallardo-Bustos C, Iglesia RDL, Vargas IT (2022) Microbial electrochemical technologies for sustainable nitrogen removal in marine and coastal environments. Multidiscip Digit Publ Inst 19(4):2411–2411. https://doi.org/10.3390/ijerph19042411

Hewitt CD, Guglielmo F, Joussaume S, Bessembinder J, Christel I, Doblas-Reyes FJ, Djurdjevic V, Garrett N, Kjellström E, Krzic A, Costa MM (2021) Recommendations for future research priorities for climate modeling and climate services. Bull Am Meteor Soc 102(3):E578–E588. https://doi.org/10.1175/BAMS-D-20-0103.1

Hidayat H, Aisyah S, Kurniawan R, Ridwansyah I, Samir O, Haryani GS (2020) Flood pulse and aquatic habitat dynamics of the Sentarum floodplain lakes area. Indonesian J Limnol 1(1):27–37. https://doi.org/10.51264/inajl.v1i1.2

Huang J, Zhang Y, Arhonditsis GB, Gao J, Chen Q, Wu N, Dong F, Shi W (2019) How successful are the restoration efforts of China's lakes and reservoirs? Environ Int 123:96–103. https://doi.org/10.1016/j.envint.2018.11.048

Indonesia WMT (2021) Pelaksanaan Teknologi Modifikasi Cuaca (TMC) Di Daerah Tangkapan Air Danau Toba Provinsi Sumatera Utara Tahun 2021—Tahap I (Periode 1–29 April 2021)

Jose AM, Cruz NA (1999) Climate change impacts and responses in the Philippines: water resources. Clim Res 12(2–3 SPEC. ISS. 6):77–84. https://doi.org/10.3354/cr012077

Lalinde LL, Maierhofer C (2020) Creating a culture of shared responsibility for climate action in Guatemala through education. Springer, Cham, pp 85–112. https://doi.org/10.1007/978-3-030-57927-2_3

Li S, Jiang C, Ma Y, Li C (2023) Spatiotemporal analysis of hydrometeorological factors in the source region of the Dongting lake basin. China Atmos 14(12):1793. https://doi.org/10.3390/atmos14121793

Lind L, Eckstein RL, Relyea RA (2022) Direct and indirect effects of climate change on distribution and community composition of macrophytes in lentic systems. Biol Rev 97(4):1677–1690. https://doi.org/10.1111/brv.12858

Lun R, Chham A, Chea S, Khem V, Ou K, Samreth S (2022) The study on water balance in mlech reservoir under climate change effects Kampot Province. Indonesian J Limnol 3(1):34–46. https://doi.org/10.51264/inajl.v3i1.25

Muralikrishna IV, Manickam V (2009) Global climatic changes and weather modification technology. BS Publications, Hyderabad

May L, Olszewska J, Gunn I, Meis S, Spears BM (2020) Eutrophication and restoration in temperate lakes. IOP Publ 535(1):012001–012001. https://doi.org/10.1088/1755-1315/535/1/012001

Obianyo JI (2019) Effect of salinity on evaporation and the water cycle. Emerg Sci J 3(4):255–262. https://doi.org/10.28991/esj-2019-01188

Özbayram EG, Çamur D, Köker L, Çam AO, Akçaalan R, Albay M (2022) The effects of climate change on aquatic ecosystems in relation to human health. Aquat Sci Eng 37(3):123–128. https://doi.org/10.26650/ASE20221057160

Paerl HW, Gardner WS, Havens KE, Joyner AR, McCarthy MJ, Newell SE, Qin B, Scott JT (2016) Mitigating cyanobacterial harmful algal blooms in aquatic ecosystems impacted by climate change and anthropogenic nutrients. Harmful Algae 54:213–222. https://doi.org/10.1016/j.hal.2015.09.009

Perreault T (2020) Climate change and climate politics: parsing the causes and effects of the drying of Lake Poopó Bolivia. Univ Tex Press 19(3):26–46. https://doi.org/10.1353/lag.2020.0070

Rendana M, Idris WMR, Rahim SA, Rahman ZA, Lihan T (2023) Predicting soil erosion potential under CMIP6 climate change scenarios in the Chini Lake Basin Malaysia. Geosci Lett 10(1):1. https://doi.org/10.1186/s40562-022-00254-7

Ridwansyah I, Setiawan F, Subehi L (2021) Identification of rainfall and inundation pattern using remotely sensed data in Lake Sentarum floodplain area. Indonesian J Limnol 2(2):9–20. https://doi.org/10.51264/inajl.v2i2.16

Roslinda E (2019) Economic valuation of the Danau sentarum national park, West Kalimantan Indonesia. Biodiversitas 20(7):1983–1989. https://doi.org/10.13057/biodiv/d200726

Sayre R, Luijendijk E, Survey, USG (2023) Increasing seasonal variation in the extent of rivers and lakes from 1984 to 2022, pp 1–14

Singh G, Singh A, Singh P, Mishra VK (2021) Impact of climate change on freshwater ecosystem. In: Water conservation in the era of global climate change. Elsevier. https://doi.org/10.1016/B978-0-12-820200-5.00017-8

Sohns A (2023) River definitions reciprocally define us. In: River research and applications, pp 1–7. https://doi.org/10.1002/rra.4195

Strassburg BB, Iribarrem A, Beyer HL, Cordeiro CL, Crouzeilles R, Jakovac CC, Braga Junqueira A, Lacerda E, Latawiec AE, Balmford A, Brooks TM (2020) Global priority areas for ecosystem restoration. Nature 586(7831):724–729. https://doi.org/10.1038/s41586-020-2784-9

Vincent WF (2009) Effects of climate change on lakes. Elsevier BV, pp 55–60. https://doi.org/10.1016/b978-012370626-3.00233-7

Wahab NA, Kamarudin MKA, Toriman ME, Juahir H, Samah MAA, Azinuddin M, Saudi ASM, Hoe LI, Saad MHM, Sunardi S (2023) The assessment of sedimentation problems in Kenyir hydropower reservoir, Malaysia. Water 15(13):2375. https://doi.org/10.3390/w15132375

Waltham NJ, Elliott M, Lee SY, Lovelock C, Duarte CM, Buelow C, Simenstad C, Nagelkerken I, Claassens L, Wen CK, Barletta M (2020) UN Decade on ecosystem restoration 2021–2030—what chance for success in restoring coastal ecosystems? Front Mar Sci 7:71. https://doi.org/10.3389/fmars.2020.00071

Williamson CE, Saros JE, Vincent WF, Smol JP (2009) Lakes and reservoirs as sentinels, integrators, and regulators of climate change. Limnol Oceanogr 54:2273–2282. https://doi.org/10.4319/lo.2009.54.6_part_2.2273

Woolway RI, Weyhenmeyer GA, Schmid M, Dokulil MT, de Eyto E, Maberly SC, May L, Merchant CJ (2019) Substantial increase in minimum lake surface temperatures under climate change. Clim Change 155:81–94. https://doi.org/10.1007/s10584-019-02465-y

Yamamoto K, Sayama T, Apip (2021) Impact of climate change on flood inundation in a tropical river basin in Indonesia. Prog Earth Planet Sci 8(1). https://doi.org/10.1186/s40645-020-00386-4

Yao F, Livneh B, Rajagopalan B, Wang J, Crétaux JF, Wada Y, Berge-Nguyen M (2023) Satellites reveal widespread decline in global lake water storage. Science 380(6646):743–749. https://doi.org/10.1126/SCIENCE.ABO2812

Yunana DA, Shittu AA, Ayuba S, Bassah EJ, Joshua WK (2017) Climate change and lake water resources in Sub-Saharan Africa: case study of Lake Chad and Lake Victoria. Niger J Technol 36(2):648–654. https://doi.org/10.4314/njt.v36i2.42

Zalewski M, Urbaniak M, Negussie YZ (2010) Ecohydrological systemic solutions for reduction of siltation, eutophication and dioxin-induced toxicity. The pilot study of the Asella BioFarm Park lake, Ethiopia. Ecohydrol Hydrobiol 10(2–4):363–368. https://doi.org/10.2478/v10104-011-0027-2

Zubieta R, Molina-Carpio J, Laqui W, Sulca J, Ilbay-Yupa M (2021) Comparative analysis of climate change impacts on meteorological, hydrological, and agricultural droughts in the Lake Titicaca Basin. Multidiscip Digit Publ Inst 13(2):175–175. https://doi.org/10.3390/w13020175

Chapter 6
Emerging Pollutants in Inland Waters

Denalis Rohaningsih, Imroatushshoolikhah, Nurul Setiadewi, and Relita Novianti

Abstract Emerging Pollutants (EPs) are gaining concern due to its increasing hazard efficacy for human and the environment along with human population growth, industrial development, and increasingly diverse urban activities. Meanwhile, studies regarding EPs occurrence in Southeast Asian countries with its tropical climates are still quite limited, China has become the country with the largest number of EPs research in Asia continent. As a developed country, China has elaborated on the presence of EPs in various environmental matrices. On the other side, research in Southeast Asia region is limited to several well-known EPs, i.e., several pollutants from the pharmaceuticals and pesticide groups.

Emerging Pollutants (EPs) are gaining concern due to its increasing hazard efficacy for human and the environment along with human population growth, industrial development, and increasingly diverse urban activities. Meanwhile, studies regarding EPs occurrence in Southeast Asian countries with its tropical climates are still quite limited, China has become the country with the largest number of EPs research in Asia continent. As a developed country, China has elaborated on the presence of EPs in various environmental matrices. On the other side, research in Southeast Asia region is limited to several well-known EPs, i.e., several pollutants from the pharmaceuticals and pesticide groups. For the past decade, the most studied EPs in the Southeast Asia region is about the occurrence or pollution status (Lee et al. 2021). However, specific research which elaborates the occurrence of the EPs in the inland water of Southeast Asian countries is rarely found.

D. Rohaningsih (✉) · Imroatushshoolikhah · N. Setiadewi · R. Novianti
Research Center for Limnology and Water Resources, National Research and Innovation Agency (BRIN), Jakarta, Indonesia
e-mail: denalis.rohaningsih@brin.go.id

© The Author(s), under exclusive license to Springer Nature Singapore Pte Ltd. 2025

A. C. Santos-Borja et al. (eds.), *Progress on Ecosystem Restoration of Tropical Inland Waters*, https://doi.org/10.1007/978-981-96-2284-9_6

6.1 Understanding the Emerging Pollutants and Its Classification

Water is the most precious resource, yet, the status of inland water has been threatened due to multiple stressors from the environment. The significant rising of emerging pollutants is one of stressors that should be considered (Geissen et al. 2015; Tagorti and Kaya 2022). Emerging Pollutants (EPs) also named as contaminant of emerging concern or emerging contaminant can be described as the group of contaminants which are not newly present group in the inland waters system. They have been long existing in the environment but less recognized for its effects. However, it has been seriously discussed in the past decades internationally. Based on NORMAN (network of reference laboratories, research, and related organizations for monitoring of emerging environmental substance), currently more than 120 thousand emerging chemicals which are divided into 30 classes of emerging substances based on their origin (listed in the database https://www.norman-network.com/nds/susdat/susdatSearchShow) that increased from 20 classes or about 700 contaminants had been detected in 2015 (Geissen et al. 2015).

EPs can be defined as chemicals both natural and synthetical that pass into water and presumably have an impact on ecosystem and human health. However, they are not commonly regulated and monitored which makes their occurrence, fate, behavior, and toxicological effect poorly understood and recognized (Diaz-Sosa et al. 2020; Geissen et al. 2015). The sources of EPs can be from everywhere. The foremost common groups which are noted for the status as EPs are pesticides, pharmaceuticals, personal care products, surfactants, flame retardants, endocrine disruptors compounds, and industrial products (Geissen et al. 2015).

The agriculture sector involves some products such as fertilizer and pesticides in most of its activities. Both fertilizer and new generations of pesticides widely spread in the aquatic ecosystems as the examples are imidacloprid (Huang 2022), flupyradifurone (Huang 2022; Mangold-Döring et al. 2022), chlorpyriphos (Lydy et al. 1999), neo-nicotinoid, and its derivatives. Meanwhile, pharmaceuticals or drug waste comes from effluent and excretion of human beings, animals, and plants (Mishra et al. 2023). Some drug waste can be in the form of antibiotics, antidepressants, or endocrinal/synthetic hormones which some of them had been studied for its impact and contaminated to aquatic ecosystem (Anh et al. 2021; Schuijt 2023). While the residue of drugs can be from livestock or healthcare waste, medical waste, or domestic waste (Schuijt 2023). Flame retardant for example the group of organophosphate flame retardants (OPFRs) which usually available as prevention for fires also be recognized as emerging flame retardant as well as functioned as plasticizers (Yang et al. 2019).

The increase amount of EPs in the aquatic ecosystem is due to the increase in their demand for human consumption. As a results, inland water both surface and groundwater have been severely polluted. The focus on emerging pollutants has been started since the effect of DDT first emerged.

This review objective is to provide an overview of the trend of emerging pollutants in the inland water ecosystem of Southeast Asian countries.

6.2 Occurrence of the EPs in the Inland Water

The Silent Spring, written by Rachel Carson, which revealed the impact of excessive use of DDT and chemicals on the environment and humans, was the beginning of the people's awareness of the EPs. Aside from pesticides, abundant new pollutants from various sectors are starting to emerge. Human need for products that support their daily activities has spurred development of numerous types of EPs.

Research examining the occurrence of EPs in the Asian continent is not evenly distributed. Globally, EPs research is mostly carried out by developing countries, such as Europe, America, Australia, and China. Only a small number of developing countries have conducted research on this topic, including in the Southeast Asian region. As of this article, there are still many EPs compounds that have not been studied. Hence, the information regarding the presence of EPs in the environment is still limited. Requirement of specific yet non-standardized analysis methods, lack of adequate instruments for analysis, and relatively high cost of analysis make EPs research challenging to carry out (Preisendanz et al. 2021; Waleng and Nomngongo 2022). In addition, the dynamics of EPs types are changing fairly rapid. This might happen due to EPs fields being dynamic and constantly evolving adjusting to the social and environmental need (Feng et al. 2023).

Almost all countries in Southeast Asia do not have specific regulations to limit the level of EPs usage. Regulations generally apply for commonly found chemical compounds, e.g., DDT, its derivatives, and several other types of pesticides. The absence of specific regulation, lack of law enforcement to manage the waste, and unawareness of EP's impact on the environment resulted in management of EPs becoming quite challenging in Southeast Asia countries (Ngoc et al. 2009 in Lee et al. 2021). Therefore, it certainly affects the water quality degradation and poses a danger to living creatures around it.

6.2.1 EPs in River Flow

The existence of EPs in river is waters mostly derived from anthropogenic activities, including household, agricultural, and industrial activities. Each of these activities produces very diverse EPs. Table 6.1 shows the types of EPs that have been found in river waters in Southeast Asian countries.

There are several specific types of compounds from each group of EPs that are distributed in Southeast Asian countries. The description in Table 6.1 refers to the major compounds that were studied or recorded. Agriculture and aquaculture are the main economic sectors of Southeast Asian countries, except Singapore and Brunei

Table 6.1 Occurrence of the EPs in rivers of Southeast Asian countries

EPs type	Compound	Range (ng L^{-1})	Location	Source	Reference(s)
Pesticides	DDT	40.42–102.31	Klang River Estuary (Malaysia)	Residential area, fisheries, industry, steel factory, dumping site, tourism	Omar et al. (2019)
	Chlordane	9–10	Ma River (Vietnam)	Agriculture, domestic	Le et al. (2023)
	HCHs	n.d.–25.03	Surface water of Cameron Highland (Malaysia)	Agriculture	Saadati et al. (2012)
Pharmaceuticals	SMX	790 ± 170	Sunter River (Indonesia)	Domestic wastewater, hospital, industry	Nozaki et al. (2023)
		< 0.05–0.86	Klang River Estuary (Malaysia)	Residential area, fisheries, industry, steel factory, dumping site, tourism	Omar et al. (2019)
		19.26–114.24	Lui, Selangor, Gombak River (Malaysia)	Agricultural, residential area	Praveena et al. (2018)
		~ 82	Pasig River (Philippines)	WWTP, domestic wastewater	Shimizu et al. (2013)
		6.0–15.9	City River (Thailand)	Veterinary wastewater, domestic wastewater	Suzuki et al. (2021)
		6.347–66.65	Cau River (Vietnam)	Domestic wastewater, agriculture run-off	Ngo et al. (2021)
		310–15,591	To Lich River, Kim Nguu River-Red River Tributaries (Vietnam)	Domestic wastewater, hospital effluents, and stormwater run-off	Tran et al. (2019)
	CAF	53–58	Upper Mekong River (Cambodia)	WWTP outlet	Sang et al. (2022)

(continued)

Table 6.1 (continued)

EPs type	Compound	Range ($ng\ L^{-1}$)	Location	Source	Reference(s)
		360–8900	Ciliwung River, Citarum River, Cisadane River (Indonesia)	Domestic wastewater	Dsikowitzky et al. (2016)
		16.27–36.75	Lui, Selangor, Gombak River (Malaysia)	Agriculture, domestic waste	Praveena et al. (2018)
		~ 2250	Chao Phraya River (Thailand)	WWTP outlet	Tewari et al. (2013)
		46.34–444	Cau River (Vietnam)	Domestic wastewater and agriculture run-off	Ngo et al. (2021)
	DCF	67–332	Upper Mekong River (Cambodia)	WWTP outlet	Sang et al. (2022)
		210 ± 33	Sunter River (Indonesia)	Domestic wastewater, hospital, industry	Nozaki et al. (2023)
		0.47–10.80	Klang River Estuary (Malaysia)	Residential area, fisheries, industry, dumping site, tourism, agriculture	Omar et al. (2019)
		3.14–20.80	Lui River (Malaysia)	Domestic, agriculture	Praveena et al. (2018)
		58–367	Chao Phraya River (Thailand)	WWTP outlet	Tewari et al. (2013)
		1.08–1.4	Cau River (Vietnam)	Domestic wastewater and agriculture run-off	Ngo et al. (2021)
	APAP	43–435	Chao Phraya River (Thailand)	WWTP outlet	Tewari et al. (2013)
		~ 61,250	Cagayan de Oro River Basin (Philippines)	Hospital WWTP outlet	Mariano et al. (2023)

(continued)

Table 6.1 (continued)

EPs type	Compound	Range (ng L^{-1})	Location	Source	Reference(s)
	Ibuprofen	60–419	Chao Phraya River (Thailand)	WWTP outlet	Tewari et al. (2013)
PCPs	DEET	13–155	Langat River Basin (Malaysia)	Sewage treatment plant effluent	Tan et al. (2015)
PFAS	PFOS	~ 14.6	Sai Gon River (Vietnam)	Industrial wastewater, WWTP discharge	Lam et al. (2017)
	PFOA	~ 40.2	Sai Gon River (Vietnam)	Industrial wastewater, WWTP discharge	Lam et al. (2017)
	PFNA	~ 53.3	Sai Gon River (Vietnam)	Industrial wastewater, WWTP discharge	Lam et al. (2017)
Industrial products	PCBs	0.13–7.67	Selangor River Basin (Malaysia)	Industrial wastewater, plastic factory	Sakai et al. (2016)
	BPA	9.76–1070.22	Bengawan Solo River (Indonesia)	Domestic, aquaculture	Ismanto et al. (2022)
		~ 37,000	Pahang River (Malaysia)	Domestic, industry, agriculture	Nazifa et al. (2020)
		50.67 ± 4.19	Chak-mak Canal (Thailand)	Industrial wastewater run-off	Ocharoen et al. (2018)

DDT dichlorodiphenyltrichloroethane; *HCHs* hexachlorocyclohexane compound; *SMX* sulfamethoxazole; *CAF* caffeine; *DCF* diclofenac; *APAP* acetaminophen; *DEET* diethyltoluamide; *PFOS* perfluorooctanesulfonate; *PFOA* perfluorooctanoic acid; *PFNA* perfluorononanoate; *PCBs* polychlorinated biphenyls; *BPA* bisphenol-a
n.d.: not detected/below the method limit detection; n.a.: not available

Darussalam (Akber et al. 2020; Lee et al. 2021). Fertilizer usage becomes a major source of EPs in surface waters. The foremost type of pesticide that is widely used in Southeast Asia are organochlorine pesticides, namely dichlorodiphenyltricholoroethane (DDT) and its derivatives, chlordane, and hexachlorocyclohexane compound (HCH) (Lee et al. 2021). DDT is dominating the fertilizer type with the highest concentration in river water, 102.31 ng L^{-1}.

Pharmaceuticals compounds are dominated by sulfamethoxazole (SMX), caffeine (CAF), diclofenac (DCF), and acetaminophen (APAP) with maximum concentrations of 15,591 ng L^{-1} in Red River tributaries of Vietnam, 8900 ng L^{-1} in the Indonesian River, 367 ng L^{-1} in the Chao Phraya River of Thailand, and 61,250 ng L^{-1} in the Cagayan de Oro River Basin of Philippines (Dsikowitzky et al. 2016; Mariano et al. 2023; Tewari et al. 2013; Tran et al. 2019). These compounds are commonly used by the community, so the high market demand has an impact on its high residue in inland waters. The compounds listed in Table 6.1 mostly come from domestic waste and WWTP, especially hospital WWTP.

Unlike pesticides and pharmaceuticals, a few research has been studied on the presence of PCPs compounds in inland waters. PCPs are the most studied group of EPs compared to other groups (Lee et al. 2021). Of the various types of PCPs found in the aquatic environment, there are 13–155 ng L^{-1} in the Langat River Basin of Malaysia which comes from industrial WWTP effluent (Tan et al. 2015). Compared with PCPs, several studies on PFAS compounds and industrial products have been carried out. PFNA or perfluorononanoate is a PFAS compound with the highest concentration found with a concentration of up to 53.3 ng L^{-1} in the Sai Gon River of Vietnam (Lam et al. 2017). Compared to the current concentration of PFAS compounds which is still below 60 ng L^{-1}, the concentration of industrial product compounds is relatively much higher, reaching 37,000 ng L^{-1} (Ismanto et al. 2022; Nazifa et al. 2020; Ocharoen et al. 2018; Sakai et al. 2016). However, the occurrence distribution data for both PCPs and industrial products is not evenly distributed. So the condition of these compounds in the environment is not yet known precisely.

Apart from PCBs and BPA, the other industrial products which currently gaining concern are microplastics (MPs). MPs cannot be degraded but fragmented and pose a danger to the health of humans and other living creatures. It can be found in various environmental matrices, both biotic and abiotic. Among Southeast Asia countries, Malaysia and Indonesia are the countries that study MPs the most (Hassan et al. 2024). However, specific studies regarding the number/concentration of MPs particles in their inland water environment are still limited.

MPs found in water sample from Cherating River and Klang River estuary of Malaysia with an abundance of 0.0070 ± 0.0033 particles m^{-3} and 500–4500 particles m^{-3} (Pariatamby et al. 2020; Zaki et al. 2021). A study conducted in the Citarum River of Indonesia shows that the average abundance of MPs in the water samples taken is 0.0574 ± 0.025 particles m^{-3} (Sembiring et al. 2020). The MPs in these locations come from urban wastewater run-off, industrial WWTP, and agricultural activities.

The abundance of MPs in urban canals and sub-river Mark-Hiao River of Lao PDR differs in the dry and rainy season, namely 0.38–4.61 items m^{-3} and 1.23–7.10 items m^{-3}, respectively, with various types of MPs (Noudeng et al. 2023). The difference in MPs abundance in both seasons may be caused by the difference in water level and water flow. River water flows more when in rainy season compared to in dry season so the volume of waste carried by the water flow will increase, and vice versa.

Aside from these countries, study on MPs was carried out in Myanmar by analyzing sediment samples from the river in the cities of Yangon and Pathein. Both are cities with dense populations. Hence, even though also come from domestic activities like in Malaysia and Indonesia river, the abundance of MPs is quite high as much as 211–2002 pcs kg^{-1} dry weight for Yangon water sample and 0–8426 pcs kg^{-1} dry weight for Pathein water sample (Tun et al. 2024).

From several studies that have been carried out, the types of MPs found to be dominated the inland waters include polypropylene (PP), polyethylene (PE), polyethylene terephthalate (PET), and polystiren (PS). The ratio of abundance of which varies depending on the type of activities in the environment.

6.2.2 *EPs in Lake*

Contaminants that flow with river water will end up in the groundwater, lake, or sea. Lakes can act as sinks, which accommodate various chemical materials produced by anthropogenic activities and channeled through rivers. Eventually, lakes are one of the main sinks of persistent organic pollutants—POPs (Li et al. 2023). Urban lakes surrounded by anthropogenic activities will be vulnerable to pollution by EPs and other pollutants. The presence of EPs in lake waters in Southeast Asia countries is described in Table 6.2.

EPs contaminants in lakes tend to be higher than in rivers. For DDT contaminants, the concentration range in rivers is only hundreds, whereas lakes can reach over 1000 ng L^{-1} for similar pollutant sources. Moreover, this is in line with the lake role as a sink, where some of the contaminants that pass through the lake pathway will eventually accumulate even for small amount.

Similar to river's condition, EPs in lakes are also dominated by pesticide and pharmaceutical contaminants. The maximum pesticide concentration recorded in Sembrong Lake Basin of Malaysia was as much as 1873.60 ng L^{-1} (Sharip et al. 2017) which was sourced from the agricultural activities. Meanwhile, the pharmaceutical product with the highest concentration is ibuprofen as much as 7.9–1500 ng L^{-1} in several Indonesian urban lakes (Nozaki et al. 2023) which its source comes from domestic waste streams.

In contrast to developing countries in this region, Singapore, as a developed country has different types of dominant contaminants. In addition, differences in major economic activities in these countries influence the types of EPs that enter the water bodies. The EPs type with maximum concentration found in Singapore was PFAS/PFOA with concentration range of 2.4–1060 ng L^{-1} (Hu et al. 2011). Aside from Singapore, PFAS contaminants were also found in the Laguna Lake of the Philippines in more diverse types. In both Singapore and Philippines, PFAS is known to originate mainly from industrial activities as PFAS is usually found around industrial estates (Kunacheva et al. 2009).

MPs as a product of industrial activity are also found in several lakes in Southeast Asia. MPs in Indonesian lakes were found in Lake Singkarak with an abundance of

Table 6.2 Occurrence of the EPs in lakes of Southeast Asian countries

EPs type	Compound	Range ($ng\ L^{-1}$)	Location	Source	Reference(s)
Pesticides	DDT	0–1873.60	Sembrong Lake Basin (Malaysia)	Agriculture	Sharip et al. (2017)
	Chlordane	0–23.5	Sembrong Lake Basin (Malaysia)	Agriculture	Sharip et al. (2017)
	HCHs	0–23	Sembrong Lake Basin (Malaysia)	Agriculture	Sharip et al. (2017)
Pharmaceuticals	SMX	5.8–660	Lake Sunter, Kemayoran Golf Lake, Situ Gintung Lake, Situ Rawa Cipondoh (Indonesia)	Domestic wastewater	Nozaki et al. (2023)
		n.d.–180	Lake Sampoloc (Philippines)		Dimzon et al. (2018)
		110 ± 40	Yen So Lake (Vietnam)	Domestic, agriculture	Nozaki et al. (2023)
	DCF	1.0–210	Lake Sunter, Kemayoran Golf Lake, Situ Gintung Lake, Situ Rawa Cipondoh (Indonesia)	Domestic wastewater	Nozaki et al. (2023)
		6.5 ± 5.1	Yen So Lake (Vietnam)		
	Ibuprofen	7.9–1500	Lake Sunter, Kemayoran Golf Lake, Situ Gintung Lake, Situ Rawa Cipondoh (Indonesia)	Domestic wastewater	Nozaki et al. (2023)
		62 ± 15	Yen So Lake (Vietnam)		
PCPs	DEET	12–240	Lake Sampoloc (Philippines)	Domestic, agriculture	Dimzon et al. (2018)

(continued)

Table 6.2 (continued)

EPs type	Compound	Range (ng L^{-1})	Location	Source	Reference(s)
PFAS	PFOS	0.40–2.85	Laguna Lake, Victoria Lake (Philippines)		Guardian et al. (2020)
		2.2–87.3	Reservoir (Singapore)	Industrial WWTP, domestic waste	Hu et al. (2011)
	PFOA	3.22–8.43	Laguna Lake, Victoria Lake (Philippines)		Guardian et al. (2020)
		2.4–1060	Reservoir (Singapore)	Industrial WWTP, domestic waste	Hu et al. (2011)
	PFNA	0.44–1.54	Laguna Lake, Victoria Lake (Philippines)		Guardian et al. (2020)
Industrial products	BPA	0.71–47.40	Laguna Lake (Philippines)		Katrina and Espino (2020)

DDT dichlorodiphenyltrichloroethane; *HCHs* hexachlorocyclohexane compound; *SMX* sulfamethoxazole; *DCF* diclofenac; *DEET* diethyltoluamide; *PFOS* perfluorooctanesulfonate; *PFOA* perfluorooctanoic acid; *PFNA* perfluorononanoate; *BPA* bisphenol-a; *MPs* microplastic
n.d.: not detected/below the method limit detection; n.a.: not available

2–9 items m^{-3} which are dominated by foam and fragment types (Henny et al. 2022) as the source of pollution comes from waste from anthropogenic activities, namely plastic film, food bag packaging, hard plastic fragments, also bottle and cup packaging (Rohaningsih et al. 2022). Just like in Lake Singkarak, MPs in Vietnam's lakes are also dominated by PET which is produced from packaging materials. Apart from that, rayon polymer also dominates the pollution. Moreover, the highest abundance was recorded in the inlet of West Lake, Bay Mau Lake, and Yen So Lake of Vietnam with an abundance of 589.46 items m^{-3}, 154.92 items m^{-3}, and 139.86 items m^{-3}, respectively.

The abundance of MPs in Taungthaman Lake of Myanmar was measured from sediment samples taken from the lake with an abundance of 11,946 pieces kg^{-1} dry weight with the dominance of polyprophylene (PP), PET and polyethylene (PE) in its distribution (Tun et al. 2024). In Brunei Darussalam, MPs are found in coastal areas with the types distribution including PET, high-density polyethylene (HDPE), polyvinyl chloride (PVC), and acrylonitrile butadiene styrene (ABS) (Norshamli and Santos 2023).

Of the four Southeast Asian countries, PET dominates the MPs type that is present in lake waters. This is of course influenced by the activity types that take place around the lake and also waste types that enter the lake water body. Lake waters surrounded by denser urbanization will have a higher abundance of waste than rural areas. It appeared in the MPs distribution of two Philippines lakes. Lake Sampoloc which is surrounded by a highly urbanized area has a higher abundance of MPs compared to Lake Yamto which is a rural lake. The respective abundances of MPs in Lake Sampoloc and Lake Yambo are 483–989 items m^{-3} and 344–789 items m^{-3} (Natuel et al. 2023).

6.2.3 EPs in Deep Groundwater

Some surface water that infiltrates into the ground can become a contaminant for the groundwater. Not only surface water run-off, contaminant pathway into groundwater can also occur through leaching from contaminated soil (Zhao et al. 2022). Groundwater is the main source of clean water in several Southeast Asian countries, namely Indonesia, Vietnam, Myanmar, Lao PDR, Cambodia, and Timor Leste (Lee et al. 2021). Groundwater pollution problem is urgent and needs to be resolved immediately. However, studying pollution in groundwater is quite difficult to conduct because the sampling method and its monitoring are not as simple as method for surface water research.

There are not many references available regarding EPs pollution in groundwater in Southeast Asia countries. Several compounds from the PCPs group which function as artificial sweeteners (ARSs) were found in Singapore groundwater samples, namely acesulphame, cyclamate, and sucralose with concentration range n.d.–730 ng L^{-1}, n.d.–790 ng L^{-1}, and n.d.–780 ng L^{-1}, respectively (Tran et al. 2014). Several PFAS compounds were found in groundwater from the Hanoi City and Ho Chi Minh City areas of Vietnam, namely PFOS as much as 0.19–8.2 ng L^{-1}, PFOA 0.14–4.5 ng L^{-1}, and PFNA 0.15–4.5 ng L^{-1} (Duong et al. 2015; Lam et al. 2017). Moreover, PFOS and PFOA are also found in groundwater around the industrial waste disposal area of Thailand with maximum concentrations of 25.88 ng L^{-1} and 34.96 ng L^{-1}, respectively (Hongkachok et al. 2023).

6.3 Addressing the Challenge of the Emerging Pollutants

In recent years, the global population has faced a significant challenge posed by Emerging Pollutants (EPs). This challenge arises due to the potential risks and diverse effects they pose to both the environment and human health. For instance, the enduring presence of POPs in the ecosystem, their accumulation in human and animal tissues, and their amplification within food chains contribute to notable effects on both human well-being and the environment (Ibrahim et al. 2023; Petrovic et al. 2016).

The use of chemicals that have the potential to become pollutants is unavoidable in human daily activities. These include man-made chemicals for industry, farming, and everyday products, as well as others unintentionally created during industrial processes or burning (by-products). The primary source of emerging pollutants in water comes from urban, industrial, and agricultural wastewater (Mateo-Sagasta et al. 2017). For example, Surfactants, originating from human activities, are prominent organic pollutants with a considerable likelihood of entering the environment, mainly through wastewater discharge, due to their extensive utilization, especially in aqueous solutions (Petrovic et al. 2016).

Wastewater treatment plants (WWTPs) serve as significant pathways for the emergence of diverse micropollutants, accumulating pollutants from various sources and potentially reducing their release into aquatic environments such as lakes and rivers (Arman et al. 2021; Nguyen et al. 2023; Setiadewi et al. 2023). Although wastewater treatment plants are primarily designed to remove pathogens and solid particles, they often lack mechanisms to remove other types of micro-contaminants, leading to their effluents serving as a primary pathway for many emerging contaminants to enter water bodies (Petrovic et al. 2016; Setiadewi et al. 2023). The most problematic pollutants are enduring polar compounds with high water solubility, capable of bypassing both natural filtration processes and conventional treatment methods, posing a potential threat to drinking water sources (ibid).

The impact of emerging pollutants on aquatic ecosystems, including in lakes and rivers, raises significant concerns. Lake and river ecosystems are necessary freshwater resources closely linked to human life. EPs can easily become a pollutant to river ecosystems where poorly treated effluents from wastewater treatment plants (WWTPs) are discharged that can accumulate in sediments and river flora and fauna due to their persistence (Vasilachi et al. 2021). First example, the presence of pharmaceuticals and PCPs in drinking water is a crucial issue as it directly affects the quality of water supplies, including surface water bodies and groundwater, with ongoing debate about whether these compounds pose a risk to human health (Stefanakis and Becker 2019). Typical pharmaceutical concentrations found in drinking water, usually lower than 0.05 μg/L and below therapeutic doses, are not considered a direct threat to human health, and regulatory guidelines for these compounds are lacking (WHO 2011). While the concentrations are typically low, there is growing concern about possible long-term effects on both people and aquatic life due to ongoing exposure to these substances in the environment. Although these levels are not high enough to cause immediate harm to humans, meaning they are much lower than the recommended prescription doses, they have been observed to impact aquatic ecosystems (Petrovic et al. 2016).

In addition, Perfluorinated Compounds (PFCs) have been detected in various parts of aquatic ecosystems, such as lakes, where they were found at concentrations at levels of picogram per liter in lakes, nanogram per liter in rivers, at levels of nanogram per gram for precipitation water, soils, and sediments, and also in biota samples like fish, where concentrations reach micrograms per kilogram (Llorca et al. 2011). They demonstrate resistance to degradation and can accumulate in organisms, potentially becoming more concentrated as they move up the food chain. PFCs are introduced into the environment through various ways, such as wastewater streams as direct ways and indirect pathways from their decomposition or disposal (Petrovic et al. 2016).

Moreover, the presence of microplastics (MPs) in aquatic environments exacerbates the complexity of pollutant interactions. This occurs because MPs can interact with other contaminants present in their surround. Much like Perflourinated Compounds (PFCs), MPs serve as carriers for hazardous substances, releasing toxic compounds and accumulating in organisms. Once in the environment, microplastics tend to absorb higher levels of micropollutants and more readily infiltrate living organisms, leading to consequences that are more severe compared to larger plastic particles (Nguyen et al. 2023). MPs also have the ability to leach by process, like photooxidation and biological interaction in the aquatic environment (Priya et al. 2022). This phenomenon will influence the environmental fate and behavior of the pollutants, which is usually called bioaccumulation, biomagnification, degradation, etc., and also pose a significant impact to the aquatic ecosystem in that environment (Nguyen et al. 2023).

Emerging pollutants can also lead to widespread pollution due to various activities. The leaching from fertilizers, the run-off from pesticides applied in farming and forestry, or the deposition of industrial pollutants through the air can all have negative impacts on the purity (quality) of surface and groundwater (Petrovic et al. 2016). In fact, EPs in river water can change into less harmful substances, but they can also stick to sediment, get carried to other water bodies where sensitive plants and animals live or seep into groundwater, which is often used for drinking, and become even more toxic (Al Aukidy et al. 2012). The cross-contamination that occurred, for example, in groundwater, might be a decline in water quality as a drinking water source and the emergence of health concerns as a long-term effect. Given that drinking water treatment plants depend on water resources, the widespread occurrence of such contamination in aquatic environments, especially surface water, poses a significant challenge (Arman et al. 2021).

A big concern about the occurrence of EPs in the environment is the detrimental impacts on ecosystems, including biota and humans which can be called as toxicological effect of EPs (Arman et al. 2021). EPs have a significant effect on the health aspect of aquatic living organisms, either directly or indirectly, depending on their fate and transport in the environment. Literature showed that some Emerging Pollutants (EPs) are known to have biological effects on ecosystems and aquatic life, with aquatic organisms often exhibiting greater sensitivity than humans due to potential disruptions in the endocrine system (Stefanakis and Becker 2019). For example, both natural and synthetic steroid hormones are considered the most powerful (potent) disruptors of the endocrine system. They are expelled in mammalian urine and enter the water bodies through wastewater treatment plant discharges or untreated discharges. These strong estrogenic substances have been proven to trigger estrogenic reactions in fish even at low water concentrations (0.1–1 ng L^{-1}). Such levels are occasionally detected downstream of wastewater treatment plant discharges and have been linked to reproductive and developmental issues in fish, such as feminization, reduced fertility, or hermaphroditism (Petrovic et al. 2016).

Endocrine Disruption Chemicals (EDCs) are synthetically or naturally occurring substances that have the ability to inhibit or imitate hormones, thereby impacting the endocrine systems of living organisms (Bharagava and Chowdhary 2018). The occurrence of EDCs in the environment could be present in such as pesticides, pharmaceutical products, personal care products (PCPs), bisphenol A (BPA), or other chemical products (vom Saal and Hughes 2005). Those micro-contaminants can have an impact on fish, such as estrogenic effects, sexual traits and gender ratio alterations, egg fertilization and feminization reduction in fish, etc. (Richardson and Ternes 2018). In addition to that, Disinfection By-Products (DBPs), which are usually detected in drinking water or swimming pools, can cause some problems to the biota exposed to aquatic life, such as inhalation, ingestion, dermal contact (Gan et al. 2013).

Other EPs, like plasticizers, can also cause several impacts on health and biota metabolism. BPA can cause sexual development disorders and also the quality of organs inside, like sperm, ovary, hormonal production, and even cancer (Arman et al. 2021). Triclosan poses a threat to fish embryos, algae, crustaceans, and fish, also potentially harming phytoplankton at particular concentrations where these biota are typically found and accustomed to living in freshwater, including inland waters, such as in rivers or lakes (Srain et al. 2021). The consumption of microplastics by living organisms and their subsequent transfer through food webs can also pose significant risks to both human health and ecological stability (Nguyen et al. 2023).

The alarming potential of EPs is when it enters the food chain. Due to the tiny size of pollutants or particles of EPs, they can be easily consumed by organisms like fish, crabs, aquatic mammals, etc., including those contained in human food, like in seafood. This exposure raises a significant risk and negative effect to the health of organisms and humans. Humans are at risk of endocrine-disrupting chemicals (EDCs) through contaminated and accumulated food and beverages because humans are at the highest level of the food chain (Kumar et al. 2022). These substances may disrupt various bodily systems, such as gastrointestinal issues, respiratory problems, immune system dysfunction, infertility or reproductive organs, nervous system, changes in chromosomes, etc. (Nguyen et al. 2023; Srain et al. 2021). However, the direct impact on humans is still being explored, while the impact of emerging contaminants on animals is well documented (Kumar et al. 2022).

The major challenge today lies in assessing the potential ecological impacts associated with the emergence of contaminants in aquatic environments and establishing connections between emerging contaminant pollution and ecosystem quality. However, directly assessing the effects of environmental pollutants on ecosystems is complex due to insufficient data on the occurrence, behavior, and long-term effects of many emerging contaminants in the environment. Detection of the contaminants in the environment is quite challenging due to the required low detection limits, the complex nature of the samples, and the difficulty in separating the compounds from interferences (Snow et al. 2018). Long-term exposure to these chemicals can be harmful despite EC values in aquatic sources being found in the range of ng L^{-1} to g L^{-1} (Gheorghe et al. 2016; Liu et al. 2020; Rout et al. 2021). Furthermore, the eco-toxicological significance of numerous emerging contaminants and the impacts of combined exposure to multiple chemicals are largely unexplored, necessitating the development of new tools to evaluate the environmental risks posed by emerging contaminants.

6.4 The Road Ahead

Population growth both in developing and developed countries will eventually influence the increasing amount of pollutants in the environment. Especially for countries in Southeast Asia which are mostly included as developing country, their environment is very vulnerable to the EPs threat as the increase in population will be proportional to the increase of intensity and concentration of pollutants that enter water bodies (Lam et al. 2017).

Limited references regarding EPs in Southeast Asian countries ring the bell for insight into EPs pollution in the environment cannot be understood holistically yet. On the other hand, these new contaminants need to be controlled immediately. However, the obstacles in implementing EPs research must first be resolved. The development of technology and methods for analyzing EPs compounds using available instruments must continue to be carried out in creative way without reducing their reliability while still paying attention to environmental condition and its sustainability.

Aside, the limitations of instruments for qualitative and quantitative analysis can be overcome by establishing collaboration between researchers across the Southeast Asia region. Research collaboration will accelerate and equalize perceptions on information regarding EPs topics. Furthermore, the occurrence of EPs compound can be studied broader and monitored so that the existing database will continue to be updated. Hence, the efforts to mitigate the efficacy of EPs hazard to the environment, including humans, can be carried out using appropriate and targeted methods.

The holistic system which will be built should not only be used for determining the occurrence, distribution, fate, and impact of EPs, but also to develop a possible environmental protection system. For EPs regulation, the government's role is to put EPs into environment regulation so that the use and monitoring of presence of EPs can be monitored. In addition, law enforcement needs to be stricter that the sustainability of the water environment can always be maintained.

In the end, it is impossible to avoid the presence of EPs in the environment, but the potential hazards posed from excessive use of compounds and inappropriate waste management can still be managed by continuing to monitor their presence in the environment along with the development of appropriate holistic mitigation systems.

References

Akber MA, Aziz AA, Lovelock C (2020) Major drivers of coastal aquaculture expansion in Southeast Asia. Ocean Coast Manag 198(Aug):105364. https://doi.org/10.1016/j.ocecoaman.2020.105364

Al Aukidy M, Verlicchi P, Jelic A, Petrovic M, Barcelò D (2012) Monitoring release of pharmaceutical compounds: occurrence and environmental risk assessment of two WWTP effluents and their receiving bodies in the Po Valley, Italy. Sci Total Environ 438:15–25. https://doi.org/10.1016/j.scitotenv.2012.08.061

Anh HQ, Le TPQ, Da Le N, Lu XX, Duong TT, Garnier J, Rochelle-Newall E, Zhang S, Oh NH, Oeurng C, Ekkawatpanit C, Nguyen TD, Nguyen QT, Nguyen TD, Nguyen TN, Tran TL, Kunisue T, Tanoue R, Takahashi S, Nguyen TAH (2021) Antibiotics in surface water of East and Southeast Asian countries: a focused review on contamination status, pollution sources, potential risks, and future perspectives. Sci Total Environ 764. https://doi.org/10.1016/j.scitotenv.2020.142865

Arman NZ, Salmiati S, Aris A, Salim MR, Nazifa TH, Muhamad MS, Marpongahtun M (2021) A review on emerging pollutants in the water environment: existences, health effects and treatment processes. Water (Switz) 13(22):1–31. https://doi.org/10.3390/w13223258

Bharagava RN, Chowdhary P (2018) Emerging and eco-friendly approaches for waste management. In: Emerging and eco-friendly approaches for waste management, pp 1–435. https://doi.org/10.1007/978-981-10-8669-4

Diaz-Sosa VR, Tapia-Salazar M, Wanner J, Cardenas-Chavez DL (2020) Monitoring and ecotoxicity assessment of emerging contaminants in wastewater discharge in the city of Prague (Czech Republic). Water (Switz) 12(4). https://doi.org/10.3390/W12041079

Dimzon IKD, Morata AS, Müller J, Yanela RK, Lebertz S, Weil H, Perez TR, Müller J, Dayrit FM, Knepper TP (2018) Trace organic chemical pollutants from the lake waters of San Pablo City, Philippines by targeted and non-targeted analysis. Sci Total Environ 639:588–595. https://doi.org/10.1016/j.scitotenv.2018.05.217

Dsikowitzky L, Sträter M, Dwiyitno, Ariyani F, Irianto HE, Schwarzbauer J (2016) First comprehensive screening of lipophilic organic contaminants in surface waters of the megacity Jakarta, Indonesia. Mar Pollut Bull 110(2):654–664. https://doi.org/10.1016/j.marpolbul.2016.02.019

Duong HT, Kadokami K, Shirasaka H, Hidaka R, Chau HTC, Kong L, Nguyen TQ, Nguyen TT (2015) Occurrence of perfluoroalkyl acids in environmental waters in Vietnam. Chemosphere 122:115–124. https://doi.org/10.1016/j.chemosphere.2014.11.023

Feng W, Deng Y, Yang F, Miao Q, Ngien SK (2023) Systematic review of contaminants of emerging concern (CECs): distribution, risks, and implications for water quality and health. Water (Switz) 15(22). https://doi.org/10.3390/w15223922

Gan W, Guo W, Mo J, He Y, Liu Y, Liu W, Liang Y, Yang X (2013) The occurrence of disinfection by-products in municipal drinking water in China's Pearl River Delta and a multipathway cancer risk assessment. Sci Total Environ 447:108–115. https://doi.org/10.1016/j.scitotenv.2012.12.091

Geissen V, Mol H, Klumpp E, Umlauf G, Nadal M, van der Ploeg M, van de Zee SEATM, Ritsema CJ (2015) Emerging pollutants in the environment: a challenge for water resource management. Int Soil Water Conserv Res 3(1):57–65. https://doi.org/10.1016/j.iswcr.2015.03.002

Gheorghe S, Petre J, Lucaciu I, Stoica C, Nita-Lazar M (2016) Risk screening of pharmaceutical compounds in Romanian aquatic environment. Environ Monit Assess 188(6). https://doi.org/10.1007/s10661-016-5375-3

Guardian MGE, Boongaling EG, Bernardo-Boongaling VRR, Gamonchuang J, Boontongto T, Burakham R, Arnnok P, Aga DS (2020) Prevalence of per- and polyfluoroalkyl substances (PFASs) in drinking and source water from two Asian countries. Chemosphere 256. https://doi.org/10.1016/j.chemosphere.2020.127115

Hassan I, Sethupathi S, Bashir MJK, Munusamy Y, Chan CW (2024) A systematic review of microplastics occurrence, characteristics, identification techniques and removal methods in

ASEAN and its future prospects. J Environ Chem Eng 12(2):112305. https://doi.org/10.1016/j.jece.2024.112305

Henny C, Rohaningsih D, Suryono T, Santoso AB, Waluyo A (2022) Microplastic pollution in the surface water of Lake Singkarak, Indonesia. IOP Conf Ser Earth Environ Sci 1118(1). https://doi.org/10.1088/1755-1315/1118/1/012050

Hongkachok C, Boontanon SK, Boontanon N, Sukeesan S, Polprasert C, Fujii S (2023) Per- and polyfluoroalkyl substances in groundwater in Thailand: occurrence, source identification and spatial distribution. Int J Environ Sci Technol 20(11):12437–12448. https://doi.org/10.1007/s13762-023-04894-w

Hu J, Yu J, Tanaka S, Fujii S (2011) Perfluorooctane sulfonate (PFOS) and perfluorooctanoic acid (PFOA) in water environment of Singapore. Water Air Soil Pollut 216(1–4):179–191. https://doi.org/10.1007/s11270-010-0525-7

Huang A (2022) Inter- and intra-species sensitivity of aquatic arthropods to imidacloprid and flupyradifurone. Wageningen University. https://doi.org/10.18174/574627

Ibrahim A, Muhamad SS, Ardiwinata AN, Sugiarti, Adam MA, Fitrada W, Riky K (2023) Organochlorine and pyrethroid residue in fish and sediment of Lake Singkarak, a tropical deep lake. J Limnol Water Resour 9:356–363. https://doi.org/10.55981/limnotek.2023.2084

Ismanto A, Hadibarata T, Kristanti RA, Maslukah L, Safinatunnajah N, Sathishkumar P (2022) The abundance of endocrine-disrupting chemicals (EDCs) in downstream of the Bengawan Solo and Brantas rivers located in Indonesia. Chemosphere 297(Feb):134151. https://doi.org/10.1016/j.chemosphere.2022.134151

Katrina KM, Espino MP (2020) Occurrence and distribution of hormones and bisphenol A in Laguna Lake, Philippines. Chemosphere 256:127122. https://doi.org/10.1016/j.chemosphere.2020.127122

Kumar R, Qureshi M, Vishwakarma DK, Al-Ansari N, Kuriqi A, Elbeltagi A, Saraswat A (2022) A review on emerging water contaminants and the application of sustainable removal technologies. Case Stud Chem Environ Eng 6(May):100219. https://doi.org/10.1016/j.cscee.2022.100219

Kunacheva C, Boontanon SK, Fujii S, Tanaka S, Musirat C, Artsalee C, Wongwattana T (2009) Contamination of perfluorinated compounds (PFCs) in Chao Phraya river and Bangpakong river, Thailand. Water Sci Technol 60(4):975–982. https://doi.org/10.2166/wst.2009.462

Lam NH, Cho CR, Kannan K, Cho HS (2017) A nationwide survey of perfluorinated alkyl substances in waters, sediment and biota collected from aquatic environment in Vietnam: distributions and bioconcentration profiles. J Hazard Mater 323:116–127. https://doi.org/10.1016/j.jhazmat.2016.04.010

Le TDH, Pham LH, Dinh QT, Le TMT, Tram NTB (2023) Land use influencing the distribution of pesticides in surface water: the case of the Ma River and its tributaries in Thanh Hoa province, Vietnam. Arch Environ Contam Toxicol 85(3):290–301. https://doi.org/10.1007/s00244-023-01018-w

Lee THY, Chuah J, Snyder SA (2021) Occurrence of emerging contaminants in Southeast Asian environments: present status, challenges, and future prospects. ACS ES&T Water. https://doi.org/10.1021/acsestwater.1c00453

Li B, Xu D, Zhou X, Yin Y, Feng L, Liu Y, Zhang, L (2023) Environmental behaviors of emerging contaminants in freshwater ecosystem dominated by submerged plants: a review. Environ Res 227(Nov 2022). https://doi.org/10.1016/j.envres.2023.115709

Liu N, Jin X, Feng C, Wang Z, Wu F, Johnson AC, Xiao H, Hollert H, Giesy JP (2020) Ecological risk assessment of fifty pharmaceuticals and personal care products (PPCPs) in Chinese surface waters: a proposed multiple-level system. Environ Int 136(8):105454. https://doi.org/10.1016/j.envint.2019.105454

Llorca M, Farré M, Picó Y, Barceló D (2011) Analysis of perfluorinated compounds in sewage sludge by pressurized solvent extraction followed by liquid chromatography-mass spectrometry. J Chromatogr A 1218(30):4840–4846. https://doi.org/10.1016/j.chroma.2011.01.085

Lydy MJ, Belden JB, Ternes MA (1999) Effects of temperature on the toxicity of M-parathion, chlorpyrifos, and pentachlorobenzene to *Chironomus tentans* environmental contamination and toxicology. J Lydy Arch Environ Contam Toxicol 37:542–547

Mangold-Döring A, Huang A, Van Nes EH, Focks A, Van Den Brink PJ (2022) Explicit consideration of temperature improves predictions of toxicokinetic-toxicodynamic models for flupyradifurone and imidacloprid in *Gammarus pulex*. Environ Sci Technol 56(22):15920–15929. https://doi.org/10.1021/acs.est.2c04085

Mariano SMF, Angeles LF, Aga DS, Villanoy CL, Jaraula CMB (2023) Emerging pharmaceutical contaminants in key aquatic environments of the Philippines. Front Earth Sci 11(Sept):1–15. https://doi.org/10.3389/feart.2023.1124313

Mateo-Sagasta J, Zadeh SM, Turral H, Burke J (2017) Water pollution from agriculture: a global review. The Food and Agricultural Organization, pp 1–35

Mishra RK, Mentha SS, Misra Y, Dwivedi N (2023) Emerging pollutants of severe environmental concern in water and wastewater: a comprehensive review on current developments and future research. Water-Energy Nexus 6:74–95. https://doi.org/10.1016/j.wen.2023.08.002

Natuel FA, Magcale-Macandog DB, Faustino-Eslava DV, Cui LE, Hotes S (2023) Microplastic occurrence in rural and urban surface waters: the cases of Lake Sampaloc and Lake Yambo in San Pablo City, Laguna, Philippines. Sci Eng J 16

Nazifa TH, Kristanti RA, Ike M, Kuroda M, Hadibarata T (2020) Occurrence and distribution of estrogenic chemicals in river waters of Malaysia. Toxicol Environ Heal Sci 12(1):65–74. https://doi.org/10.1007/s13530-020-00036-8

Ngo TH, Van DA, Tran HL, Nakada N, Tanaka H, Huynh TH (2021) Occurrence of pharmaceutical and personal care products in Cau river, Vietnam. Environ Sci Pollut Res 28(10):12082–12091. https://doi.org/10.1007/s11356-020-09195-0

Nguyen MK, Lin C, Nguyen HL, Le VR, Priya KL, Singh J, Chang SW, Um MJ, Nguyen DD (2023) Emergence of microplastics in the aquatic ecosystem and their potential effects on health risks: the insights into Vietnam. J Environ Manag 344(June). https://doi.org/10.1016/j.jenvman.2023.118499

Norshamli AZ, Santos JH (2023) Occurrence and distribution of microplastics on coastal intertidal zones of Brunei Darussalam. Eur Chem Bull 12(8):2694–2712

Noudeng V, Van Quan N, Xuan TD, Vathanamixay C, Souvanna P (2023) Risk assessment of microplastics dispersion and accumulation in urban canals to the water environment in Vientiane capital, Laos. Water Air Soil Pollut 234(9). https://doi.org/10.1007/s11270-023-06587-z

Nozaki K, Tanoue R, Kunisue T, Tue NM, Fujii S, Sudo N, Isobe T, Nakayama K, Sudaryanto A, Subramanian A, Bulbule KA, Parthasarathy P, Tuyen LH, Viet PH, Kondo M, Tanabe S, Nomiyama K (2023) Pharmaceuticals and personal care products (PPCPs) in surface water and fish from three Asian countries: species-specific bioaccumulation and potential ecological risks. Sci Total Environ 866(Sept 2022):161258. https://doi.org/10.1016/j.scitotenv.2022.161258

Ocharoen Y, Boonphakdee C, Boonphakdee T, Shinn AP, Moonmangmee S (2018) High levels of the endocrine disruptors bisphenol-A and 17β-estradiol detected in populations of green mussel, *Perna viridis*, cultured in the Gulf of Thailand. Aquaculture 497(July):348–356. https://doi.org/10.1016/j.aquaculture.2018.07.057

Omar TFT, Aris AZ, Yusoff FM, Mustafa S (2019) Risk assessment of pharmaceutically active compounds (PhACs) in the Klang river estuary, Malaysia. Environ Geochem Health 41(1):211–223. https://doi.org/10.1007/s10653-018-0157-1

Pariatamby A, Hamid FS, Bhatti MS, Anuar N, Anuar N (2020) Status of microplastic pollution in aquatic ecosystem with a case study on Cherating river, Malaysia. J Eng Technol Sci 52(2):222–241. https://doi.org/10.5614/j.eng.technol.sci.2020.52.2.7

Petrovic M, Elosegi A, Sabater S, Barcelo D (2016) Emerging contaminants in river ecosystems. In: The handbook of environmental chemistry. https://doi.org/10.1007/978-3-319-29376-9

Praveena SM, Shaifuddin SNM, Sukiman S, Nasir FAM, Hanafi Z, Kamarudin N, Ismail THT, Aris AZ (2018) Pharmaceuticals residues in selected tropical surface water bodies from Selangor

(Malaysia): occurrence and potential risk assessments. Sci Total Environ 642:230–240. https://doi.org/10.1016/j.scitotenv.2018.06.058

Preisendanz HE, Barnes RG, Mashtare ML, Lintern A, Mina O, Williams C, Elliott HA (2021) The emergence, trajectory, and impacts of emerging contaminants publications. J Environ Qual 50(6):1339–1346. https://doi.org/10.1002/jeq2.20299

Priya AK, Jalil AA, Dutta K, Rajendran S, Vasseghian Y, Qin J, Soto-Moscoso M (2022) Microplastics in the environment: recent developments in characteristic, occurrence, identification and ecological risk. Chemosphere 298(Feb):134161. https://doi.org/10.1016/j.chemosphere.2022.134161

Richardson SD, Ternes TA (2018) Water analysis: emerging contaminants and current issues. Anal Chem 90(1):398–428. https://doi.org/10.1021/acs.analchem.7b04577

Rohaningsih D, Henny C, Suryono T, Santoso AB (2022) Macroplastic abundance at Lake Singkarak riparian, West Sumatera. IOP Conf Ser Earth Environ Sci 1062(1). https://doi.org/10.1088/1755-1315/1062/1/012025

Rout PR, Zhang TC, Bhunia P, Surampalli RY (2021) Treatment technologies for emerging contaminants in wastewater treatment plants: a review. Sci Total Environ 753:141990. https://doi.org/10.1016/j.scitotenv.2020.141990

Saadati N, Abdullah MP, Zakaria Z, Rezayi M, Hosseinizare N (2012) Distribution and fate of HCH isomers and DDT metabolites in a tropical environment-case study Cameron Highlands-Malaysia. Chem Cent J 6(1). https://doi.org/10.1186/1752-153X-6-130

Sakai N, Dayana E, Abu Bakar A, Yoneda M, Nik Sulaiman NM, Ali Mohd M (2016) Occurrence, distribution, and dechlorination of polychlorinated biphenyls and health risk assessment in Selangor river basin. Environ Monit Assess 188(10). https://doi.org/10.1007/s10661-016-5595-6

Sang D, Cimetiere N, Giraudet S, Tan R, Wolbert D, Cloirec PL (2022) Online SPE-UPLC-MS/MS for herbicides and pharmaceuticals compounds' determination in water environment: a case study in France and Cambodia. Environ Adv 8(Feb):100212. https://doi.org/10.1016/j.envadv.2022.100212

Schuijt LM (2023) Aquatic life on drugs: assessing the ecological impacts of pharmaceuticals on aquatic ecosystems. Wageningen University. https://doi.org/10.18174/633024

Sembiring E, Fareza AA, Suendo V, Reza M (2020) The presence of microplastics in water, sediment, and milkfish (*Chanos chanos*) at the downstream area of Citarum river, Indonesia. Water Air Soil Pollut 231(7). https://doi.org/10.1007/s11270-020-04710-y

Setiadewi N, Henny C, Rohaningsih D, Waluyo A, Soewondo P (2023) Microplastic occurrence and characteristics in a municipal wastewater treatment plant in Jakarta. IOP Conf Ser Environ Earth Sci 1201. https://doi.org/10.1088/1755-1315/1201/1/012053

Sharip Z, Hashim N, Suratman S (2017) Occurrence of organochlorine pesticides in a tropical lake basin. Environ Monit Assess 189(11). https://doi.org/10.1007/s10661-017-6274-y

Shimizu A, Takada H, Koike T, Takeshita A, Saha M, Rinawati NN, Murata A, Suzuki T, Suzuki S, Chiem NH, Tuyen BC, Viet PH, Siringan MA, Kwan C, Zakaria MP, Reungsang A (2013) Ubiquitous occurrence of sulfonamides in tropical Asian waters. Sci Total Environ 452–453:108–115. https://doi.org/10.1016/j.scitotenv.2013.02.027

Snow DD, Cassada DA, Biswas S, Shafieifini M, Li X, D'Alessio M, Carter L, Sallach JB (2018) Detection, occurrence and fate of emerging contaminants in agricultural environments. Water Environ Res 90(10):1348–1370. https://doi.org/10.2175/106143018x15289915807254

Srain HS, Beazley KF, Walker TR (2021) Pharmaceuticals and personal care products and their sublethal and lethal effects in aquatic organisms. Environ Rev 29(2):142–181. https://doi.org/10.1139/er-2020-0054

Stefanakis AI, Becker JA (2019) A review of emerging contaminants in water: classification, sources, and potential risks. In: Waste management: concepts, methodologies, tools, and applications, pp 177–202. https://doi.org/10.4018/978-1-7998-1210-4.ch008

Suzuki S, Ogo M, Takada H, Seki K, Mizukawa K, Kadoya A, Yokokawa T, Sugimoto Y, Sato-Takabe Y, Boonla C, Anomasiri W, Sukpanyatham N (2021) Contamination of antibiotics and sul and tet(M) genes in veterinary wastewater, river, and coastal sea in Thailand. Sci Total Environ 791:148423. https://doi.org/10.1016/j.scitotenv.2021.148423

Tagorti G, Kaya B (2022) Genotoxic effect of microplastics and COVID-19: The hidden threat. Chemosphere 286(P3):131898. https://doi.org/10.1016/j.chemosphere.2021.131898

Tan ESS, Ho YB, Zakaria MP, Latif PA, Saari N (2015) Simultaneous extraction and determination of pharmaceuticals and personal care products (PPCPs) in river water and sewage by solid-phase extraction and liquid chromatography-tandem mass spectrometry. Int J Environ Anal Chem 95(9):816–832. https://doi.org/10.1080/03067319.2015.1058929

Tewari S, Jindal R, Kho YL, Eo S, Choi K (2013) Major pharmaceutical residues in wastewater treatment plants and receiving waters in Bangkok, Thailand, and associated ecological risks. Chemosphere 91(5):697–704. https://doi.org/10.1016/j.chemosphere.2012.12.042

Tran NH, Li J, Hu J, Ong SL (2014) Occurrence and suitability of pharmaceuticals and personal care products as molecular markers for raw wastewater contamination in surface water and groundwater. Environ Sci Pollut Res 21(6):4727–4740. https://doi.org/10.1007/s11356-013-2428-9

Tran NH, Hoang L, Nghiem LD, Nguyen NMH, Ngo HH, Guo W, Trinh QT, Mai NH, Chen H, Nguyen DD, Ta TT, Gin KYH (2019) Occurrence and risk assessment of multiple classes of antibiotics in urban canals and lakes in Hanoi, Vietnam. Sci Total Environ 692:157–174. https://doi.org/10.1016/j.scitotenv.2019.07.092

Tun TZ, Mon EE, Nakata H (2024) Microplastics distribution in sediments collected from Myanmar. Arch Environ Contam Toxicol 86(1):1–12. https://doi.org/10.1007/s00244-023-01042-w

Vasilachi IC, Asiminicesei DM, Fertu DI, Gavrilescu M (2021) Occurrence and fate of emerging pollutants in water environment and options for their removal. Water (Switz) 13(2):1–34. https://doi.org/10.3390/w13020181

vom Saal FS, Hughes C (2005) An extensive new literature concerning low-dose effects of bisphenol A shows the need for a new risk assessment. Environ Health Perspect 113(8):926–933. https://doi.org/10.1289/ehp.7713

Waleng NJ, Nomngongo PN (2022) Occurrence of pharmaceuticals in the environmental waters: African and Asian perspectives. Environ Chem Ecotoxicol 4(Nov 2021):50–66. https://doi.org/10.1016/j.enceco.2021.11.00

WHO (2011) Guidelines for drinking-water quality, 4th edn. World Health Organization

Yang J, Zhao Y, Li M, Du M, Li X, Li Y (2019) A review of a class of emerging contaminants: the classification, distribution, intensity of consumption, synthesis routes, environmental effects and expectation of pollution abatement to organophosphate flame retardants (OPFRs). Int J Mol Sci 20(12). https://doi.org/10.3390/ijms20122874

Zaki MRM, Ying PX, Zainuddin AH, Razak MR, Aris AZ (2021) Occurrence, abundance, and distribution of microplastics pollution: an evidence in surface tropical water of Klang River estuary, Malaysia. Environ Geochem Health 43(9):3733–3748. https://doi.org/10.1007/s10653-021-00872-8

Zhao Z, Li J, Zhang X, Wang L, Wang J, Lin T (2022) Perfluoroalkyl and polyfluoroalkyl substances (PFASs) in groundwater: current understandings and challenges to overcome. Environ Sci Pollut Res 29(33):49513–49533. https://doi.org/10.1007/s11356-022-20755-4

Chapter 7
Biotic Communities: Endemic and Invasive Fish

Gadis Sri Haryani, Livia Rossila Tanjung, Djamhuriyah S. Said, Syahroma Husni Nasution, Rahmi Dina, Rudhy Gustiano, and Triyanto

Life within aquatic ecosystems represents a delicate balance, where every species plays a unique role in preserving environmental harmony. One crucial element of biotic diversity in aquatic environments is the fish population, often serving as a primary indicator of ecosystem health. However, significant challenges arise from two distinct groups of fish that play different roles: endemic fish, naturally involved in local ecological cycles, and invasive fish, introducing new imbalances that threaten biotic sustainability. In many waters worldwide, we witness the presence of endemic fish that have been integral to local ecosystems for centuries. Their existence contributes to species diversity and supports the sustainability of food cycles and the harmonious balance of nature. On the flip side, the invasion of introduced fish poses a severe threat, carrying consequences that extend far beyond normal ecological boundaries. Issues arise when invasive fish replace endemic species, significantly altering the dynamics of biotic communities.

The threats caused by invasive fish to endemic species occur in almost every country, including ASEAN members such as Malaysia, the Philippines, and Indonesia. This article details the threats, problems, and impacts caused by the presence of invasive species on endemic fish in the biotic communities of freshwater ecosystems, particularly those in Indonesia. Indonesia is the second highest in the world for freshwater fish diversity, trailing only behind Brazil.

Hubert et al. (2015) reported that Indonesia has 1218 species of freshwater fish belonging to 84 families. Of these, there were 1172 native species from 79 families, among which 630 species were endemic. Further identification and documentation of

G. S. Haryani (✉) · L. R. Tanjung · D. S. Said · S. H. Nasution · R. Dina · Triyanto
Research Center for Limnology and Water Resources, National Research and Innovation Agency (BRIN), Jakarta, Indonesia
e-mail: gadi001@brin.go.id

R. Gustiano
Research Center for Biosystematics and Evolution, National Research and Innovation Agency (BRIN), Jakarta, Indonesia

© The Author(s), under exclusive license to Springer Nature Singapore Pte Ltd. 2025
A. C. Santos-Borja et al. (eds.), *Progress on Ecosystem Restoration of Tropical Inland Waters*, https://doi.org/10.1007/978-981-96-2284-9_7

the diversity of fish species showed that Indonesia has approximately 1700 freshwater species (Gustiano et al. 2021). However, FishBase recorded 1272 freshwater fish species in Indonesia, including 141 endemic species (Froese and Pauly 2023). The remaining 428 species (25%) may not have been reported. Given that there were 630 endemic species in 2015, the unreported 428 species were more likely to be endemic.

If the unreported 428 species were added to the recorded 141 species, the number of endemic species reaches 569, with a difference of 61 species. The decline of 61 species illustrated the extinctions that may have occurred over the past eight years. The increase in the number of recorded species and the decline in the number of endemic species during the eight years are believed to be related to the discovery of new species accompanied by external factors that caused a decline in these endemic species, showing the richness and vulnerability of aquatic ecosystems. However, this noteworthy diversity continues to face threats from various human activities that have the potential to hasten the extinction of endemic fish species. Factors that threaten and cause the decline in fish biodiversity in inland waters include habitat degradation, overexploitation, flow modification, water pollution, and invasion of introduced species (Dudgeon et al. 2006). Therefore, special consideration must be taken since most Indonesian freshwater fish threatened with extinction are generally endemic species (Umar et al. 2015).

Through a profound understanding of these two groups (endemic and invasive species), necessary protection and management measures can be identified to preserve the sustainability of aquatic ecosystems. Additionally, this article is crucial for providing in-depth insights into the ecological implications of interactions between endemic and invasive fish. Based on research results and various references, this review forms the basis for effective conservation policies and management actions to protect biotic diversity in freshwater ecosystems.

7.1 Endemic Fishes in Inland Aquatic Ecosystems: Status and Threat

Endemic fish refer to species naturally found exclusively in a specific geographical region, such as a lake, river, or water system. These fish are restricted to a particular habitat and are not naturally found elsewhere. This definition emphasizes the limited and specific geographic distribution of a fish species.

One of the critical characteristics of endemic fish is their highly restricted distribution. They are typically confined to specific water systems and do not naturally extend into other regions. Endemic fish often exhibit specialized physical and behavioral adaptations that enable them to thrive in their specific environments. These adaptations may involve coloration or body shapes tailored to local water conditions or feeding patterns suited to local food resources. At times, endemic fish also form

Table 7.1 The number of endemic fish species in the five biggest Indonesian Islands

Island	Endemic species	Reference
Sumatra	73	Dahruddin et al. (2021)
Java	16	Jubaedah and Sayida (2022)
Borneo	294	Hubert et al. (2015)
Sulawesi	68	Hadiaty (2018)
Papua	124	Hubert et al. (2015)

unique relationships with other species within the aquatic ecosystem. This relationship can include dependence on other organisms for food or play crucial roles in the local food chain.

Due to ecological and biological barriers, endemic fish are frequently found in isolated or remote ecosystems, such as specific lakes or small rivers. These conditions contribute to their limited distribution. Unique characteristics may encompass dependence on specific environmental factors, such as particular water temperatures, acidity levels, or habitat structures. Due to their specialized adaptations for these habitats, they are susceptible to threats originating from external sources.

Asia is the site of some 3500 recorded freshwater fish species. Nearly a third of these fishes are found in the archipelago islands such as Sumatra, Java, and Borneo, and more than half are endemic to the country's freshwaters (Kottelat and Witten 1996). Hubert et al. (2015) showed that the 630 endemic fish species consist of 431 species in the Sunda Shelf, 82 in the Wallacea region, and 125 in the Sahul Shelf.

Recent data showed that 73 endemic fish species have been identified in the fresh waters of Sumatra (Dahruddin et al. 2021). In the freshwater ecosystem of Java, 16 endemic fish species have been recorded (Jubaedah and Sayida 2022). This data (Table 7.1) indicates a 48% decrease in the population of endemic fish species over seven years compared to Hubert et al. (2015). In Borneo, Hubert et al. (2015) identified 294 endemic fish species, almost double the previously identified by Kottelat et al. (1993). Hadiaty (2018) categorized the endemic ichthyofauna in Sulawesi's freshwater ecosystem into 45 species that inhabit lakes and 23 species that inhabit rivers. In West Papua, 124 freshwater endemic fish species were recorded (Hubert et al. 2015). However, the recent data regarding endemic freshwater fish identified on the island is nonexistent, so the current number of endemic fish is unknown (Table 7.1).

7.1.1 Sumatra

The inland waters of Sumatra, including Lake Laut Tawar, Lake Toba, Lake Maninjau, Lake Singkarak, Batanghari River, and Musi River, are home to various endemic fish.

Lake Laut Tawar, located in Central Aceh of Aceh Province, is the habitat of two species of endemic small fish, *Poropuntius tawarensis* (Weber and de Beaufort 1916) and *Rasbora tawarensis* (Weber and de Beaufort 1916), also known as Kawan and Depik in vernacular (Muchlisin et al. 2010). *Poropuntius tawarensis* (Fig. 7.1) attains a maximum length of 13.7 cm (Kottelat et al. 1993) and is classified as Critically Endangered (Lumbantobing 2020f).

R. tawarensis is present throughout Lake Laut Tawar (Fig. 7.2) but is more prevalent in the shallow waters near the shore (Muchlisin 2013). So, fishermen tend to catch fish in this area, which leads to overfishing and causes the size of the fish caught to decrease. Conversely, fish living in offshore deep waters tend to be larger but less abundant. The fish can grow up to 12 cm in length (Kottelat et al. 1993) and is classified as Critically Endangered (Lumbantobing 2019d). However, commercial fishing of this fish is still going strong as a profitable commodity (Hanafiah 2021; Setiawan 2022; Wibisono 2021; Figs. 7.3 and 7.4).

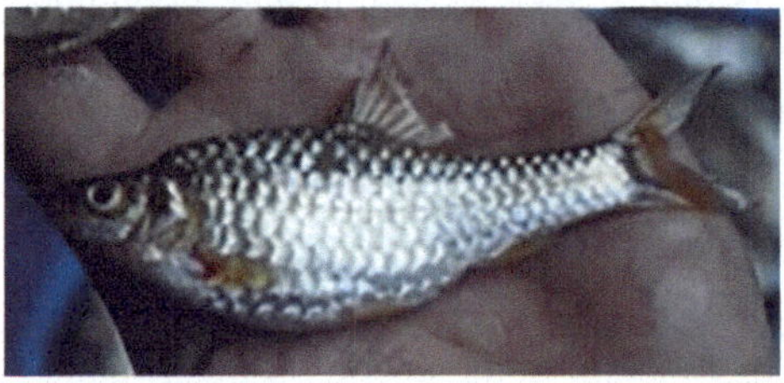

Fig. 7.1 *Poropuntius tawarensis* (Nurasiahriza 2016)

Fig. 7.2 *Rasbora tawarensis* (Rahmi et al. 2021)

Fig. 7.3 Fresh *R. tawarensis* traded at the Takengon market, Aceh (Setiawan 2022; Wibisono 2021)

Fig. 7.4 The daily amount of Depik fish being dried in the sun indicates overfishing (Hanafiah 2021)

Factors causing the decline in Depik and Kawan fish populations include:

- Environmental degradation around the lake, such as forest clearing and illegal logging. The tree roots, which function as water traps in the rainy season, continuously supply water to the lake in the dry season. The environmental conditions around the lake can be seen as bare hills with a few pine trees that grow sparsely. The water level of the lake is decreasing from year to year. The reduced water supply has caused several small rivers around the lake, which are fish spawning areas, to dry up. Deforestation also causes a decrease in water clarity due to siltation, which reduces sunlight penetration into the water, affecting the growth and reproduction of plankton as food for Depik fish.
- Intentional or accidental introduction of fish into Lake Laut Tawar has resulted in a population reduction and the potential extinction of endemic species.
- Destructive fishing techniques practiced in the lake include the use of potash poison or potassium cyanide (KCN) and traditional traps called Dedesen, which are installed at the Depik fish migration area. This trap has a hole as a way for fish to enter, and fish that enter can no longer get out. The fish caught are generally the spawning broodstock.
- Pollution occurs in lake waters from household waste, agriculture, hotels, restaurants, and fish farming in floating net cages due to excessive feeding.

Lake Toba, located in North Sumatra, is the largest lake in Indonesia and the world's largest volcanic lake. This lake is the habitat of three endemic fish species: Pora-Pora, *Neolissochilus thienemanni* Ahl 1933, known locally as Ihan, and *Rasbora tobana* Ahl 1934, known as Wader in its vernacular.

Neolissochilus thienemanni (Fig. 7.5) is found in Lake Toba and the Upper Asahan River (Rachmad et al. 2019). This medium-sized fish, belonging to the Cyprinidae family, reaches a maximum length of 21.1 cm (Kottelat et al. 1993), is vulnerable to extinction (Lumbantobing 2020a), and is extremely hard to locate in their natural habitat. On the other hand, another endemic species in Lake Toba, *Rasbora tobana* (Fig. 7.6), is listed as Least Concern (Lumbantobing 2020b) and not threatened with

Fig. 7.5 *Neolissochilus thienemanni* (Rachmad et al. 2019)

Fig. 7.6 *Rasbora tobana* (Lumbantobing 2010, with permission from the author)

extinction. This species can achieve a maximum length of 13.7 cm (Lumbantobing 2010).

Lake Toba once had an endemic fish called Pora-Pora. Since the 1990s, this fish has grown increasingly scarce. Therefore, in 2003, the government massively introduced a fish species that resembles Pora-Pora to Lake Toba. The introduced species is called Bilih *Mystacoleucus padangensis* Bleeker, which is endemic to Lake Singkarak. Due to their similarities, residents around Lake Toba still call this introduced fish as Pora-Pora (Sumut Pos 2020).

The average size of the new Pora-Pora is 12.0 cm, although it can grow up to 17.5 cm long (Ginting et al. 2013). After restocking in 2003 (the government claimed it was Pora-Pora restocking, not introducing), the number of Pora-Pora in Lake Toba increased considerably to the point that fishermen caught tons of fish daily and traded outside the region (Kelana 2015). Eventually, the fish disappeared from the lake's waters for almost a decade and only reappeared in mid-2022. However, the fish caught are much smaller than before (Argus 2022). Figures 7.7 and 7.8 portray local fishermen using nets to catch Pora-Pora in 2012 when the fish population was still abundant. The fish is classified as vulnerable (Lumbantobing 2020c).

The growth of water hyacinth is almost uncontrolled throughout the lake's shores, indicating pollution by nutrient elements rich in organic matter, nitrogen, and phosphorus. The waste that is directly dumped into the lake by restaurant and hotel proprietors, the unconsumed fish pellets thrown into the lake from fish farming activities in floating net cages scattered throughout the lake, and detergents used by nearby

Fig. 7.7 Pora-Pora fishing uses nets on the shore of Lake Toba, showing the fish abundance in Tongging Village, Merek District, Karo Regency, North Sumatra (Sani 2012, with permission from the author)

Fig. 7.8 Freshly caught Pora-Pora (Sani 2012, with permission from the author)

residents for washing clothes in the lake are believed to be one of the causes of the decline in lake water quality and endemic fish populations.

Lake Maninjau, located in the Agam District of West Sumatra, is home to two endemic fish species, namely Rinuak and *Rasbora maninjau* Lumbantobing 2014, both of which are consumption fish. Rinuak is a small fish (about 2 cm long) with a transparent body and lives in schools (Fig. 7.9). In the local dialect, 'Rinuak' means tiny, similar to the Indonesian word 'renik', meaning minuscule and delicate. The taxonomic classification of this fish is unknown.

Prior to 2016, Rinuak populations were abundant in the lake. The fish placed in large basins were sold by many traders in traditional markets around the lake (Fig. 7.10), indicating that fishing continued as long as the fish were still present. In 2016–2017, we noticed that these fish were no longer being traded in the market due to the absence of catch.

The recurrence of mass mortality of caged fish in November 2022 led to complaints from fishermen and small traders over the increasing scarcity and difficulty of capturing Rinuak (Maulana 2023). We suspect that these repeated disappearances are due mainly to habitat degradation, overfishing, and most likely to the introduction of invasive predatory fishes into the lake.

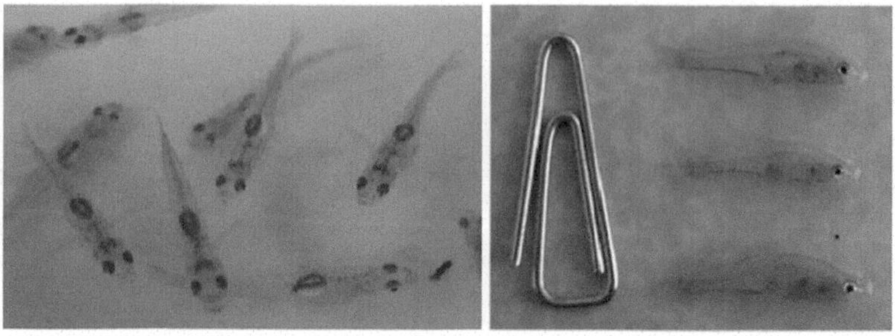

Fig. 7.9 Live Rinuak from Lake Maninjau. Photo credit: LR Tanjung, Triyanto

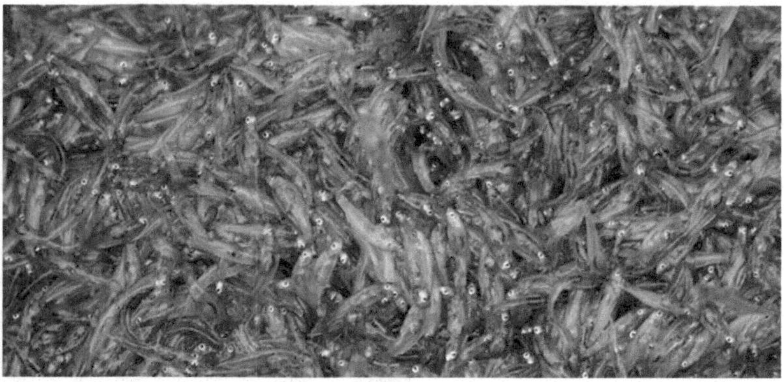

Fig. 7.10 Fresh Rinuak sold in the market near the lake (Sastra 2021)

The scientific names used for this fish in the published manuscript were *Psilopsis* sp. (Astuti et al. 2016; Fendri et al. 2019) and *Gobiopterus brachypterus* Bleeker 1855 (Yoga and Samir 2020). However, the genus *Psilopsis* is not present in the FishBase data, and the body color and the shape of the caudal fin of *Gobiopterus brachypterus* in FishBase and Kottelat et al. (1993) are distinct from the Rinuak we caught in Lake Maninjau (Fig. 7.9). We believe that the classification of this fish is still uncertain, supported by the results of the mtDNA analysis of this fish by Roesma et al. (2020). However, Rinuak may constitute a new species from the genus *Gobiopterus* that has not yet been identified.

Bada Maninjau, the local name for *Rasbora maninjau* (Fig. 7.11), is classified as endangered according to IUCN Red List Status (Lumbantobing 2020g). This fish is one of the five *Rasbora* species found in the lake, and all five species are locally referred to as Bada fish. Bada has been overfished due to its economic value as consumption fish, where the catch in 2017 decreased by 80% compared to 2008. The population is constantly decreasing, as is the size of the fish caught. The substantial decrease in the Rasbora population may be blamed on overfishing, deteriorating habitat and water quality (Dina et al. 2019).

Fig. 7.11 *Rasbora maninjau* (Lumbantobing 2014 with permission from the author)

Fig. 7.12 *Mystacoleucus padangensis* of Lake Singkarak. Photo credit: SH Nasution

Lake Singkarak is one of the largest National Priority Lakes in West Sumatra and is part of the Lake Singkarak National Park. The lake is the habitat for an endemic fish with high economic value, known locally as Bilih (*Mystacoleucus padangensis* Bleeker 1852, Fig. 7.12). This fish is classified as vulnerable (Lumbantobing 2020c). Even though this fish is protected, its population and the size of the fish caught are decreasing. The length of fish caught in 1996 reached 14.9 cm (Syandri 1996), compared to 9.5 cm in 2021 (Nasution et al. 2021). A decline in population numbers and the length of fish caught indicates overfishing.

The non-environmentally friendly fishing gear used by many fishermen in Lake Singkarak is one of the causes that threaten the sustainability of Bilih. Local fishermen continue to employ destructive fishing gear, including explosives (fish bombs), poison, and electricity. Gill nets with extremely small mesh sizes (1–1.5 cm) are the most commonly employed, reaching 544 units with a length of 100 m and a depth of 10 m for each unit (Coubout 2014), indicating massive catches of Bilih which capture all fish from juvenile to adult. In addition to overfishing, inhabitants around Lake Singkarak are endangering the sustainability of Bilih by disposing of various types of waste from agricultural, domestic, and market activities. Should this persist, the degradation of the ecology surrounding the lake would worsen and eventually endanger the lake's biota, especially Bilih, and cause harm to all the surrounding populations.

7.1.2 Java and Bali

Inland waters in Java are home to 132 species of native and endemic fish. Of these, 116 species (88%) are native, and the remaining 16 species (12%) are endemic (Jubaedah and Sayida 2022). The fish found in many aquatic ecosystems and have the highest level of endemism on the island of Java are represented by two genera of Cypriniformes, namely *Rasbora* (Cyprinidae) and *Nemacheilus* (Nemacheilidae), which are symbolic endemic lineages on the Java Island and the Lesser Sunda Islands consisting of the islands of Bali and Lombok (Hubert et al. 2019).

The four species of *Rasbora* that are still located are (a) *Rasbora aprotaenia* Hubbs & Brittan 1954 (SL 47 mm), found in Cisiih River, Banten, (b) *Rasbora argyrotaenia* Bleeker 1849 (SL 33 mm), found in Cilacap, Central Java, (c) *Rasbora lateristriata* Bleeker 1854 (SL 87 mm), found in Kali Dauwan, Mojokerto, East Java, and (d) *Rasbora baliensis* Hubbs & Brittan 1954 (SL 72 mm), found in Jembrana, West Bali (Fig. 7.13). The two *Nemacheilus* species successfully identified are *Nemacheilus fasciatus* Valenciennes 1846 (SL 42–45 mm), collected from (a) Ciasem River, Purwakarta, West Java and (b) Ciheulang River, Central Java, and *Nemacheilus chrysolaimos* Valenciennes 1846 (SL 55–58 mm), collected from (c) Ngerjo, Blitar, East Java and (d) Bicoro, Lumajang, East Java, shown in Fig. 7.13 (Hubert et al. 2019).

When first recorded, *Rasbora baliensis* Hubbs & Brittan 1954 was considered endemic to Lake Bratan, Bali Island. Over time, this fish was found in East Java and introduced to the Lesser Sunda Islands (Hubert et al. 2019). *Rasbora baliensis* has also been reported in rivers of Central and East Java (Fig. 7.14; Hasan et al.

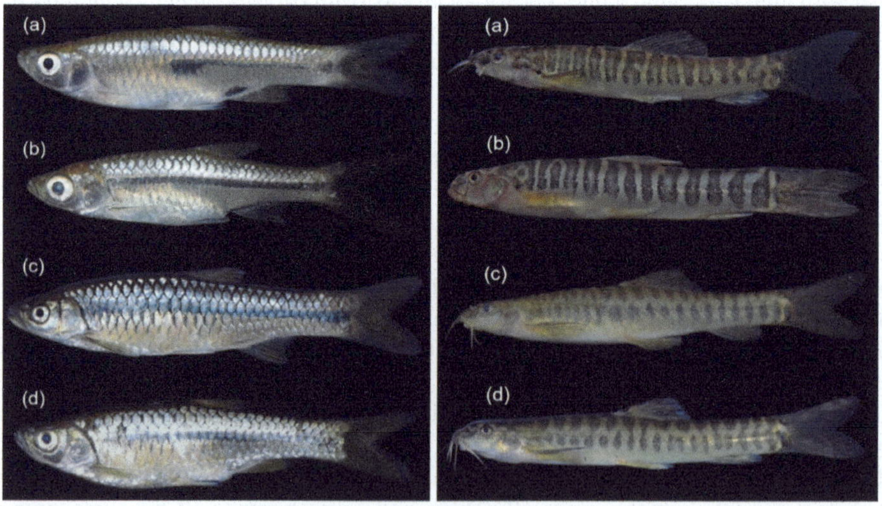

Fig. 7.13 Four endemic *Rasbora* species (left) and two *Nemacheilus* species (right) were found in Java (Hubert et al. 2019, with permission from the author)

Fig. 7.14 Specimen of *Rasbora baliensis* obtained from the Solo River, Central Java. Photo: H Cahyono in Hasan et al. (2021a)

2021a, b). This species is not threatened with extinction since it is still extensively distributed throughout its natural habitat, stretching from East Java to East Nusa Tenggara (Lumbantobing 2021).

In addition to *R. baliensis*, another endemic fish species found in Bali is *Lentipes whittenorum* Watson & Kottelat 1994. This species has only been documented from the islands of Bali and Lombok, and other information is unknown, so the species is classified as Data Deficient to accurately reflect the lack of information regarding its biology, ecology, population, and potential threat impacts (Jaafar 2019).

Lobocheilos falcifer Valenciennes 1842 is another endemic fish, also called Mangut in Java (Fig. 7.15). In West Java, this fish used to be commonly found in the Cisadane River, Ciliwung River, and Citarum River. This fish species was recently observed in the Wadaslintang Reservoir and Lake Rawa Pening, Central Java Province (Hasan et al. 2019, 2021b).

Mangut is becoming harder to locate these days. The natural habitat of this fish is clear and unpolluted water. High environmental pressure has caused the fish habitat to be surrounded by settlements and industries, resulting in polluted rivers making this species vulnerable to extinction. Mangut is listed as vulnerable, which indicates its declining population (Lumbantobing 2019b).

Mangut is extremely rare and nearly extinct in the Cisadane and Ciliwung Rivers, but it is still rarely present in the Serayu River. However, the fish is still present in the upstream section of Lake Rawa Pening (Widianto 2021). The population of Mangut is decreasing, but there are currently only 25 fish species in the Brantas River, down from 60 species in 1990, while in the Bengawan Solo River, less than ten fish species remain after twenty species became extinct.

Fig. 7.15 Specimen of *Lobocheilos falcifer* collected from the outlet of Lake Rawa Pening, Semarang Regency, Central Java. Photo: V Hasan in Hasan et al. (2021b)

Numerous fish, including Leat fish (*Lobocheilos lehat* Bleeker 1858), Belida (*Chitala lopis* Bleeker 1851), *Acrochordonichthys ischnosoma* Bleeker 1858, *Acrochordonichthys rugosus* Bleeker 1846, and *Bagarius lica* Volz 1903 are extinct in the Citarum River. The species *Lobocheilos lehat* was first reported in West Java in 1858, but this species has not been found again until now. *Bagarius lica* was once also prevalent in the Bengawan Solo River but is now extinct throughout Java. Ciliwung River was inhabited by 187 species of fish in the 1910s. Within a century, there were just 20 species in 2009, five of which were introduced. In the last ten years, three species of fish throughout Java Island have become extinct and rare, including Jambal or *Pangasius macronema* Bleeker 1850, Bala or *Luciosoma setigerum* Valenciennes 1842, and *Homalopteroides wassinkii* Bleeker 1853 (Gustiano 2009; Widianto 2021).

Major rivers in Java are severely polluted. Endemic fish continue to experience extinction. Several species of fish in Javanese rivers have become extinct. The remaining species, for example, Mangut fish, are on the verge of extinction. In addition to agricultural and domestic wastes dumped directly into rivers, industrial liquid waste also contributes to river pollution and has seriously harmed fish populations in the rivers of Java by disrupting the food chain.

The government does not place a high priority on reducing pollution in river water. The damage done to rivers in Java demonstrates the government's inability to manage rivers. Sadly, even though five institutions, including the River Basin Center/Ministry of Public Works and Public Housing, the Ministry of Environment and Forestry, the state-owned company called Perum Jasa Tirta that operates in the water resources management sector, the Governor, and the Mayor/Regent, are involved in river management, the state of the rivers is still far from improving.

Textile factories and recycled paper manufacturers make use of the Brantas River (East Java), Bengawan Solo River (Central Java–East Java), Citarum River (West Java), and Ciujung River (Banten) as liquid waste disposal sites. The liquid waste that is completely processed is directly thrown into the river. Recycled paper, mainly imported from the European Union, still contains high amounts of plastic waste. Removing plastic impurities from recycled papers produces liquid waste, which, if not completely processed and is still black, foul-smelling, and poisonous, still contains hazardous chemical waste and microplastics that will pollute waterways.

Industries that do not want to cover the costs of comprehensive waste processing will throw their waste into rivers by installing pipes below the water surface to make the waste pipes invisible. Apart from polluting the air, rivers, and land, this waste also contains microplastics originating from recycled paper, thereby increasing health risks for residents living around the company's locations. Heavy industrial waste pollution in rivers will impact not only the fish and other biota that inhabit the river but also the drinking water companies that process the river water as raw material for drinking water.

Government regulations that explicitly control the cross-ministerial assignments in charge of river management in Java are necessary to eliminate pollution and restore river quality. Clear and effective enforcement of the regulations in the field is required for both the restrictions and the penalties imposed on those who violate them.

Until now, the government has not seriously monitored the disposal of industrial liquid waste. Unscrupulous industries continue to dispose of untreated waste at night. The government also easily issues permits for liquid waste disposal, ignoring the rivers' carrying capacity and pollution load. In addition, there are no strict sanctions against these perpetrators, so this illegal conduct continues.

7.1.3 Borneo

Lake Sentarum in West Borneo Province consists of small and large lakes, peat swamp forests, and freshwater swamp forests, forming a vast natural reservoir (Hidayat et al. 2020). Kottelat and Widjanarti (2005) recorded 212 fish species in the Kapuas Lakes area (Western Borneo), of which 146 species came from the Lake Sentarum National Park (Taman Nasional Danau Sentarum, TNDS). This lake represents the most comprehensive freshwater fish habitat in the world. Species diversity can be best explained by water level, hydrology, habitat, aquatic productivity, and migration nexus, as exemplified by the 'flood–pulse' concept (Haryani et al. 2020).

Among the diverse species, the smallest fish in Lake Sentarum, known locally as Linut (*Sundasalanx microps* Roberts 1981), measures 1–2 cm and has a transparent glass-like body. On the other end of the spectrum, the biggest fish in this lake is called Tapah (*Wallago leeri* Bleeker 1851), reaching 150 cm long. Some ornamental fish enthusiasts prefer large fish, like Arowana from Lake Sentarum (Haryani and Said 2011). Two ornamental fish species worth looking into include Botia (*Chromobotia macracranthus* Bleeker 1852) and Arowana (*Scleropages formosus* Müller & Schlegel 1840). With a total length of 30.5 cm, Botia is not threatened with extinction (Least Concern). In contrast, Arowana, with a maximum length of 90 cm, is categorized as endangered according to the IUCN Red List (Larson and Vidthayanon 2019). Figure 7.16 shows the super red Arowana from Lake Sentarum.

Several *Lobocheilos* species from Borneo, for example, *Lobocheilos bo* Popta 1904, *L. kajanensis* Popta 1904, and *L. tenura* Kottelat & Tan 2008, show a population trend that is currently still stable and not threatened with extinction (Kottelat and Tan 2008). However, the species *Lobocheilos erinaceus* Kottelat & Tan 2008, described

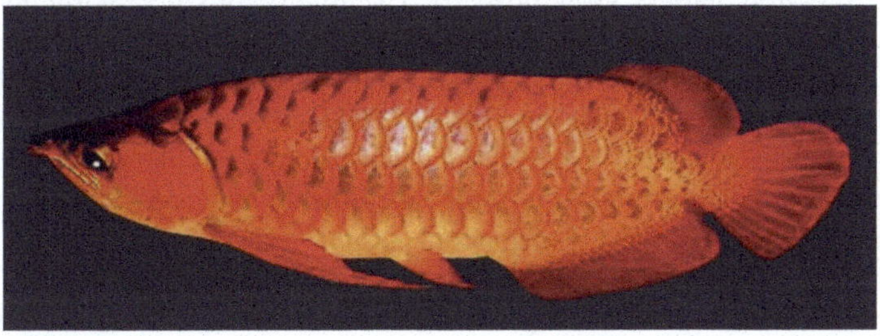

Fig. 7.16 Super red Arowana from Lake Sentarum (DJPB KKP 2023)

from northwestern Borneo by Kottelat and Tan (2008), has a highly fragmented population and is nearly threatened by the continuing decline in the number of adult individuals (Lumbantobing 2020d).

Rasbora borneensis Bleeker 1860 is collected only in the Kapuas mainstream and reported to enter the lakes during the wet season (Kottelat and Widjanarti 2005). This fish measures up to 8 cm long (Kottelat et al. 1993). The population trend of this species is uncertain; however, it is Least Concerned or not threatened with extinction (Lumbantobing 2020e), as are *Rasbora myersi* Brittan 1954 and *R. vaillantii* Popta 1905.

7.1.4 Sulawesi

Sulawesi Island has 68 endemic species from seven families and four orders (Hadiaty 2018). Endemic fishes on Sulawesi are spread across several lakes and rivers such as Lake Tondano (Minahasa Regency, North Sulawesi), Lake Lindu (Sigi Regency, Central Sulawesi), Lake Matano (East Luwu Regency, South Sulawesi), Lake Towuti (Luwu Regency, South Sulawesi), Lake Poso (Poso Regency, Central Sulawesi), and Maros River (Maros Regency, South Sulawesi).

Lake Tondano in North Sulawesi is inhabited by an endemic fish *Tondanichthys kottelati* Collette 1995. The population trends and size of this species are unknown. Widespread threats include high levels of agricultural pollution, sedimentation, heavy metals, and infrastructure damming of the rivers' main course, and these continue to result in a decline in habitat quality throughout the lake. Therefore, this species is rated Critically Endangered (Daniels 2020).

Oryzias sarasinorum Popta 1905, formerly *Xenopoecilus sarasinorum* (Parenti 2008), also called Rono locally, is an endemic species to Lake Lindu, which is in critically endangered condition according to the IUCN red list of threatened species (Mokodongan 2019). The primary diet of Rono is diatom (phytoplankton) from the genus *Melosira*. Lake Lindu is classified as oligotrophic, meaning that the source of nutrients and organic material is insufficient to support producer organisms. This insufficiency indicates that phytoplankton that serve as food for Rono and aquatic plant *Phragmites karka* Retz, which functions as a place for Rono to feed, spawn, and seek protection, cannot grow properly. Eventually, phytoplankton and aquatic plant populations will decline. As a result, the Rono population in the area will also decrease. In addition, endemic fish in Lake Lindu face competition or predation from introduced fish species that have long been introduced to the lake. For endemic fish to remain in existence and survive, foreign fish introduction must cease, and any fish that have already been introduced into the lake must be eradicated.

Endemic fish in Lake Matano generally cannot survive outside their natural habitat, particularly those belonging to the genus *Telmatherina*. However, *Telmatherina celebensis* Boulenger 1897 (Fig. 7.17), a fish endemic to the Malili Lake Basin, recorded in Lakes Towuti, Mahalona, and Lontoa, is widely traded commercially.

Fig. 7.17 *Telmatherina celebensis* (Univers Aquatique 2024)

The population of *Telmatherina celebensis* is decreasing, and this species is in the Near-Threatened category (Lumbantobing 2019c). With a maximum length of 6.8 cm (Kottelat et al. 1993), this fish is included in the small fish group. Therefore, due to the fish's small size, the inhabitants in the Lake Towuti area name this fish Pangkilang, while in the Lake Matano area, the small fish are called Opudi (Nasution 2017).

Oryzias marmoratus Aurich 1935 is an endemic species in Lakes Towuti, Mahalona, Lontoa, and some adjacent streams and creeks. This species reaches a maximum length of 4.1 cm (Herder and Chapuis 2010), shown in Fig. 7.18.

This species is classified as Near-Threatened (Lumbantobing 2019e) based on its limited distribution area and the various ongoing threats to this species, particularly high levels of sedimentation in several estuaries and invasive species causing the number of subpopulations and habitat quality to decline.

Tominanga sanguicauda Kottelat 1990 is a fish endemic to Lake Towuti (Fig. 7.19) with a maximum length of 5.4 cm (Kottelat et al. 1993). The fish is listed as Near-Threatened based on its highly restricted distribution range and multiple ongoing threats to the species, especially the high levels of sedimentation in several estuarine areas within the species range that have caused the ongoing decline in habitat quality (Lumbantobing 2019a).

Fig. 7.18 *Oryzias marmoratus*, Medaka of Lake Towuti (Said and Hidayat 2015)

Fig. 7.19 *Tominanga sanguicauda* (Said and Hidayat 2015)

Residents in the Lake Towuti area use dip nets (bagan) to catch this species, especially for consumption. However, this fish is also traded and exported to other regions as ornamental fish due to its attractive color.

Significant threats to the endemic aquatic fauna in Lakes Malili include water regulation rights granted to foreign mining companies that convert natural rivers into artificial channels, erosion due to increased land use and deforestation by illegal logging, damming of Lake Towuti, and mining with massive earth-moving. The government must not give foreign companies the right to upset the spatial planning of natural waters. Habitat destruction and environmental degradation occurring in Lake Towuti is also caused by sawdust from sawmills processing logs cut all around the lake, overflows into natural habitats, and shifts the shoreline by tens of meters throughout former inshore habitats (Hoese et al. 2015).

Additional threats come from population increase and habitat loss due to ever-expanding mining activities. For example, streams in the former mining area have been found either not to exist anymore or to have changed to biologically dead remains filled with mud, oil, and waste.

The Maros River is one of the habitats for *Marosatherina ladigesi* Ahl 1936, also known as Celebes Rainbowfish (Fig. 7.20). The fish has a maximum length of 8 cm (Kottelat et al. 1993). The fish is listed as vulnerable as the species faces major threats throughout its range, including overfishing for the international aquarium trade and local pesticide runoff from rice fields, which together have resulted in a decline in wild populations and a decline in sex ratios (Daniels and Palmer-Newton 2019).

Fig. 7.20 *Marosatherina ladigesi*, endemic of Maros River (Said and Hidayat 2015)

Most of Sulawesi's endemic species are within protected areas, which comprise the waterbodies of Lakes Towuti and Mahalona with the status of Nature Park, administered altogether by the Conservation Forest Management Unit of Towuti (KPHK Towuti). In addition, many of the land areas around the two lakes have Protected Forest status, administered by the Protection Forest Management Unit of Larona Malili (KPHL Larona Malili). Despite the status of these areas for conservation and protection, the implementation of strict regulations that include a sustainably managed zonation system for various human activities along with nature conservation programs appears to be very limited or even lacking throughout Lakes Towuti and Mahalona.

7.1.5 Papua

In Papua, endemic fish inhabit lakes, rivers, and tributaries. The most prevalent group in West Papua and Papua New Guinea is the Rainbowfishes group, which comprises the Melanotaeniidae and Pseudomugilidae families. Five of the seven genera of the Melanotaeniidae family, *Chilatherina*, *Iriatherina*, *Glossolepis*, *Melanotaenia*, and *Pelangia*, are found in West Papua, Indonesia. This group of fish is well-known for being ornamental fish due to its beautiful appearance and small size, up to 17 cm in total length. Many local endemic fish are at risk of extinction, so they are listed as endangered by the IUCN.

In West Papua, 124 freshwater endemic fish species were recorded in 2015, including all fish species within the family Melanotaeniidae (Hubert et al. 2015; Tappin 2018). Lake Sentani, located in West Papua, is home to two endemic fish species, *Glossolepis incisus* Weber 1907 and *Chilatherina sentaniensis* Weber 1907. Red Rainbowfish or *Glossolepis incisus* also inhabit the tributaries of the lake. The adult fish measures up to 12 cm in length (Fig. 7.21). This species was exceedingly abundant during the 1990s. It was abundant in 2015, and the population trend is considered stable. However, this species is listed as endangered in the IUCN Redlist (Allen and Kadarusman 2020a).

Fig. 7.21 *Glossolepis incisus* from Lake Sentani (Said and Hidayat 2015)

Fig. 7.22 Male and female *Chilatherina sentaniensis*. Photo: Buddy Jonkers in Tappin (2018)

Chilatherina sentaniensis has a maximum size of around 8.5 cm (Fig. 7.22) and is only known from Lake Sentani and small tributaries that flow into the lake near Jayapura City. It is now thought to exist only at the eastern end of the lake. This species is listed as critically endangered since its habitat is continuing to decline, brought on by deforestation around the lake, water pollution from neighboring communities, and the threat posed by invasive and predatory species introduced into the lake, in addition to gold mining, which negatively affects the western end of the lake (Tappin 2018; Allen et al. 2020a).

Melanotaenia ajamaruensis Allen & Cross 1980 have a metallic blue to yellowish green color, with longitudinal bands of orange and yellow and dark scale edges (Fig. 7.23). Males are usually deeper bodied than females, and they reach a length of around 11 cm. The name *Melanotaenia ajamaruensis* comes from Lake Ajamaru, the only known collection site and type locality for this species up to 1980. The type specimens of *M. Ajamaruensis*, as described by Allen and Cross in 1980, were collected in an upstream section of the Kaliwensi River, about 5 km west of Lake Ajamaru, the only known location of *M. ajamaruensis* until 2013. Lake Ajamaru is not connected to Kaliwensi River, an old outlet to the western side of the lake, but it does not flow into the lake. In 2013, live specimens of *M. ajamaruensis* were collected for the aquarium hobby by Hans-Georg Evers and Jeffrey Christian from both the Kaliwensi and Soum Rivers (Tappin 2018).

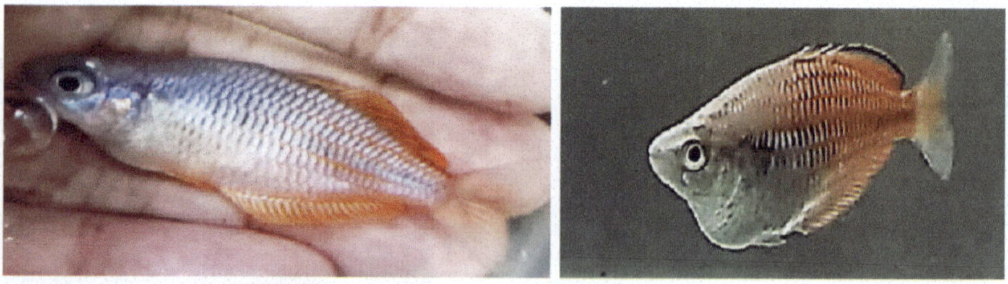

Fig. 7.23 *Melanotaenia ajamaruensis*. Photo credit: DS Said

Fig. 7.24 *Melanotaenia boesemani* (Said and Hidayat 2015)

This species is assessed as Critically Endangered as it is undergoing a continuing decline in mature individuals due to predation and competition with invasive species (Kadarusman 2020).

Melanotaenia boesemani Allen & Cross 1980 is remarkably similar to *M. ajamaruensis*. When fully matured, the male *Melanotaenia boesemani* shows a half-and-half coloration (Fig. 7.24). The head and front portion of the body are a brilliant blue-gray, while the fins and rear portion of the body are bright red–orange. This species typically grows to a maximum size of less than 10 cm, but it can reach 12 cm. *M. boesemani* has been mainly collected in Lake Ajamaru and a few surrounding tributaries but has also been reported from Lake Hain, Lake Aitinjo, and Lake Uter in West Papua Province, Indonesia (Tappin 2018).

Massive declines in habitat and the number of adult individuals related to anthropogenic activities, sedimentation, invasive plant and fish species, and bird predation are causing the species' population trend to decline. As a result, this species is listed as Endangered (Allen and Kadarusman 2020b).

Melanotaenia praecox Weber & de Beaufort 1922 has a vivid, neon blue body with red dorsal, anal, and caudal fins. They are usually smaller than 5 cm SL but can reach 8 cm in size (Fig. 7.25).

The initial collection of *M. praecox* was taken from a tributary of the Mamberamo River in West Papua in 1910. Subsequently, in 1991, 2000, and 2008, specimens were collected from locations near the airstrip at Dabra and Iritoi in the middle part of the Mamberamo River, from a tributary of Tiri River about 4 km west of Dabra on the edge of the Mamberamo Plain, and a tributary of Mamberamo River in Pagai village (Tappin 2018).

Fig. 7.25 *Melanotaenia praecox* (Said and Hidayat 2015)

Fig. 7.26 *Pseudomugil reticulatus*. Photo: GR Allen in Tappin (2018)

The area where this species lives is largely undisturbed, and the species is in the Little Concern category on the IUCN red list (Allen and Kadarusman 2020c). Although the population trend is currently stable, future threats, including regional development plans, such as the construction of forest concessions, oil palm plantations, coal mines, the proposed Mamberamo Dam, and invasive species, are endangering the biodiversity of this region.

Pseudomugil reticulatus Allen & Ivantsoff 1986 is a small, slender, and laterally compressed species, growing to a length of about 3–4 cm SL (Fig. 7.26). The body color is yellow-brown to light brown, the stomach area is silvery, and the eyes are intensely blue (Tappin 2018). The sole known specimens of the species were taken from a rainforest stream flowing into Lake Ajamaru in West Papua, Indonesia. This species is rated Critically Endangered and continues to experience declines in adult numbers due to primary threats of predation by invasive species (Allen et al. 2020b).

7.1.6 Malaysia

There are 470 strictly freshwater fish species in Malaysia. These species occupy habitats that include streams, rivers, lakes, ponds, paddy fields, peat swamps, and other freshwater swamps. Five of the freshwater fish highly threatened (HT) species came from peat swamps, while the remaining eleven were from streams and upper sections of rivers. Freshwater fish that are considered locally extinct (LE) in the country are two species of Cyprinids, namely *Balantiocheilos melanopterus* Bleeker 1850 and *Neobarynotus microlepis* Bleeker 1851 (Chong et al. 2010).

Freshwater habitats encompass the highest percentage of threatened fish species (87%) followed by estuarine habitats (66%). Of the 32 species of highly threatened (HT), 16 are freshwater species. Fish extinctions in Malaysia are confined to two freshwater species, but both freshwater and marine species are being increasingly threatened by habitat loss or modification (76%), overfishing (27%), and bycatch

Fig. 7.27 *Neolissochilus hendersoni* from Teluk Bahang River in Penang Island. Photo: A Ahmad, with permission from the author; https://fishbase.mnhn.fr/photos/PicturesSummary.php?resultPage=1&ID=54692&what=species

(23%). The most crucial threat to freshwater fishes is habitat modification and overfishing, while 35 species are threatened due to their endemism (Chong et al. 2010).

In Peninsular Malaysia, *Neolissochilus* is locally known as Tengas fish or Copper Mahseer. This fish lives naturally in rivers with clean water and rapid currents. One of the species of *Neolissochilus*, which is endemic in Malaysia, is *Neolissochilus hendersoni* Herre 1940 (Fig. 7.27). This fish is found in Penang, Langkawi, Kedah, and Kelantan. Until now, the spread of this species has only been known from Peninsular Malaysia, and there have been no reports of its existence in other Southeast Asian areas (Khaironizam et al. 2015).

Neolissochilus hendersoni in Penang and Langkawi is under critical threat since it is restricted to specific areas. This fish has been listed as a vulnerable species by the IUCN Red List of Threatened Species (Ho and Ahmad 2019). This endangering condition has been attributed to the rapid development of these islands and the pollution in the river system, which includes several streams in Batu Feringgi, Teluk Bahang, and Balik Pulau Penang. Environmental alterations are anticipated to imperil this species in Peninsular Malaysia (Khaironizam et al. 2015).

The bearded gudgeon fish *Pogoneleotris heterolepis* Günther 1869 was initially discovered in estuaries and the middle parts of rivers in Sarawak, Malaysia. Larson and Hui (2019) redescribed the fish based on the fifteen specimens obtained from fish markets and local fishers near Simunjan River in Sarawak between 1982 and 2017. Recently, Valen et al. (2022) reported the presence of *P. heterolepis* with TL around 19 cm in the Kapuas River, West Borneo, Indonesia (Fig. 7.28). This rare finding is the first known record of this species in Indonesian Borneo. This fish was spotted in the river mouth area approximately 5 km from the estuary. This discovery demonstrates that this species' distribution range has expanded in Borneo Rivers.

This species has been found in four river systems in Sarawak. It appears to be a rare occurrence as these river systems remain understudied due to lack of access, logistical assistance, and concerns about saltwater crocodiles (*Crocodylus porosus*) in the areas. The population trend remains uncertain, and threats to this species are

Fig. 7.28 *Pogoneleotris heterolepis* caught in Kapuas River in West Borneo. Photo: Marcel Alveri Adis in Valen et al. (2022)

unknown. However, the area where the species lives is susceptible to residential and commercial development, mangrove forest destruction, and the possibility of water pollution from domestic and agricultural sources. Local fishermen catch this species as a bycatch, although the effort is minimal, and the impact is unknown. Therefore, this fish is on the IUCN Red List of Threatened Species as a Data Deficient species (Larson 2019).

7.1.7 The Philippines

Fish endemic to the Philippines are slowly disappearing, which seems normal. It is no exception in many rivers and lakes. Even protected biodiversity hotspots such as Lake Taal, home to the unique freshwater sardines known as Tawilis, are not exempted from the influx of introduced fish imported to supply the food needs of a growing population and the aquarium trade.

The freshwater sardine (*Sardinella tawilis* Herre 1927) is endemic to Lake Taal (formerly Lake Bombon) in Luzon, Philippines. This fish is the only *Sardinella* species inhabiting fresh water (Castillo et al. 1975). The current population trend of this species is declining, and Tawilis is classified as endangered by the IUCN Red List due to many factors, which include pollution, predation, and competition with exotic species (Santos et al. 2019) (Figs. 7.29 and 7.30).

Fig. 7.29 *Sardinella tawilis* from Batangas in the southern part of Luzon, Philippines. Photo: NFRDI-NFBC, https://fishbase.mnhn.fr/photos/UploadedBy.php?autoctr=22014&win=uploaded

Fig. 7.30 *Sardinella tawilis* from Lake Taal (https://en.wikipedia.org/wiki/Sardinella_tawilis#/media/File:Tawilis_(Sardinella_tawilis),_an_endemic_fish_that_can_only_be_found_in_Taal_Lake.jpg)

Another endemic fish to Luzon Island is the Silver Therapon (*Leiopotherapon plumbeus* Kner 1864), also known as Ayungin in Tagalog, and is an important freshwater food species that are mostly caught from Laguna de Bay (Fig. 7.31). It was introduced to other lakes such as Sampaloc Lake and Taal Lake in the late 1950s and early 1970s, respectively. This fish is considered one of the tastiest freshwater fish, so it is sold fresh or dried at high market prices (Fig. 7.32). Accounting for approximately 70% of the total fishery catch in Laguna de Bay, it was once the most abundant fish caught in the lake and was used primarily for domestic human consumption and as a component of duck feed. Demand for this fish species remains high, causing the wild stock to decrease. This fish has been added to the IUCN Red List as a vulnerable species (Torres et al. 2021).

According to the Philippine Statistics Authority, capture for this fish has declined from 4765 metric tons in 2002 to 1408 metric tons in 2018. Heavy fishing pressure on wild fisheries stocks, habitat degradation due to pollution and sedimentation, and the introduction of invasive alien species in Laguna de Bay and surrounding inland waters have resulted in the decline in the population of this endemic fish in their natural habitat. However, this has also raised interest in the domestication and conservation of this species (Aya and Garcia 2020).

Overall, environmental changes impact endemic fish, which can vary depending on the nature of the alterations. Some common impacts include:

Fig. 7.31 Hatchery-bred *Leiopotherapon plumbeus* (Aya and Garcia 2020, with permission from the authors and SEAFDEC Aquaculture Department)

Fig. 7.32 Fresh (**a**) and dried (**b**) Silver Therapon sold as a market commodity (Aya and Garcia 2020, with permission from the authors and SEAFDEC Aquaculture Department)

- Habitat Loss: Environmental changes such as deforestation, urbanization, or construction projects can lead to the loss of natural habitats for endemic fish. This habitat loss can reduce breeding grounds, shelter, and food resources.
- Water Quality Changes: Water pollution from industrial, agricultural, or urban runoff can affect the water quality in areas where endemic fish reside. These changes may negatively impact the fish's health, reproduction, and survival.
- Water Temperature Changes: Global warming and climate change can cause shifts in water temperatures in the habitats of endemic fish. Some fish species may be more sensitive to temperature changes than others, affecting their life cycles.
- Alien Species: Introducing non-native fish species to the environment can compete with endemic fish for resources and habitats. These alien species may also carry diseases harmful to endemic fish.
- Weather Pattern Changes: Extreme weather events or climate fluctuations can alter environmental conditions in the habitats of endemic fish. These changes can lead to changes in food distribution, water levels, and the availability of other resources.
- Overfishing: Unsustainable or excessive fishing practices can threaten the survival of endemic fish. If the number of fish caught exceeds the reproductive and population growth rates, the endemic fish population can decline dramatically.

These impacts are interconnected and complex, posing threats to the survival of endemic fish. Protecting natural habitats, implementing sustainable fisheries management, and engaging in environmental conservation efforts are vital in mitigating the impacts of environmental changes on endemic fish and ensuring their survival.

In addition, some organisms function as primary producers in aquatic ecosystems, using natural resources and generating biomass that supports food webs. Conversely, particular endemic fish species act as primary consumers, regulating populations of other organisms to prevent uncontrolled expansion.

As integral components of the food web, endemic fish contribute to trophic balance by controlling organism populations at various levels. They prevent the overpopulation of specific species that could disrupt the ecosystem. Some endemic fish may act as natural predators for certain organisms in aquatic ecosystems. For example, *Telmatherina celebensis* Boulenger 1897 and *Neolissochilus sumatranus* Weber & de Beaufort 1916 tend to consume aquatic insects (Sulistiono et al. 2006; Mayasari et al. 2020).

Endemic fish also play diverse roles in supporting the ecological balance and sustainability of their habitat. Participating in ecological processes, some endemic fish such as *Oryzias sarasinorum* Popta 1905, *Tor douronensis* Valenciennes 1842, and *Osteochilus vittatus* Valenciennes 1842 contribute to maintaining water quality by consuming phytoplankton or aquatic plant, such as Lemna, to reduce the potential for excessive growth of algae or aquatic plants (Gani et al. 2015; Mayasari et al. 2020; Said et al. 2021).

7.2 Invasive Species in Inland Aquatic Ecosystems

Invasive fish can threaten freshwater ecosystems and existing fish communities, often negatively impacting natural resources and aquatic ecology. According to the IUCN (2003), as cited in Umar et al. (2015), Invasive Alien Species are foreign species capable of establishing themselves in natural or semi-natural ecosystems, initiating changes, and posing a threat to local/native biodiversity.

7.2.1 Indonesia

Invasive fish species have spread across various inland waters in Indonesia, including rivers, lakes, and reservoirs. Floating net cages in lakes are thought to contribute to the spread of unwanted non-native fishes inadvertently transported into the lake that compete with endemic species for food sources and habitat. In addition to spreading disease, introduced species may be more resilient to thriving in deteriorating aquatic environmental conditions, unsuitable for some endemic fish, multiplying, becoming invasive, dominating, and disrupting aquatic systems. The most vital concern is the introduction of predatory fish.

Many fish species that have been introduced for aquaculture and are now widely cultivated in some priority lakes with increasingly declining conditions are truly invasive species, such as Nile Tilapia (*Oreochromis niloticus* Linnaeus 1758), Mozambique Tilapia (*Oreochromis mossambicus* Peters 1852), Common Carp (*Cyprinus carpio* Linnaeus 1758), and North African Catfish (*Clarias gariepinus* Burchell 1822). However, referring to the Minister of Marine Affairs and Fisheries Regulation Number 19/Permen-KP/2020, these fish species are not included in the group of invasive fish species.

Invasive fish observed in Indonesian lakes generally comprise the same species in all invaded lakes. For example, invasive fish species found in Lake Laut Tawar include *Oreochromis mossambicus*, *O. niloticus*, *Clarias gariepinus*, *Cyprinus carpio*, Puffer fish (*Xiphophorus* sp.), and Amazon sailfin catfish (*Pterygoplichthys pardalis* Castelnau 1855). Grass carp (*Ctenopharyngodon idella* Valenciennes 1844) have been introduced by the local government to control the development of water hyacinths in the lake. However, this effort is considered unsuccessful because Lake Laut Tawar is still overgrown with water hyacinth, while the presence of introduced fish has become a competitor for endemic fish (Nurasiahriza 2016).

Cichlid species initially introduced as ornamental fish have now flourished in public waters. An example is the Red Devil (*Amphilophus labiatus* Günther 1864). This fish was introduced to Indonesia in the early 1990s. Currently, this species has become an invasive pest in several lakes and reservoirs (Fig. 7.33), such as Lake Matano, Lake Batur, Lake Toba, Kedungombo Reservoir (Central Java), Sermo Reservoir in Kulon Progo Regency (Yogyakarta), Cirata and Jatiluhur Reservoirs (West Java). Since the introduction of Red Devil into these waters, local fishermen's catches have declined, with over 75% of the catch now consisting of Red Devil cichlids (Wahyudewantoro and Rachmatika 2016).

The community around Lake Sentani in Papua has begun exploiting the invasive Red Devil fish as a food source and selling it in marketplaces alongside other caught fish species like Tilapia (Fig. 7.34).

Suckermouth Catfish or Amazon sailfin catfish (*Pterygoplichthys pardalis*) belongs to a family of South American catfishes characterized by the presence of the ventral mouth and bony plates covering the body. They are popular aquarium fishes used for cleaning aquarium glasses. Two species have been introduced to Indonesian freshwaters from aquaria or breeding ponds, particularly around Jakarta. Kottelat et al. (1993) have revised the genus name '*Pterygoplichthys*' and tentatively identified the introduced species as *Liposarcus pardalis.*

This species can reach a length of 70 cm, and its food is composed of algae, benthic algae, detritus, worms, and several types of insect larvae (Özdilek 2007). Chaichana and Jongphadungkiet (2013) stated that this fish also preys on small fish and other fish eggs. Therefore, once it becomes invasive, this fish will pose severe threats to native and endemic fishes. Figure 7.35 shows live and dead *Pterygoplichthys pardalis* after being captured from Lake Matano.

Fig. 7.33 Red Devil from Lake Batur, Bali. Photo credit: GS Haryani and R Gustiano

Fig. 7.34 Red Devil sold in the Sentani market in Papua. Photo credit: GS Haryani

Fig. 7.35 *Pterygoplichthys pardalis*. Photo credit: SH Nasution and Safiuddin

The two species of fish mentioned above (Figs. 7.33 and 7.35) were originally ornamental fish commodities, but both have become invasive fish with the most extensive distribution in Indonesian inland waters (Dina et al. 2022).

7.2.2 Malaysia

Malaysian inland waters, like Indonesian waters, have a long history of introduced fish, whether intentional or accidental, dating back to the early twentieth century. The number of introduced fish in Malaysian waters is also increasing, with 42 species now inhabiting a variety of inland natural waters, including highland and isolated streams, rivers, rice fields, swamps, drainage areas, dams and reservoirs, small lakes created from mining activities, and estuaries.

The fishes were introduced into Malaysian waters for a variety of purposes, including aquaculture (64.3%), ornamental fish (19.0%), recreational fishing (14.3%), and biological control (2.4%). Table 7.2 shows different species of introduced freshwater fish found in Malaysian waters (Rahim et al. 2013).

The presence of introduced fish in Malaysian waters has affected the structure of the native ichthyofauna. Serious intrusions of invasive fish species that threaten the survival of several endemic fishes have been reported in various rivers in Malaysia. These invasive species include Silver Barb (*Barbonymus gonionotus* Bleeker 1849), Mosquitofish (*Gambusia affinis* Baird & Girard 1853), Nile Tilapia (*Oreochromis niloticus*), Amazon Sailfin Catfish (*Pterygoplichthys pardalis*), Asian Red Tail Catfish (*Hemibagrus wyckioides* Fang & Chaux 1949), and Flowerhorn fish which is a hybrid of different species and breeds of Neotropical cichlids that originated in Central America. Among these invasive fishes, Nile Tilapia is the most dominant species in many rivers (Ahmad et al. 2020).

Until 2013, introduced fish were not restricted from spreading into Malaysian rivers and lakes (Rahim et al. 2013). However, the Malaysian government has recently prepared a National Action Plan on Invasive Alien Species 2021–2025, which includes invasive alien fish (Jabatan Pertanian Malaysia 2021). Nine invasive introduced fish have been declared, shown in Table 7.3, which also includes three recently identified invasive alien fish: *Amphilophus citrinellus*, *Phractocephalus hemioliopterus*, and *Hemibagrus wyckioides*.

The Malaysian government has defined three goals in the national plan regarding invasive alien species in general, which are:

- Improvement of understanding and public awareness about invasive alien species (IAS):
  - Increase research in all aspects of IAS management, including understanding their pathways and ecological impacts and developing technologies to contain and control IAS.
  - Educate the public on the threats of IAS, the problems posed by smuggling wild animals and plants, accidental escapes from imports by legitimate enterprises, and the pet trade.
  - Strengthen information
  - Intensify extension services to stakeholders so that prevention, control, eradication, and mitigation efforts on IAS can be effectively implemented.

Table 7.2 The occurrence of introduced freshwater fish species in Malaysia

Family	Species	Common name	Previous locality	Purpose
Arapaimidae	*Arapaima gigas*	'Pirarucu'	S. America	Aquarium
Belontiidae	*Betta splendens*	Siamese fighting fish	Thailand	Aquarium
	Trichogaster pectoralis	Skinned gourami	Thailand	Aquaculture
Centrarchidae	*Micropterus salmoides*	Largemouth bass	Florida	Recreational
Characidae	*Colossoma* sp.	Red promfet	Taiwan	Aquaculture
Cichlidae	*Etroplus suranensis*	Green chromide	Sri Lanka	Recreational
	Oreochromis mossambicus	Mozambique tilapia	Indonesia	Aquaculture
	Oreochromis niloticus	Nile tilapia	Thailand	Aquaculture
	Oreochromis urolepis hornorum	Nile tilapia	N.A.	Aquaculture
	O. hornorum × *O. mussambicus*	N.A.	N.A.	Aquaculture
	Cichla ocellaris	Peacock Bass	S. America	Recreational
	Cichla monoculus	Peacock Bass	S. America	Recreational
	Red tilapia hybrid	Tilapia	Thailand	Aquaculture
	Tilapia zillii	Redbelly tilapia	N.A.	Aquaculture
Clariidae	*Clarias macrocephalus*	Broadhead catfish	Thailand	Aquaculture
	Clarias gariepinus	African catfish	N.A.	Aquaculture
Cyprinidae	*Barbonymus gonionotus*	Java barb	Indonesia	Aquaculture
	Carassius auratus auratus	Goldfish	China	Aquaculture
	Ctenopharyngodon idella	Grass carp	China	Aquaculture
	Hypophthalmichthys nobilis	Bighead carp	China	Aquaculture
	Hypophthalmichthys molitrix	Silver carp	China	Aquaculture
	Cyprinus carpio	Common carp	China	Aquaculture
	Cyprinus carpio specularis	Mirror carp	China	Aquaculture
	Cirrhinus chienensis	Chinese mud carp	China	Aquaculture
	Mylopheryngodon piceus	Black carp	China	Aquaculture
	Carassius sp. *x Cyprinus* sp.	Japanese carp	China	Aquaculture
	Catla catla	Catla	India	Aquaculture

(continued)

Table 7.2 (continued)

Family	Species	Common name	Previous locality	Purpose
Cichlidae	*Lobeo rohita*	Rohu	India	Aquaculture
	Cirrihinus cirrhosus	Mrigal	India	Aquaculture
Gobiidae	*Rhinogobius giurinus*	Barcheek Goby	China	Aquarium
Helostomatidae	*Helostoma temminckii*	Kissing gourami	Thailand	Aquarium
Lepisostidae	*Lepisostius spatula*	Alligator gar	USA	Aquarium
Loricariidae	*Hypostomus plecostomus*	Armored catfish	S. America	Aquarium
	Pterygoplichthys pardalis	Amazon sailfin catfish	S. America	Aquarium
Salmonidae	*Oncorhynchus mykiss*	Rainbow trout	Scotland	Recreational
	Salmo gairdneri	Rainbow trout	New Zealand	Recreational
Pangasiidae	*Pangasius pangasius*	Yellowtail catfish	N.A.	Aquaculture
	Pangasius hypophthalmus	Striped catfish	Thailand	Aquaculture
Poleciliidae	*Poecilia reticulate*	Guppy	N.A.	Aquarium
	Poecilia sphenops	Molly	N.A.	Aquarium
	Gambusia affinis	Mosquitofish	N.A.	Biological control
Terapontidae	*Scortum barcoo*	Jade perch	Australia	Aquaculture

N.A.: data not available

Table 7.3 Invasive introduced fish in Malaysia

No.	Common name	Scientific name
1	African Catfish	*Clarias gariepinus*
2	Amazon Redtail Catfish	*Phractocephalus hemioliopterus* Bloch & Schneider 1801
3	Asian Redtail Catfish	*Hemibagrus wyckioides*
4	Black Tilapia	*Oreochromis mossambicus*
5	Black Tilapia	*Oreochromis niloticus*
6	Earth-eater	*Geophagus surinamensis* Bloch 1791
7	Midas Cichlid	*Amphilophus citrinellus* Günther 1864
8	Peacock Bass	*Cichla* sp.
9	Zebra Tilapia	*Heterotilapia buettikoferi* Hubrecht 1881

- Develop and implement a work program to control and manage the risk of IAS from ship ballast water and sediments.

- Risk assessment on all introduced species before their release:
  - Establish a formal and transparent mechanism to ensure all alien species are subject to a rigorous risk assessment.
  - Establish response plans and the capacity to contain and eradicate potential invasive species.

- Enhanced quarantine inspection and enforcement at entry points and international borders:
  - Enhance legislation enforcement on importing, trading, and keeping wild animals, fish, and plants.
  - Implement provisions of the International Maritime Organisation (IMO) Ballast Water Management Convention to safeguard against marine invasive species.
  - Enhance quarantine facilities and improve the skills and capabilities of quarantine, customs, and other border officials.
  - Strengthen and monitor the implementation of the National Action Plan for the Prevention, Eradication, Containment, and Control of IAS.

7.2.3 The Philippines

The latest data showed that 51 freshwater fish species have been introduced into the Philippines (Froese and Pauly 2019). However, Guerrero III (2014) pointed out that 62 freshwater fish species were introduced in the country from 1905 to 2013. A further evaluation showed that 48 species of the introduced fishes have been beneficial in terms of contributing to fish production (aquaculture) in the country for ornamental purposes, recreational fishing, and biological control of mosquitoes. Ten introduced fish species are considered invasive for causing ecological or economic loss, and four species are considered potentially invasive. Ornamental fish and aquaculture species comprise the majority of species brought into the country, and although intended to be restricted to aquariums and aquaculture facilities, some have escaped into natural waters, reproduced, and become invasive with devastating impacts to endemic species and ecosystems.

Several fish species introduced to the Philippines have contributed to the country's food security by providing a cheaper source of protein. Among these are the cultured tilapias, such as Nile tilapia (*Oreochromis niloticus*), carps, mainly the bighead carp (*Hypophthalmichthys nobilis* Richardson 1845), and catfishes (*Clarias gariepinus* and *Pangasianodon hypophthalmus* Sauvage 1878). Four species introduced for cultivation became invasive, including *Channa striata* Bloch 1793, *Clarias batrachus* Linnaeus 1758, *Monopterus albus* Zuiew 1793, and *Oreochromis mossambicus*. The mudfish (*Channa striata*) is declared a pest that preys on the young of cultured fish such as Nile Tilapia. The Asiatic catfish imported from Thailand (*Clarias batrachus*) has entirely dominated and ecologically displaced the native catfish (*Clarias macrocephalus* Günther 1864) in Laguna de Bay and other water bodies in Luzon, where it has been introduced. *Clarias batrachus* now dominates natural populations in lakes and rivers, and the indigenous *Clarias macrocephalus* can hardly be found in the markets nowadays, although it has better qualities as a food fish than *Clarias batrachus* (Juliano et al. 1989).

The rice paddy eel (*Monopterus albus*) has been documented to infest rice paddies in the Cagayan Valley. It is known locally as Kiwet and feeds on frogs, snail eggs, small fish, and shrimp, and is becoming a threat to native fish species. This invasive species, which grows between 25 and 40 cm, causes damage to rice fields by burrowing into the soil, loosening its composition, and disturbing irrigation (Perez 2013). The introduced Mossambique Tilapia (*Oreochromis mossambicus*) caused a disaster for brackishwater farms, as it competed for food with cultivated milkfish (*Chanos chanos* Fabricius 1775). However, this species is currently well-established in brackishwater aquaculture throughout the country (Juliano et al. 1989).

Some introduced fishes for ornamental purposes have become invasive after escaping from commercial or private keepers into open waters. These invasive fishes include Janitor Fish (*Pterygoplichthys* disjunctivus Weber 1991 and *P. pardalis*), Jaguar Guapote (*Parachromis managuensis* Günther 1867), Clown Featherback (*Chitala ornate* Gray 1831), Giant Snakehead (*Channa micropeltes* Cuvier 1831), and Blackchin Tilapia (*Sarotherodon melanotheron* Rüppell 1852) (Guerrero III 2014).

Invasive fish have colonized many waterways in the Philippines, including the Marikina River, which flows through the eastern part of Metropolitan Manila, the Philippines' capital city. Most invasive fishes come from Africa, South America, and Indochina and have effectively adapted to Philippine waters, which is terrible news for endemic species. Local fishermen can readily capture Nile Tilapia, African Catfish, Janitor Fish, Pangasius, and Cream Dory, as well as smaller fish like Guppies and Mollies, which feed on mosquito larvae (Lacsamana 2021).

Invasive fish have caused adverse environmental impacts in the Philippines' waters. Eleven species should be included in the species list with significant ecological, environmental, and socioeconomic implications (Casal et al. 2007). The movement of this species throughout the country must be carefully assessed and monitored. These species include Goldfish (*Carassius auratus* Linnaeus 1758), Walking catfish (*Clarias batrachus*), Grass carp (*Ctenopharyngodon idella*), Common carp (*Cyprinus carpio*), Mosquitofish (*Gambusia affinis*), Silver carp (*Hypophthalmichthys molitrix* Valenciennes 1844), Bluegill (*Lepomis macrochirus* Rafinesque 1819), Largemouth black bass (*Micropterus salmoides* Lacepède 1802), Mozambique tilapia (*Oreochromis mossambicus*), Nile tilapia (*Oreochromis niloticus*), and Guppy (*Poecilia reticulata* Peters 1859).

Some mechanisms by which invasive fish colonize and spread in freshwater ecosystems include human involvement, transport methods, adaptations supporting dispersion, and environmental factors facilitating invasion.

- Human Factors and Transportation Systems: Invasive fish can enter new ecosystems intentionally, such as by importing fish for aquaculture purposes or accidentally, for example, by delivering ballast water from ships or transporting goods by ship.
- Water Channels and Hydrological Connections: Water channels connecting different water systems, such as rivers or canals, can serve as migration routes for invasive fish and facilitate their movement from one region to another.
- Climate Change: Increased water temperatures and changes in rainfall patterns caused by climate change may create a more favorable environment for invasive fish and open up previously inaccessible areas.
- Water Quality Resilience: Poor water quality, characterized by high nutrient levels, high pollutant levels, or low oxygen, can increase the competitiveness of invasive fish that are more tolerant of these conditions.
- Habitat Structure Changes: Destruction of aquatic vegetation or river modifications that constitute changes in habitat structure may provide an advantage for invasive fish that can adapt to altered habitats.
- Fragmented Habitat Utilization: Invasive fish capable of thriving in fragmented habitats can exploit habitat structures modified by human activities.
- Adequate Spawning and Hatching Sites: Sufficient spawning and hatching sites can facilitate invasive fish's reproduction and biological success in new environments. Invasive fish with broad spawning preferences and adaptability to various substrates have a greater chance of establishing and reproducing successfully.
- International Trade: The international trade of fish, whether for aquarium or commercial fisheries purposes, can lead to the release of invasive fish in new areas, accelerating the invasion process. Fish released by aquarium traders or escaping from cultivation facilities can spread to natural aquatic ecosystems.

Negative impacts resulting from the invasion of alien fish can be ecological, for example, reduced abundance or extinction of endemic fish species, alteration of food chains, and ecological balance. Alien fish invasions can change the biotic community structure by replacing local species or creating new ecological pressures.

Negative impacts due to the invasion of introduced fish can also be economic. It is estimated that the economic cost of biodiversity loss is ten to one hundred times higher than protection costs (Gómez-Baggethun and Martín-López 2015). Some economic impacts include:

- Decline in Local Fisheries Yield: Invasion of alien fish species can threaten local fish populations with high economic value. Reduction in local fisheries yields, potentially harming fishermen and the fishing industry.
- Damage to Fishing Equipment: Some alien fish species tend to damage traditional fishing equipment, resulting in additional costs for local fishermen.
- Changes in Migration Patterns and Fishing Seasons: The invasion of alien fish species can alter the migration patterns and fishing seasons of local species. The difficulty local fishermen have in capturing their target species reduces the potential economic yield.

- Costs of Control and Rehabilitation: Efforts to control and rehabilitate ecosystems to address the invasion of alien fish species require significant financial investment. Local communities and governments bear the cost of combating the negative impacts of the invasion of alien fish.

7.3 Strategies for Management and Conservation

Understanding the environmental and human factors that reinforce the invasion of introduced fish species is crucial for designing effective management strategies. Strict monitoring of human activities that may accelerate invasions and protecting vulnerable aquatic environments against changes can help mitigate the impacts of introduced fish invasions on ecosystems. Awareness and preventive actions are crucial to safeguarding the sustainability of aquatic ecosystems from the threats posed by the invasion of introduced fish species.

Practical steps in management and conservation strategies include real solutions to combat the threats faced by endemic fish due to the invasion of introduced species. Management of Invasive Species:

1. Early Detection and Monitoring:

    - Enhance monitoring systems to detect invasive fish species as early as possible and develop monitoring programs involving local communities, researchers, and other stakeholders. Collaboration between the government, local civil society, academic institutions, and policymakers is essential to address threats to endemic species and their habitat, establish fish conservation zones, and conduct monitoring. Conservation zones are managed by the local community, which also oversees efforts to prevent illegal fishing and safeguard crucial habitat areas for endemic fish species. This collaborative approach aims to prevent the introduction of potentially invasive species while promoting the sustainable conservation of both the species and its habitat.

2. Population Control and Management:

    - Influential and sustainable population control methods should be implemented, including natural predators, artificial predators, or other biological control methods.
    - Natural predators can be employed as a means to control invasive fish populations. This strategy involves introducing or promoting the presence of indigenous or introduced species that are natural predators of the invasive fish to regulate their numbers. Using natural predators is a potential biological control method to restore ecological balance by leveraging predator–prey relationships within the ecosystem. Implement management strategies to control and reduce the population of invasive fish, for example, by holding invasive fishing competitions that provide prizes. However, implementing such strategies necessitates thorough ecological assessments and considerations to

prevent unintended consequences and ensure the effectiveness of the control measures.

- Using invasive fish as a food source or as livestock feed involves exploring various methods to manage and control the impact of invasive species on local ecosystems. Harvesting and incorporating invasive fish into human consumption or animal feed can be considered as a potential strategy. This strategy seeks to mitigate the adverse effects of invasive species by turning them into a valuable resource. However, careful consideration must be given to ecological, health, and regulatory aspects to ensure sustainable and safe practices. Implementing such strategies may provide economic benefits and contribute to reducing the invasive species population while addressing food or feed needs.

3. Community Education and Awareness:

   - Educate local communities to truly understand the impact of invasive fish species on endemic fish and aquatic ecosystems and encourage community participation in conservation and natural resource management activities.
   - Educate local communities about the dangers of discarding unwanted aquarium fish into public waters because these fish are at risk of becoming invasive and threatening native and endemic fish.
   - Increase local community awareness of the importance of protecting endemic species and preventing the introduction of invasive fish, thereby ensuring the long-term sustainability of both endemic fish and their habitat. Integrating local and scientific knowledge is crucial, including raising awareness of the importance of endemic fish conservation in local communities and the threats posed by invasive species to endemic species and their habitats. Local communities will gain greater knowledge about endemic and invasive species and actively work to ensure the long-term survival of fish communities in their environment.

4. Trade and Transportation Control:

   - Tighten rules and regulations related to the trade of potentially invasive fish, and monitor and manage fish transportation rigorously to prevent invasive spread.

5. Ecosystem Rehabilitation, Restoration, and Conservation:

   - Implement ecosystem rehabilitation and restoration programs affected by invasive fish and take actions to support the recovery of affected endemic fish populations, such as managing a controlled increase in the number of endemic or native fish populations outside their natural habitat, which is then used for restocking.

6. International Conservation Collaboration:

   - Numerous international collaborative efforts are currently related to the conservation of endemic fish and the management of introduced fish invasions.

By engaging in collaborations such as research and information exchange, policy and guideline development, training and education initiatives, and collaborative projects, the efforts to protect endemic fish and their ecosystems and control the threats posed by invasive fish can be more effective and successful.

7. Research and Innovation:

   - Support for scientific research into the behavior and dynamics of invasive fish species. Studying endemic fish species and their habitats and ecosystems is equally essential.
   - Innovative technologies and new management methods that include advances in monitoring technology, sensors, or the application of new control methods need to be developed to overcome problems related to invasive fish species. For example, an environmentally friendly, low-cost, vision-based fish recognition and separation system. The system includes an underwater imaging system that captures fish images for processing, a fish species recognition algorithm that identifies invasive fish species, and a mechanical system that guides fish movement and restrains invasive fish species from removal (Chai et al. 2021).

8. Strict Regulations and Penalties:

   - The Indonesian government implements prevention and control measures for invasive fish through policies issued by the Ministry of Marine Affairs and Fisheries, specifically through Law Number 32 of 2004 on fisheries, which has been amended into Law Number 45 of 2009. Additionally, the prohibition on the entry of Hazardous Fish into Indonesia is regulated by Ministerial Regulation Number 41/Peraturan Menteri-KP/2014 and Regulation of the Minister of Marine Affairs and Fisheries Number 19/PERMEN-KP/2020 Year 2020 concerning Prohibition of Entry, Cultivation, Circulation, and Exit of Fish Species Harmful and/or Detrimental into and from the Territory of the Fisheries Management Area of the Republic of Indonesia. The Fisheries Quarantine, Quality Control, and Safety of Fisheries Products Agency (BKIPM) under the Ministry of Marine Affairs and Fisheries also releases a database listing fish species with the potential to become invasive alien species in Indonesia.
   - Establish strict regulations regarding the ownership, trade, and release of invasive fish.
   - Sanctions and penalties must be strictly imposed for any violation of regulations regarding invasive fish species.

Integrating these approaches is essential to achieving effective and sustainable management of invasive fish species and protecting endemic fish and aquatic ecosystems. The collaboration between institutions, communities, and countries is essential for resolving the challenges posed by invasive species and ensuring the conservation of freshwater ecosystems.

Remarks

In our ongoing efforts toward sustainability, we continuously seek solutions to protect the countless lives within our aquatic ecosystems. Through this writing, we have opened a window into the complexity of the relationships between endemic and invasive fish and, thus, our responsibility to maintain the fragile biotic balance.

We must remember that every action we take has an impact. Every policy, every small step to raise awareness, and every collective effort is a foothold toward a sustainable future. We can find significant solutions to preserve freshwater biota communities in these various endeavors.

Conservation challenges ahead of us may be significant, but we can bring about positive change through collaboration, innovation, and dedication. We are not just guardians of these aquatic ecosystems but the authors of their future. We have the power to preserve biodiversity and ensure that future generations can also enjoy the wonders beneath the water's surface.

In conclusion, let us safeguard our freshwater together as an invaluable heritage. Understanding and respecting this precious biotic balance protects endemic fish and aquatic ecosystems and safeguards our nation's future.

References

Ahmad AK, Nur-Hazwani MNR, Omar SAS, Aweng ER, Taweel A (2020) Preliminary study on invasive fish species diffusion in selected Malaysian freshwater ecosystems. Pakistan J Biol Sci 23(11):1374–1379. https://doi.org/10.3923/pjbs.2020.1374.1379

Allen GR, Kadarusman (2020a) *Glossolepis incisus*. In: IUCN red list threat. Species. https://www.iucnredlist.org/species/9268/147681075. Accessed 16 Feb 2024

Allen GR, Kadarusman (2020b) *Melanotaenia boesemani*. In: IUCN red list threat. Species. https://www.iucnredlist.org/species/13058/147682045. Accessed 17 Feb 2024

Allen GR, Kadarusman (2020c) *Melanotaenia praecox*. In: IUCN red list threat. Species. https://www.iucnredlist.org/species/13074/147688071. Accessed 17 Feb 2024

Allen GR, Hammer M, Kadarusman (2020a) *Chilatherina sentaniensis*. In: IUCN red list threat. Species. https://www.iucnredlist.org/species/4631/147680762. Accessed 16 Feb 2024

Allen GR, Kadarusman, Unmack P (2020b) *Pseudomugil reticulatus*, Vogelkop Blue-Eye. In: IUCN red list threat. Species. https://www.iucnredlist.org/species/161077487/161077492. Accessed 17 Feb 2024

Argus AA (2022) Di Tengah Gempuran Ikan Iblis Merah, Ikan Pora-pora Kembali Muncul, Tapi Ukurannya Makin Kecil. In: tribun-medan.com. https://medan.tribunnews.com/2022/07/08/di-tengah-gempuran-ikan-iblis-merah-ikan-pora-pora-kembali-muncul-tapi-ukurannya-makin-kecil. Accessed 10 Feb 2024

Astuti T, Yusra, Mardiah A (2016) Studi Mutu Ikan Rinuak (*Psilopsis* sp.) Olahan di Danau Maninjau, Kecamatan Tanjung Raya Kabupaten Agam Sumatera Barat. J Katalisator 1(1):10. https://doi.org/10.22216/jk.v1i1.982

Aya FA, Garcia LMB (2020) Biology and hatchery rearing of the silver therapon *Leiopotherapon plumbeus*. Southeast Asian Fisheries Development Center Aquaculture Department, Tigbauan, Iloilo, Philippines

Casal CMV, Luna S, Froese R, Bailly N, Atanacio R, Agbayani E (2007) Alien fish species in the Philippines: pathways, biological characteristics, establishment and invasiveness. J Environ Sci Manag 10(1):1–9

Castillo BB, Castillo AS, Gonzales CL (1975) Tawilis fishery resources investigation of Taal Lake. In: Philippine Council for Agriculture and Resources Research Fisheries Forum. Philippine Council for Agriculture and Resources Research, Los Baños, Laguna, Philippines, pp 75–98

Chai J, Lee DJ, Tippetts B, Lillywhite K (2021) Implementation of an award-winning invasive fish recognition and separation system. Electron 10(17):1–13. https://doi.org/10.3390/electronics10172182

Chaichana R, Jongphadungkiet S (2013) Assessment of the invasive catfish *Pterygoplichthys pardalis* (Castelnau, 1855) in Thailand: ecological impacts and biological control alternatives. Trop Zool 25(4):173–182. https://doi.org/10.1080/03946975.2012.738494

Chong VC, Lee PKY, Lau CM (2010) Diversity, extinction risk and conservation of Malaysian fishes. J Fish Biol 76(9):2009–2066. https://doi.org/10.1111/j.1095-8649.2010.02685.x

Coubout R (2014) Ikan Bilih, Ikan Endemik Danau Singkarak Yang Terancam Punah. Kenapa? In: Mongabay. https://www.mongabay.co.id/2014/10/30/ikan-bilih-ikan-endemik-danau-singkarak-yanag-terancam-punah-kenapa/. Accessed 12 Feb 2024

Dahruddin H, Haryono WG, Rusdianto UIV, Marwayana ON, Mokodongan DF, Sauri S, Priyatna Y (2021) Ikan Endemik Indonesia Seri Sumatra: Biologi dan Pemanfaatannya. PT Penerbit IPB Press, Bogor

Daniels A (2020) *Tondanichthys kottelati*. In: IUCN red list threat. Species. https://www.iucnredlist.org/species/21989/90982079. Accessed 18 Feb 2024

Daniels A, Palmer-Newton A (2019) *Marosatherina ladigesi*, Celebes Rainbow. In: IUCN red list threat. Species. https://www.iucnredlist.org/species/21574/126023418. Accessed 15 Feb 2024

Dina R, Samir O, Lukman, Haryani GS, Nasution SH (2019) Fish and fisheries of Bada (*Rasbora* spp.) in Lake Maninjau, West Sumatra. IOP Conf Ser Earth Environ Sci 306, 124. https://doi.org/10.1088/1755-1315/306/1/012004

Dina R, Wahyudewantoro G, Larashati S, Aisyah S, Lukman, Sulastri, Imroatushshoolikhah, Sauri S (2022) Distributional mapping and impacts of invasive alien fish in Indonesia: an alert to inland waters sustainability. Sains Malaysiana 51(8):2377–2401. https://doi.org/10.17576/jsm-2022-5108-04

DJPB KKP (2023) Jenis-Jenis Arwana yang Memiliki Nilai Ekonomis Tinggi. In: Kementeri. Kelaut. dan Perikan. Republik Indones. https://kkp.go.id/djprl/bpsplpadang/page/3021-jenis-jenis-arwana-yang-memiliki-nilai-ekonomis-tinggi. Accessed 14 Feb 2024

Dudgeon D, Arthington AH, Gessner MO, Kawabata ZI, Knowler DJ, Lévêque C, Naiman RJ, Prieur-Richard AH, Soto D, Stiassny MLJ, Sullivan CA (2006) Freshwater biodiversity: importance, threats, status and conservation challenges. Biol Rev Camb Philos Soc 81(2):163–182. https://doi.org/10.1017/S1464793105006950

Fendri STJ, Ifmaily I, Syarti SR (2019) Analisis Protein Pada Rinuak, Pensi dan Langkitang dengan Spektrofotometri UV-Vis. J Katalisator 4(2):119–124. https://doi.org/10.22216/jk.v4i2.4425

Froese R, Pauly D (2019) List of introduced fishes reported from the Philippines. In: Fishbase. https://fishbase.mnhn.fr/country/CountryChecklist.php?c_code=608&vhabitat=introduced&csub_code. Accessed 19 Mar 2024

Froese R, Pauly D (2023) List of freshwater fishes reported from Indonesia. In: FishBase. http://fishbase.org/country/CountryChecklist.php?resultPage=17&c_code=360&vhabitat=fresh. Accessed 6 Feb 2024

Gani A, Nilawati J, Rizal A (2015) Studi Habitat dan Kebiasan Makan (Food Habit) Ikan Rono Lindu (Oryzias sarasinorum Popta, 1905). J Sains dan Teknol Tadulako 4(3):9–18

Ginting DW, Purnomo PW, Ghofar A (2013) Potensi dan Pengelolaan Sumber Daya Ikan Pora-Pora (*Mystacoleucus padangensis* Bleeker) di Danau Toba, Sumatra Utara. Manag Aquat Resour J 2(4):28–37. https://doi.org/10.14710/marj.v2i4.4265

Gómez-Baggethun E, Martín-López B (2015) Ecological economics perspectives on ecosystem services valuation. In: Handbook of ecological economics, pp 260–282

Guerrero RD III (2014) Impacts of introduced freshwater fishes in the Philippines (1905–2013): a review and recommendations. Philipp J Sci 143(1):49–59

Gustiano R (2009) Pangasiid Catfishes of Indonesia. Buletin Plasma Nutfah 15(2):91–100

Gustiano R, Kurniawan K, Haryono H (2021) Optimizing the utilization of genetic resources of Indonesian native freshwater fish. Asian J Conserv Biol 10(2):189–196. https://doi.org/10.53562/ajcb.67022

Hadiaty RK (2018) Status taksonomi iktiofauna endemik perairan tawar Sulawesi. J Iktiologi Indones 18(2):175–190. https://doi.org/10.32491/jii.v18i2.428

Hanafiah J (2021) Ikan Depik yang Tidak Lagi Bahagia di Danau Lut Tawar. In: Mongabay. https://www.mongabay.co.id/2021/12/14/ikan-depik-yang-tidak-lagi-bahagia-di-danau-lut-tawar/. Accessed 8 Feb 2024

Haryani GS, Said DS (2011) Pola Keanekaragaman Ikan Hias Endemik/Asli Indonesia Berdasarkan Garis Wallacea. Prosiding Forum Perairan Umum Indonesia 8:252–260

Haryani GS, Hidayat, Samir O (2020) Diversity of fish caught using gill nets in Lake Sentarum, West Kalimantan-Indonesia. In: IOP conference series: earth and environmental science

Hasan V, Soemarno, Widodo MS, Wiadnya DGR, Mukti AT, Irawan B (2019) Distribution extension and first record of *Lobocheilos falcifer* (Cypriniformes, cyprinidae) in Central Java Province, Indonesia. Ecol Environ Conserv 25:S158–S161

Hasan V, Samitra D, Widodo MS, Islam I, Ottoni FP (2021a) An updated checklist of *Rasbora baliensis* (Hubbs & Brittan 1954) (Cypriniformes: Cyprinidae) in Indonesia. Malaysian J Sci 40(3):107–113. https://doi.org/10.22452/MJS.VOL40NO3.8

Hasan V, Vieira L de O, Ottoni FP, Masithah ED (2021b) Two new localities for *Lobocheilos falcifer* (Valenciennes, 1842) (Teleostei: Cyprinidae), a rare and vulnerable freshwater fish species of Java, Indonesia. Int J Aquat Biol 9(4):244–247. https://doi.org/10.22034/ijab.v9i4.1285

Herder F, Chapuis S (2010) *Oryzias hadiatyae*, a new species of ricefish (Atherinomorpha: Beloniformes: Adrianichthyidae) endemic to Lake Masapi, Central Sulawesi Indonesia. Raffles Bull Zool 58(2):269–280

Hidayat H, Aisyah S, Kurniawan R, Ridwansyah I, Samir O, Haryani GS (2020) Flood pulse and aquatic habitat dynamics of the Sentarum Floodplain Lakes area. Indones J Limnol 1(1):27–37. https://doi.org/10.51264/inajl.v1i1.2

Ho JKI, Ahmad AB (2019) *Neolissochilus hendersoni*. In: IUCN red list threat. Species. https://www.iucnredlist.org/species/181189/1707888#geographic-range. Accessed 15 Mar 2024

Hoese DF, Hadiaty RK, Herder F (2015) Review of the dwarf *Glossogobius* lacking head pores from the Malili lakes, Sulawesi, with a discussion of the definition of the genus. Raffles Bull Zool 63:14–26

Hubert N, Kadarusman WA, Busson F, Caruso D, Sulandari S, Nafiqoh N, Pouyaud L, Rüber L, Avarre J-C, Herder F, Hanner R, Keith P, Hadiaty RK (2015) DNA barcoding Indonesian freshwater fishes: challenges and prospects. DNA Barcodes 3(1):144–169. https://doi.org/10.1515/dna-2015-0018

Hubert N, Lumbantobing D, Sholihah A, Dahruddin H, Delrieu-Trottin E, Busson F, Sauri S, Hadiaty R, Keith P (2019) Revisiting species boundaries and distribution ranges of *Nemacheilus* spp. (Cypriniformes: Nemacheilidae) and Rasbora spp. (Cypriniformes: Cyprinidae) in Java, Bali and Lombok through DNA barcodes: implications for conservation in a biodiversity hotspot. Conserv Genet 20(3):517–529. https://doi.org/10.1007/s10592-019-01152-w

Jaafar Z (2019) *Lentipes whittenorum*. In: IUCN red list threat. Species. https://www.iucnredlist.org/species/11502/91080832. Accessed 12 Feb 2024

Jabatan Pertanian Malaysia (2021) National action plan on invasive alien species 2021–2025. Jabatan Pertanian Malaysia, Putrajaya

Jubaedah I, Sayida BR (2022) Ikan Asli dan Endemik di Perairan Umum Daratan Pulau Jawa: Biologi, Pemanfaatan dan Upaya Pelestariannya. In: Prosiding Seminar Nasional Ikan XI, pp 40–53

Juliano RO, Guerrero III R, Ronquillo I (1989) The Introduction of exotic aquatic species in the Philippines. In: De Silva SS (ed) Proceedings of a workshop on introduction of exotic aquatic organisms in Asia, pp 83–90

Kadarusman (2020) *Melanotaenia ajamaruensis*. In: IUCN red list threat. Species. https://www.iucnredlist.org/species/13055/147681943. Accessed 17 Feb 2024

Kelana I (2015) Ikan Pora-Pora Jadi Andalan Nelayan Pesisir Danau Toba. In: republika.co.id. https://ekonomi.republika.co.id/berita/ntac7p374/ikan-porapora-jadi-andalan-nelayan-pesisir-danau-toba. Accessed 10 Feb 2024

Khaironizam MZ, Zakaria-Ismail M, Armbruster JW (2015) Cyprinid fishes of the genus *Neolissochilus* in Peninsular Malaysia. Zootaxa 3962(1):139–157. https://doi.org/10.11646/zootaxa.3962.1.7

Kottelat M, Tan HH (2008) A synopsis of the genus *Lobocheilos* in Java, Sumatra and Borneo, with descriptions of six new species (Teleostei: Cyprinidae). Ichthyol Explor Freshw 19(1):27–58

Kottelat M, Whitten AJ, Kartikasari SN, Wirjoatmodjo S (1993) Freshwater fishes of Western Indonesia and Sulawesi. Periplus Editions (HK)

Kottelat M, Widjanarti E (2005) The fishes of Danau Sentarum National Park and the Kapuas Lakes Area, Kalimantan Barat Indonesia. Raffles Bull Zool 13(13):139–173

Kottelat M, Witten T (1996) Freshwater biodiversity in Asia with special reference to fish. Washington, DC

Lacsamana BH (2021) PHL waters being taken over by invasive fish. In: Business world online. https://www.bworldonline.com/agribusiness/2021/09/20/397595/phl-waters-being-taken-over-by-invasive-fish/. Accessed 14 Mar 2024

Larson H (2019) *Pogoneleotris heterolepis*, Bearded Gudgeon. In: IUCN red list threat. Species. https://www.iucnredlist.org/species/91088359/91088382. Accessed 15 Mar 2024

Larson H, Vidthayanon C (2019) *Scleropages formosus*. In: IUCN red list threat. Species 2013. https://doi.org/10.2305/IUCN.UK.2011-1.RLTS.T20034A9137739.en. Accessed 14 Feb 2024

Larson HK, Hui TH (2019) A redescription of the bearded gudgeon, *Pogoneleotris heterolepis* (Günther, 1869) (Teleostei: Gobioidei: Butidae). Raffles Bull Zool 67:385–390. https://doi.org/10.26107/RBZ-2019-0029

Lumbantobing D (2019a) *Tominanga sanguicauda*. In: Iucn red list threat. Species. https://www.iucnredlist.org/species/21980/90980344. Accessed 15 Feb 2024

Lumbantobing D (2019b) *Lobocheilos falcifer*. In: IUCN red list threat. Species. https://www.iucnredlist.org/species/91005744/91005782. Accessed 12 Feb 2024

Lumbantobing D (2019c) *Telmatherina celebensis*. In: IUCN red list threat. Species. https://www.iucnredlist.org/species/21573/90980085. Accessed 14 Feb 2024

Lumbantobing D (2019d) *Rasbora tawarensis*. In: IUCN red list threat. Species

Lumbantobing D (2019e) *Oryzias marmoratus*, Marmorated Ricefish. In: IUCN red list threat. Species. https://www.iucnredlist.org/species/15574/90980644. Accessed 15 Feb 2024

Lumbantobing D (2020a) *Neolissochilus thienemanni*, Ihan. In: IUCN red list threat. Species

Lumbantobing D (2020b) *Rasbora tobana*. In: IUCN red list threat. Species. https://www.iucnredlist.org/species/91074229/91074237. Accessed 9 Feb 2024

Lumbantobing D (2020c) *Mystacoleucus padangensis*. In: IUCN red list threat. Species. https://www.iucnredlist.org/species/91059695/91059767. Accessed 9 Feb 2024

Lumbantobing D (2020d) *Lobocheilos erinaceus*. In: IUCN red list threat. Species. https://www.iucnredlist.org/species/91005692/91005701. Accessed 14 Feb 2024

Lumbantobing D (2020e) *Rasbora borneensis*. In: IUCN red list threat. Species. https://www.iucnredlist.org/species/91071245/91071249. Accessed 14 Feb 2024

Lumbantobing D (2020f) *Poropuntius tawarensis*, Kawan. In: IUCN red list threat. Species. https://www.iucnredlist.org/species/18067/91069249. Accessed 9 Feb 2024

Lumbantobing D (2020g) *Rasbora maninjau*, Bada Maninjau. In: IUCN red list threat. Species. https://www.iucnredlist.org/species/143695241/143695247. Accessed 11 Feb 2024

Lumbantobing D (2021) *Rasbora baliensis*. In: IUCN red list threat. Species. https://www.iucnredlist.org/species/19315/91070323. Accessed 12 Feb 2024

Lumbantobing DN (2010) Four new species of the *Rasbora trifasciata*-group (Teleostei: Cyprinidae) from Northwestern Sumatra Indonesia. Copeia 4:644–670. https://doi.org/10.1643/CI-09-155

Lumbantobing DN (2014) Four new species of Rasbora of the Sumatrana group (Teleostei: Cyprinidae) from northern Sumatra, Indonesia. Zootaxa 3764(1):1–25. https://doi.org/10.11646/zootaxa.3764.1.1

Maulana A (2023) Ikan rinuak jadi langka di Danau Maninjau akibat air tercemar. In: Antara

Mayasari N, Said DS, Sulawesty F (2020) Preferensi Pakan Ikan Batak (*Neolissochilus sumatranus* dan *Tor douronensis*) Asal Sungai Bonan Dolok, Samosir, Sumatra Utara. In: Ikan Batak Bonan Dolok, Kabupaten Samosir, Sumatra Utara, p 20

Mokodongan DF (2019) *Xenopoecilus sarasinorum*. In: IUCN red list threat. Species. https://www.iucnredlist.org/species/23123/90980952. Accessed 14 Feb 2024

Muchlisin ZA (2013) Distributions of the endemic and threatened freshwater fish depik, *Rasbora tawarensis* Weber & de Beaufort, 1916 in Lake Laut Tawar, Aceh Province Indonesia. Songklanakarin J Sci Technol 35(4):483–488

Muchlisin ZA, Musman M, Siti Azizah MN (2010) Length-weight relationships and condition factors of two threatened fishes, *Rasbora tawarensis* and *Poropuntius tawarensis*, endemic to Lake Laut Tawar, Aceh Province Indonesia. J Appl Ichthyol 26(6):949–953. https://doi.org/10.1111/j.1439-0426.2010.01524.x

Nasution SH (2017) Biodiversity and conservation of endemic fish species in some lakes of Sulawesi. In: Maghfiroh M, Dianto A, Jasalesmana T, Melat I, Samir O, Kurniawan R (eds) Proceedings of the 16th world lake conference. Research Center for Limnology, Indonesian Institute of Sciences, p 9

Nasution SH, Akhdiana I, Syawal MS, Nurhidayat A (2021) Length-weight relationship and condition factor of endemic fish Bilih (*Mystacoleucus padangensis* Blkr.) in Lake Singkarak, West Sumatra, Indonesia. In: E3S web of conferences, pp 1–16

Nurasiahriza (2016) Depik (*Rasbora tawarensis*): Ikan Endemik di Danau Laut Tawar yang Terancam Punah. https://nurasiahriza042.wordpress.com/2016/11/10/depik-rasbora-tawarensis-ikan-endemik-di-danau-laut-tawar-yang-terancam-punah/. Accessed 8 Feb 2024

Özdilek ŞY (2007) Possible threat for middle east inland water: an exotic and invasive species, *Pterygoplichthys disjunctivus* (Weber, 1991) in Asi River, Turkey (Pisces: Loricariidae). EÜ J Fish Aquat Sci 24(May):303–306

Parenti LR (2008) A phylogenetic analysis and taxonomic revision of ricefishes, Oryzias and relatives (Beloniformes, Adrianichthyidae). Zool J Linn Soc 154(3):494–610. https://doi.org/10.1111/j.1096-3642.2008.00417.x

Perez A (2013) Pesky rice eel now an export commodity. In: Philipp. Star. https://www.philstar.com/business/2013/08/11/1075891/pesky-rice-eel-now-export-commodity. Accessed 19 Mar 2024

Rachmad B, Sihombing E, Sabariyah N (2019) Pengelolaan Ikan Batak (*Neolissochilus thienemanni*, Ahl 1933) di Perairan Umum Daratan, Danau Toba, Provinsi Sumatra Utara. J Kelaut Dan Perikan Terap 2(2):73–83

Rahim KAA, Esa Y, Arshad A (2013) The influence of alien fish species on native fish community structure in Malaysian waters. Kuroshio Sci 7(1):81–93

Rahmi F, Susanti Z, Nilda C, Muzaifa M (2021) A mini review on characteristic of depik fish (*Rasbora tawarensis*) and its processed products. In: IOP conference series: earth and environmental science

Roesma DI, Tjong DH, Aidil DR (2020) Phylogenetic analysis of transparent gobies in three sumatran lakes, inferred from mitochondrial cytochrome oxidase I (COI) gene. Biodiversitas 21(1):43–48. https://doi.org/10.13057/biodiv/d210107

Said DS, Hidayat (2015) 101 Ikan HIas Air Tawar Nusantara. LIPI Press, Jakarta

Said DS, Mayasari N, Febrianti D, Chrismadha T (2021) Kinerja pertumbuhan dan sintasan ikan nilem *Osteochilus vittatus* (Valenciennes, 1842) pada pemeliharaan dengan pakan kombinasi tumbuhan lemna (*Lemna perpusilla* Torr) dan pakan komersial [Growth performance of bonylip barb *Osteochilus vittatus* Vale]. J Iktiologi Indones 21(2):151–165

Sani H (2012) Ada Apa dengan Ikan Pora-Pora [Endemic Fish in Lake Toba]. In: steemit.com. https://steemit.com/travel/@kakilasak/ada-dengan-ikan-pora-pora-endemic-fish-in-lake-toba. Accessed 9 Feb 2024

Santos M, Munroe TA, Di Dario F, Hata H, Torres F, Quilang JP (2019) *Sardinella tawilis*, Bombon Sardine. In: IUCN Red List Threat. Species. https://www.iucnredlist.org/species/98836352/143839946. Accessed 14 Mar 2024

Sastra Y (2021) Menangguk Rinuak, Ikan Endemik Danau Maninjau Yang Kini Muncul Lagi Setelah Terancam Punah. In: kompas.com. https://www.kompas.id/baca/nusantara/2021/04/20/menangguk-rinuak-ikan-endemik-danau-maninjau-yang-kini-muncul-lagi-setelah-terancam-punah. Accessed 10 Feb 2024

Setiawan B (2022) Mahalnya Depik Ikan Endemik Danau Lut Tawar, Punya Nilai Ekonomi Tinggi. In: serambinews.com. https://aceh.tribunnews.com/2022/03/28/mahalnya-depik-ikan-endemik-danau-lut-tawar-punya-nilai-ekonomi-tinggi. Accessed 8 Feb 2024

Sulistiono FA, Affandi R (2006) Studi isi lambung ikan opudi (*Telmatherina celebensis*) di Danau Towuti, Sulawesi Selatan. J Akuakultur Indones 5(2):149–156

Sumut Pos (2020) Ikan Bilih, si Kecil Pembersih Air Danau. In: sumutpos.com. https://sumutpos.jawapos.com/daerah/10/11/2020/ikan-bilih-si-kecil-pembersih-air-danau/. Accessed 16 Feb 2024

Syandri H (1996) Aspek Reproduksi Ikan Bilih (*Mystacoleucus padangensis* Blkr) dan Kemungkinan Pembenihannya di Danau Singkarak. Institut Pertanian Bogor

Tappin AR (2018) Rainbowfishes. https://rainbowfish.angfaqld.org.au/Melano.htm. Accessed 10 Jan 2024

Torres AG, Leander NJS, Kesner-Reyes K, Ballad EL, Ame EC, Palla HP, Angeles IJP, Gonzalez JCT, Reyes AT, Gaerlan RSP, Guino-o RSI, Cecilio MAF, Garcia MPC (2021) *Leiopotherapon plumbeus*, Ayungin. In: IUCN red list threat. Species. https://www.iucnredlist.org/species/162161619/162161646. Accessed 14 Mar 2024

Umar C, Kartamihardja ES, Aisyah A (2015) Dampak Invasif Ikan Red Devil (*Amphilophus citrinellus*) terhadap Keanekaragaman Ikan di Perairan Umum Daratan di Indonesia. J Kebijak Perikan Indones 7(1):55–61. https://doi.org/10.15578/jkpi.7.1.2015.55-61

Univers Aquatique (2024) *Telmatherina celebensis*. In: Univers aquat. https://www.universaquatique.fr/poissons-arc-en-ciel-et-apparentes/13872-telmatherina-celebensis-7-8cm.html. Accessed 14 Feb 2024

Valen FS, Hasan V, Ottoni FP, Nafisyah AL, Erwinda M, Annisa AN, Adis MA (2022) First country record of the bearded gudgeon *Pogoneleotris heterolepis* (Günther, 1869) (Teleostei: Eleotridae) from Indonesia. In: IOP Conference series: earth and environmental science

Wahyudewantoro G, Rachmatika I (2016) Jenis Ikan Introduksi dan Invasif Asing di Indonesia. LIPI Press

Weber M, de Beaufort LF (1916) The fishes of the Indo-Australian archipelago. E. J. Brill Ltd., Leiden

Wibisono A (2021) Ikan Depik, Salah Satu Ikan Endemik Makanan Khas Danau Laut Tawar. In: DJKN Kemenkeu

Widianto E (2021) Sungai-sungai di Jawa Sakit, Ikan Endemik Punah Perlahan. In: Mongabay. https://www.mongabay.co.id/2021/05/22/sungai-sungai-di-jawa-sakit-ikan-endemik-punah-perlahan/. Accessed 12 Feb 2024

Yoga GP, Samir O (2020) Ammonia toxicity to rinuak (*Gobiopterus brachypterus*) of Lake Maninjau. Indones J Limnol 1(1):11–17. https://doi.org/10.51264/inajl.v1i1.6

Chapter 8
Biotic Communities: Cyanobacteria

Sulastri Sulastri, Cynthia Henny, and Fachmijany Sulawesty

Excessive growth of cyanobacteria results in cyanobacteria blooms and responsible for harmful algal poses (HABs), which become a major problem in freshwater ecosystems globally and posed a serious threat to the ecological integrity and sustainability of inland waters. Cyanobacteria blooms adversely affect ecosystem structure and function, water quality, and ecological services due to the production and accumulation of cyanotoxins. Cyanotoxins have caused disfunction in aquatic ecosystems by degrading the quality of water utilized for fisheries, recreation, and drinking water and could pose a high risk of toxicity to humans, fish, animals, and other organisms (Huang et al. 2021; Dokulil and Teubner 2000; Chorus and Bartram 1999; Niamien-Ebrottie et al. 2015). Cyanobacteria bloom's utilization of oxygen during respiration at night and decomposition of decayed cells result in rapid oxygen depletion, which kills fish and other aquatic organisms. Additionally, decayed cells and decomposition process generate foul odors (Burford et al. 2019; Hamilton et al. 2013).

Increasing the occurrence of cyanobacteria blooms has been accelerated by nutrient both nitrogen and phosphorus enrichment. Cyanobacteria blooms occur more frequently and with greater intensity in eutrophic waters, characterized by high nutrient levels, as opposed to mesotrophic waters. Nevertheless, they can also manifest intermittently in less productive ecosystems, particularly those that have experienced substantial nutrient input and organic accumulation as a result of human activities (Burford et al. 2019; Barathan et al. 2023; O'Neil et al. 2012). The occurrence of cyanobacteria blooms is becoming far more complex in tropical aquatic ecosystems due to global warming-induced warmer conditions and irregular rainfall. The blooming event can occur periodically or persistently throughout the course of the year (Henny and Nomosatryo 2016; Sulastri et al. 2019, 2023; Sulawesty et al. 2015).

S. Sulastri (✉) · C. Henny · F. Sulawesty
Research Center for Limnology and Water Resources, Research and Innovation Agency (BRIN), Bogor, Indonesia
e-mail: sula001@brin.go.id

© The Author(s), under exclusive license to Springer Nature Singapore Pte Ltd. 2025
A. C. Santos-Borja et al. (eds.), *Progress on Ecosystem Restoration of Tropical Inland Waters*, https://doi.org/10.1007/978-981-96-2284-9_8

Cyanobacteria thrive in stagnant, warm water such as lakes and reservoirs. Compared to other algae, cyanobacteria also demonstrate high mobility both vertically and laterally within the water column. They can obtain nutrients from the colder, darker, bottom layers and utilize them to their growth in the more light, upper, warmer layers. At higher temperatures, cyanobacteria species develop more rapidly than other algae. Consequentially, an increase in water temperatures resulting from climate change is expected to contribute to an intensified proliferation and extent of cyanobacteria blooms (Burford 2019; Gobler 2020; O'Neil et al. 2012).

The assessment and management of excessive cyanobacterial growth and their impact on ecological function and human health are important for the sustainable use of inland water. This chapter provides an overarching introduction, fundamental information on cyanobacteria blooms especially in lakes and reservoirs, and environmental factors that promote excessive growth of cyanobacteria in tropical areas. We will highlight the impact of cyanobacteria blooms on the aquatic ecosystem and the underlying factors that contribute to such blooms in water bodies.

8.1 Cyanobacteria Blooms

Cyanobacteria, classified as photosynthetic prokaryotic organisms, possess the ability to synthesize chlorophyll *a*nd phycobilin production, specifically phycocyanin pigment (Whitton and Pott 2000). This pigment, which indicates their blue hue, has led to the term "blue-green algae" for cyanobacteria. Among the most prevalent genera of cyanobacteria discovered in freshwater are *Microcystis, Cylindrospermopsis, Planktothrix, Synechococcus, Oscillatoria, Anabaena, Anabaenopsis, Lyngbya, Aphanizomenon, Nostoc, and Synechocystis* (Chorus and Bartram 1999).

Cyanobacteria blooms are characterized by an excessive average density of cyanobacteria in the water. This phenomenon is also known as mass development or cyanobacteria proliferation (Ibelings et al. 2021). Cyanobacteria blooms were distinguished by the noticeable alterations in the coloration of the lake, which evolved into a vivid mucilage green (Huisman et al. 2018). Bloom conditions may be characterized by cyanobacteria cell densities exceeding 100,000 cells.mL^{-1} (EPA 2019; Chorus and Welker 2021). A range of (3–26) $\times$ 10^8 cells.mL^{-1} were observed in Hoan Kiem Lake, Thailand, during cyanobacteria blooms (Duong et al. 2014). As is the case in tropical lakes in Indonesia, the density of cyanobacteria blooms in several lakes and reservoirs displays a range of (15–70) $\times$ 10^5 cells.mL^{-1}, and in warmer conditions, very dense colonies of phytoplankton were observed, making it very hard to count the cells. The highest observed abundance of cyanobacteria is in lakes and reservoirs that are heavily exploited by massive floating cage fisheries (Sulastri et al. 2023; Sulawesty et al. 2012, 2018; Sulawesty and Yustiawati 2021; Sulawesty and Sumarni 2004; Haryani et al. 2022). During the blooms, concentrations of chlorophyll could reach as high as > 100–400 μg.L^{-1} (Sulastri et al. 2019, 2022). Table 8.1 presents chlorophyll-*a* concentrations during the cyanobacteria blooms in several lakes and reservoirs of South East Asian countries. WHO has defined threshold for chlorophyll-*a* of 10 $\mu g.L^{-1}$ increase allergy effect and chlorophyll-*a* of 50 $\mu g.L^{-1}$ increase probability of irritative symptoms and toxic impacts both in condition of cyanobacteria dominant (Ho and Michalak 2015).

One of the Indonesian lakes that is heavily exploited by massive floating cage fisheries is Lake Maninjau. The lake has suffered severe eutrophication in the past two decades due to the enrichment of nutrients, which has been worsened by the extreme climate conditions (Henny and Nomosatro 2016; Sulastri et al. 2023). Figure 8.1 indicates an extensive layer of green mucilage covering the lake surface during cyanobacteria blooms in Lake Maninjau, where *Microcystis aeruginosa* colonies proliferate and become extremely dense.

An increase in cyanobacteria bloom intensity in Lake Maninjau has been observed over the years, as displayed in Fig. 8.2, where the phytoplankton community composition is shifting to dominant cyanophyta (Sulastri et al. 2019). Increased frequency

Table 8.1 Chlorophyll-*a* concentrations during cyanobacteria blooms in lakes and reservoirs in several South Asian countries

Country	Name of waterbody	Chlorophyll-*a* concentration ($\mu g.L^{-1}$)	References
Indonesia	Lake Maninjau Urban shallow lake Lake Telaga Warna	35.3–120 89–431 16.48	Sulastri et al. (2019) Sulastri et al. (2022) Sulastri et al. (2022)
Malaysia	6 Tasik (urban ponds)	12.7–903	Sinang et al. (2015)
Philippines	Laguna Lake	20–33	Caballero and Navarro (2021)
Singapore	17 reservoirs	20–220	Lim et al. (2020)
Vietnam	3 reservoirs	46–187	Trung et al. (2018)
Thailand	Reservoirs	14.6–17.8	Fukushima et al. (2017a, b)
Cambodia	Tonle Sap Lake	20	Fukushima et al. (2017a, b)

Fig. 8.1 *Microcystis* blooms in Lake Maninjau, Indonesia, showed an extensive green mucilage layer in the water surface **a** 2011, **b** 2018 (Private collection. Photo credit: BRIN/Cynthia Henny)

of *Microcystis* blooms was also corresponded by an increase of chlorophyll-a concentration, which could escalate to 100 $\mu g.L^{-1}$ (Fig. 8.3).

What makes cyanobacteria thrive in alga blooming conditions is due to their distinctive characteristics. Cyanobacteria form gas vacuoles, which can regulate their buoyancy in the water. Gas vacuoles provide buoyancy for certain species of cyanobacteria, such as genera of *Microcystis*, to regulate their cells positions in the water column and to acquire optimal light and nutrient availability. Optimal buoyancy is primarily governed by light, although variation in the buoyancy state is also influenced by the availability of nutrients such as carbon, nitrogen, and phosphorus. At low irradiance, cyanobacteria generate more of gas vesicles, which enhances their buoyancy. Gas vesicles may experience dilution due to growth when exposed to high irradiance. Additionally, cyanobacteria could break down under high turgor pressure following prolonged light exposure (Kinsman et al. 1991). During minimal

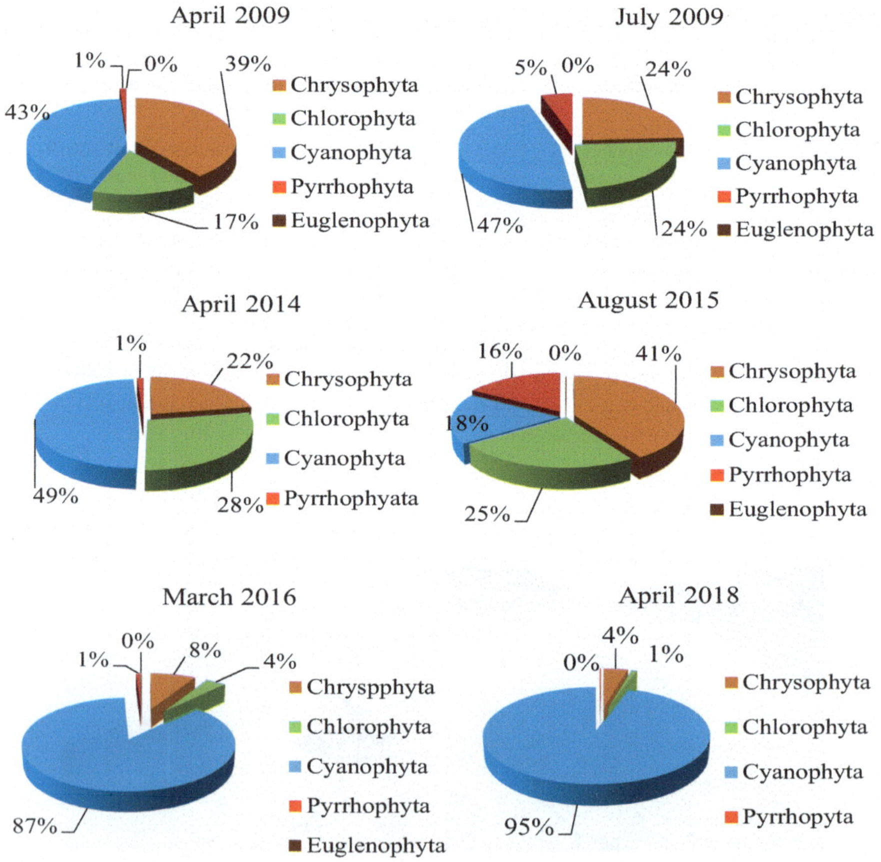

Fig. 8.2 Community structure of phytoplankton in Lake Maninjau, Indonesia, in 2009, 2014, 2015, 2016, and 2018 (Sulastri et al. 2019)

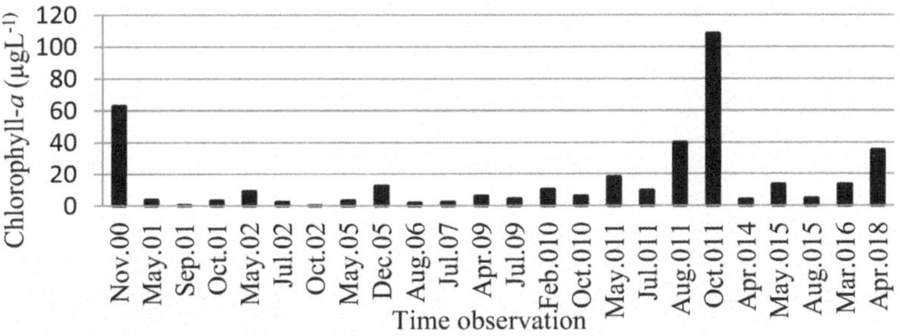

Fig. 8.3 Long-term monitoring chlorophyll-*a* concentration in Lake Maninjau, Indonesia (Sulastri et al. 2019)

wind mixing, buoyant cyanobacteria float up into the near-surface water, which leads to the formation of intense blooms (Ibelings et al. 2021; Huisman et al. 2018). A cohesive layer is formed when buoyant cyanobacteria accumulate significantly at the water's surface, often resulting in a dense layer several centimeters thick at the shore.

Another characterization of cyanobacteria is their ability to form large colonies, thereby maintaining the bloom conditions. The formation of surface bloom of cyanobacteria colonies or filament forming species such as *Microcystis, Aphanizomenon*, and *Nodularia* (Ibelings et al. 2021; Huisman et al. 2018). Cyanobacteria also have the capability of nitrogen gas (N_2) fixation from the atmosphere, which makes them thrive in the waters with nitrogen-limiting. Morphologically, these groups are filament cells with highly differentiated and special cells, namely heterocystis and akinet. Heterocystis has the function of N_2 fixation from the atmosphere or dissolved in water (Paerl 2014; Komárek et al. 2003). The genera that belong to this group are *Dolichospermum, Aphanizomenon, Anabaena, Nodularia. Lyngbya, Oscillatoria,* and *Cylindrospermopsis* (Fig. 8.4). Another filament cell with undifferentiated cells is mostly comprised of non-N_2 fixing genera such as *Oscillatoria* and *Planktothrix*; however, some of these groups have characterized N_2 fixing genera such as *Lyngbya* and *Trichodesmium* (Paerl 2014).

Cyanobacteria bloom taxa with coccoid cells that do not fix N_2 very well, such as *Synechococcus*, *Chroococcus*, and *Microcystis*. Their form ranges from solitary, such as *Chroococcus*, to colonies aggregate, such as *Microcystis*, which is found in planktonic and wide-ranging trophic status (Paerl 2014). Therefore, they are dependent on nitrogen supplies to support growth. Nitrogen-fixing (diazotrophic) cyanobacteria have a competitive advantage over non-diazotrophic cyanobacteria and eukaryotic phytoplankton in nitrogen-limited waters, and they can develop dense blooms if other nutrients, especially iron and phosphorus, are plentiful (Huisman et al. 2018).

An additional attribute that confers dominance upon cyanobacteria is their ability to store excess P as polyphosphate within the cells, which is utilized for multiple cell divisions, even when external P supplies are depleted to levels below analytical detection (Ibelings et al. 2021). Furthermore, cyanobacteria exhibit remarkable light

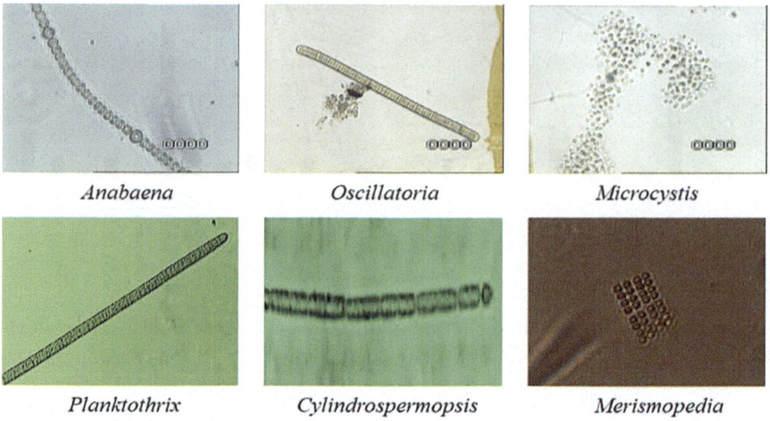

Fig. 8.4 Several dominant species of cyanobacteria found bloom in tropical lake (Indonesia) (Private collection. Photo credit: BRIN/Sulastri and Fachmijany Sulawesty)

harvesting efficiency and can typically thrive in environments with low light intensities. The majority of this group consists of filamentous forms, such as *Planktothrix* and *Cylindropermopsis* (Reynolds 2002). The balance of nitrogen and phosphorus has been the key factor causing the occurrence of cyanobacteria blooms in lakes. The lake becomes favorable for cyanobacteria blooms when the ratio of total nitrogen (TN) and total phosphorus (TP) are low. A low TN:TP ratio of 4.7 has triggered the occurrence of cyanobacteria blooms dominated by *Microcystis* blooms in Lake Maninjau, Indonesia (Sulastri et al. 2023).

Cyanobacteria are prevalent in reservoirs and lakes across the globe. The distribution of dominant cyanobacteria in South East Asia, Africa, and Latin America is detailed in Table 8.2. The most prevalent cyanobacteria found in the aquatic environments of lakes and reservoir in each region are as follows: *Microcystis*, *Anabaena*, and *Oscillatoria* in South East Asia; *Anabaena*; and *Microcystis* in Africa; *Cylindrospermopsis, Dolichospermum, and Microcystis* in Latin America. The long-term dominance of cyanobacteria is typically influenced by various factors such as nutrient concentration, lake morphometry, water temperature, light availability, mixing, and food web structure (Dokulil and Teubner 2000). However, it is often difficult to identify the cause of the dominance of one group or another because there are interacting factors involved, and the reaction will not necessarily be the same in different environments. Therefore, the strategy for overcoming it depends on the dominant species and the specific cause (Dokulil and Teubner 2000).

Table 8.2 Dominant cyanobacteria in tropical lakes and reservoirs

No.	Name of lake or reservoir	Name of dominant cyanobacteria	Sources
South East Asia			
1	Artificial urban lakes in Selangor (Malaysia)	*Microcystis* spp.	Sinang et al. (2015)
2	Lake Laguna (Philippines)	*Anabaena, Aphanizomenon, Microcystis, Oscillatoria*	Tingson-Tamayo and Zafaralla (2018)
3	Kranji Reservoir (Singapore)	*Microcystis, Anabaena*	Te and Gin (2011)
4	Hoan Kiem Lake and the NuiCoc reservoir (Vietnam)	*Microcystis aeruginosa, M. viridis, M. wesenbergii, M. ichthyoblabe, and M. novaceki*	Duoang et al. (2014)
5	Lake Phayao (Thailand)	*Oscillatoria, Microcystis, Anabaena*	Suwanpakdee et al. (2016)
6	Hanjalutung (Centtral Kalimantan, Indonesia); oxbow lake	*Anabaena, Oscillatoria, Spirulina*	Sulawesty (2018)

(continued)

Table 8.2 (continued)

No.	Name of lake or reservoir	Name of dominant cyanobacteria	Sources
7	Pondok (Banten, Indonesia), urban lake	*Spirulina, Microcystis, Oscillatoria*	Sulawesty and Sumarni (2004)
8	Rawa Kalong (West Java, Indonesia); urban lake)	*Anabaena, Oscillatoria, Aphanothece*	Sulawesty et al. (2012, 2015)
9	Limboto (Gorotalo, Indonesia); eutrophic lake	*Microcystis, Oscillatoria*	Hasim et al. (2023), Kadim et al. (2020)
10	Tempe (South Sulawesi, Indonesia); eutrophic lake	*Anabaena, Aphanocapsa, Microcystis*	Sulawesty and Yustiawati (2021)
11	Singkarak (West Sumatra)	*Anabaena*	Sulawesty (2007), Sulawesty et al. (2002)
12	Maninjau (West Sumatra, Indonesia); eutrophic lake, caldera lake	*Cylindrospermopsis, Anabaea, Aphanizomenon, Microcystis, Aphanocapsa, Oscillatoria, Chroococcus*	Sulastri et al. (2015, 2016a, b, 2020, 2023)

(continued)

Table 8.2 (continued)

No.	Name of lake or reservoir	Name of dominant cyanobacteria	Sources
13	Batur (Bali, Indonesia); eutrophic lake, caldera lake	*Anabaena, Microcystis, Merismopedia, Aphanocapsa, Oscillatoria*	Sulawesty and Awalina (2013), Haryani et al. (2022)
14	Menjer (Central Java, Indonesia); highland lake	*Aphanocapsa, Microcystis, Oscillatoria*	Sulawesty et al. (2021)
15	Segara Anak (Lombok, West Nusa Tenggara Indonesia); highland lake	*Anabaena, Lyngbya, Oscillatoria*	Arianto (2019)
African Continent			
1	Lake Victoria (Uganda)	*Anabaena, Microcystis, Gomphosphaeria, Planktolyngbya,*	Haande et al. (2011)
2	Lake Tanganyika (Africa)	*Dolichospermum flos-aquae* (*Anabaena flos-aquae*), *Chroococcus limneticus, Limnococcus, Synechococcus elongatus, Aphanothece, Aphanopapsa*	Cocquyt et al. (2021), Plisnier et al. (2023)
3	Swakoppoort Dam, Von Bach Dam (Namibia)	*Microcystis, Dolichospermum* (*Anabaena*), *Merismopedia, Cylindrospermopsis, Oscillatoria*	Sirunda et al. (2021)

(continued)

Table 8.2 (continued)

No.	Name of lake or reservoir	Name of dominant cyanobacteria	Sources
Latin America			
1	Itupararanga Reservoir (Brazil)	*Aphanizomenon, Aphanocapsa, Dolichospermum, Geitlerinema, Limnothrix, Pseudanabaena, Raphidiopsis, and Synechocystis*	Moraes et al. (2021)
2	Lakes and Reservoirs in Brazil	*Cylindrospermopsis, Dolichospermum, and Microcystis*	Soares et al. (2013)
3	Carpina, Mundau, Billings, and Guarapiranga Reservoir (Brazil)	*Cylindrospermopsis, Dolichospermum, and Microcystis*	Piccin-Santos (2012)
4	Carpina and Jucazinho Reservoirs (Brazil)	*Planktothrix agardhii, Cylindrospermopsis raciborskii,* and *Geiterinema amphibium*	De Oliveira et al. (2015)

8.2 Spatial and Temporal Cyanobacteria Blooms

Under conditions of cyanobacteria blooms, there is spatial and temporal variation in the species and density of cyanobacteria. The growth and development of cyanobacteria species are temporally and spatially determined by the dynamic nature of environmental conditions (dos Santos Silva et al. 2019). Cyanobacteria distribution and population density were also significantly impacted by the physical and chemical properties, which were significantly altered by the season. In temperate regions, cyanobacteria blooms commonly occur in summer, which is characterized by high temperatures, intense light, and a stable water column (Sivonen 1996). While in tropics, cyanobacteria bloom conditions can occur at any time or persistently throughout the year (Mowe et al. 2015).

The temporal and vertical variability of cyanobacteria within the water column are influenced by thermal structure, seasonal temperature fluctuation, and light availability (Moura et al. 2011). Lake stratification with persistent thermocline may proliferate cyanobacteria as observed in many tropical lakes and can maintain their dominance even under reduced nutrients. Cyanobacteria blooms may occur in the epilimnion, metalimnion layers or throughout the water column especially in shallow lakes (Perkins 2019). Certain cyanobacteria species are tolerant of low light intensity, which allows them to proliferate in deep water layers. *Planktothrix,* genera of cyanobacteria, commonly blooms in the metalimnion layer (Perkins 2019). Indeed, a uniform distribution of *Planktothrix agardhii* species was observed throughout the water column (Ibelings et al. 2021). Cyanobacteria belonging to the *Cylindrospermopsis* genera are frequently found in the metalimnion layer of stratified lakes, in conditions characterized by low light intensity. This species, nevertheless, migrates and accumulates at the surface as a result of water mixing (Ibelings et al. 2021).

Another factor affecting the distribution of cyanobacteria in the water column is the specific characterization of gas vacuoles, which are utilized to regulate buoyancy. Through the control of buoyancy, colonies and cells are capable of acquiring light and nutrients in the water column, giving them a competitive edge over other phytoplankton. Specific cyanobacterial species, including *Microcystis*, possess gas vacuoles that enable them to maintain a stable position in the water column and exploit advantageous environmental conditions by providing buoyancy (Wang et al. 2021).

Deep tropical lakes, such as Lake Maninjau, Indonesia, are susceptible to cyanobacteria blooms periodically or annually (Sulastri et al. 2019). High nutrient concentrations and water column stability of this lake increased the growth of *Microcystis aeruginosa* (Sulastri et al. 2019). Perennial cyanobacteria blooms have been documented in shallow lakes Indonesia (Sulastri et al. 2022). High nutrient concentration and limited light availability are critical determinants of perennial cyanobacteria blooms in tropical lakes, whereas temperature is not a limiting factor (Vanderly et al. 2021; Sulastri et al. 2022). Eutrophic tropical reservoirs characterized by high temperatures throughout the water column can cause perennial multi-species cyanobacteria blooms (Moura et al. 2011; Luthfiani et al. 2020).

8.3 Factors Promoting Cyanobacteria Bloom

8.3.1 Nutrient Enrichment

Both intensification of human activities in the catchment areas for agriculture and exploitation of the water body for cage aquaculture have increased pollutant input, especially nutrients. There is evidence of an increasing trend in nutrient concentration in global lakes and reservoirs, especially in tropical regions such as Indonesia, in the last three decades (Henny 2023; Henny and Nomosatryo 2016; Lehmusluoto et al. 1997). Increased inputs of nitrogen (N) and phosphorus (P) have caused eutrophication and subsequent harmful cyanobacteria blooms (Sulastri et al. 2023). Warm climates throughout the years and nutrient enrichment have caused major Indonesian lakes and reservoirs to experience eutrophication and cyanobacteria blooms (Henny and Nomosatryo 2016; Sulastri et al. 2023).

Major sources of external input of nutrients to lakes and reservoirs can come from various point and nonpoint sources. Discharged domestic wastewater and floating cage aquaculture (FCA) are major point sources of nutrient inputs to lakes and reservoirs (Henny and Nomosatryo 2016; Lukman et al. 2021; Piranti et al. 2017). The majority of nutrient loads that enter the lakes and reservoirs via catchment areas originate from nonpoint sources, such as urban runoff, agricultural and livestock farming, wastewater disposal, atmospheric deposition, and river inflow (Henny and Meutia 2014; Sulastri et al. 2016a, b; Gold and Sims 2005; Fong et al. 2020). Both nutrient nitrogen (N) and phosphorus (P) mainly come from fertilizers, animal manures, and other by-products such as compost and sewage (Barathan et al. 2023).

Prominent lakes in Indonesia, including Lake Toba and Lake Maninjau, are experiencing increased nutrient-enhanced algal blooms due to the substantial nutrient input originating from floating cage aquaculture (FCA) (Henny and Nomosatryo 2016; Lukman et al. 2015; Junaedi et al. 2014). An estimated ninety percent of the nutrient waste input into Lake Maninjau originated solely from FCA. Annually, the lake received around 6000 t $N.yr^{-1}$ and 350 t $P.yr^{-1}$ (Lukman et al. 2015; Marganof 2007). The anticipated P loading from the catchment area into Lake Toba was 138 $t.yr^{-1}$, whereas the estimated phosphorus loading from FCA was 570.33 $t.yr^{-1}$ (Lukman et al. 2021). Annual burdens of approximately 8000 tons of nitrogen and 2300 tons of phosphorus were estimated to have been discharged from the drainage basin into Laguna Lake, the largest lake in the Philippines, as a result of urbanization and intensive fish farming. Cyanobacteria blooms occur periodically, and the lake is perpetually impacted by nutrient-driven eutrophication (Zimmer and Bendoricchio 2001; Caballero and Navarro 2021).

Interconnected with river systems, lakes such as urban shallow lakes, floodplain lakes, and reservoirs receive substantial nutrient inputs from the catchment areas via river flow. On average, 3280 tons of nitrogen and 31 tons of phosphorus were deposited annually in Rawa Pening Lake, Indonesia, by major streams in the catchment that was heavily influenced by livestock farming and intensive agriculture. The lake has been under eutrophication for almost five decades (Piranti et al. 2017;

Sulastri et al. 2016a, b). Floodplain Tonle Sap Lake in Cambodia receives approximately 35.8 ± 12.5 kt.yr^{-1} NO_3^- and 8.7 ± 3.3 kt.yr^{-1} TP annually from the Mekong River (Sok et al. 2022).

An excess of external nutrients results in the accumulation of nitrogen (N) and phosphorus (P) in the sediment of lakes and reservoirs, which act as a sink and source of nutrients. By means of a process referred to as internal loading, both N and P may be released from the sediments into the water column. The influx of N and P from the internal loading varied in lakes and reservoirs between regions. Nitrogen (N) that has accumulated as nitrate in anoxic sediment is typically changed into N gas biologically, diffuses back into the water column, and is finally released into the atmosphere. Additionally, N gas could be fixed by phytoplankton, including N-fixing cyanobacteria (Aeriyanie et al. 2021; Lee and Oh 2018; Paerl 2014).

Unlike N compounds such as ammonia and nitrate which are more mobile in sediments, phosphate ions are typically adsorbed by clay minerals and iron (III) oxy-hydroxides. With the exception of a minor fraction that is transported out of the lake via discharge, P will cycle exclusively within the lake and primarily deposit as phosphate minerals in the sediment. Subsequently, accumulated P in sediment is recycled back into the water column via internal loading (Schindler et al. 2016; Søndergaard 2013). Lake P enrichment is significantly affected by P internal loading, resulting in persistent eutrophication.

The P internal loading is an issue of particular concern due to its challenging being controlled. Extensive research has been devoted to comprehending the mechanism that governs the internal accumulation of P. Nevertheless, research on the internal accumulation of P, particularly in tropical lakes of South East Asia, is scarce. Presumably, the rate of P release is greater in tropical lakes; in some lakes, it surpasses the external P input (Jilbert et al. 2020). The extended time span of improvements in lake water quality subsequent to controlling external nutrient loading is often attributed to internal P loading (Schindler 2012). Typically, after a reduction in external P input, P release persists for five to twenty years. This leads to an imbalance in N and P concentrations and decreases the N:P ratio, in which conditions favor the occurrence of cyanobacteria blooms (Chen et al. 2013; Ding et al. 2018; Schindler et al. 2016; Søndergaard 2007).

Many factors influence P internal loading. Anoxia is frequently associated with the internal release of nutrients from sediments, where P is released by the redox-mediated reduction of iron in phosphate-binding iron hydroxides. Minerals rich in iron (Fe) are known to be associated with phosphorus burial in lake sediments (Sondergaard et al. 2013). However, phosphate in the Fe hydroxide bond could be replaced by sulfide produced by the microbial reduction of sulfate in the anoxic sediments, leading to the formation of sulfide minerals. As a consequence, phosphate is released into the water column (Lehtoranta et al. 2009).

In different types of lakes, the recirculation of elements from the sediment to the water column affects the internal loading of nutrients. The process is dependent on the lake morphology, such as mean depth, physical (mixing and thermal stratification), and chemical (dissolved oxygen and redox potential) characteristics of lakes (Lewis 2010). Very deep lakes characterized by persistent stratification are likely to have negligible internal loading, and the lakes maintain oligotrophic conditions. In contrast to deep lakes with shallow thermocline that are susceptible to seasonal mixing and shallow lakes with full lake mixing, the lakes are highly subject to nutrient internal loading from sediments. Internal loading results in increased N gas emission into the atmosphere and abundant P in lake water, which renders the lake highly susceptible to eutrophication due to a low N:P ratio and conditions that are more favorable for the growth of cyanobacteria blooms (Crowe et al. 2008; Ding et al. 2018; Qin et al. 2020; Zhang et al. 2018).

Elevated concentrations of nitrogen and phosphorus promote the growth of toxic cyanobacteria and are associated with a high occurrence of specific cyanobacterial genera (Pham et al. 2017; Mowe et al. 2015). The study identified *Cylindrospermopsisi raciborkii* as the prevailing species in a small lake (Lake Telaga Warna), Indonesia, which had a phosphorous deficiency and a high nitrogen concentration. While the bloom *of Planktothrix agardhii* was found to be dominant in another lake (Lake Sunter, Indonesia), which was characterized by high phosphorus concentrations and nitrogen limitation (Sulastri et al. 2022).

Figure 8.5 presents the variation of nutrient concentrations over 2 decades in Lake Maninjau, Indonesia, which experienced prolonged eutrophication due to nutrient enrichment by fish cage aquaculture and frequent cyanobacteria blooms. Concentrations of total nitrogen (TN) as high as 1.117 mg.L^{-1} and total phosphorus (TP) concentrations of 0.229 mg.L^{-1} triggered cyanobacteria blooms with an abundance of 4,180.87 × 10^3 individual.L^{-1} as observed in April 2018 (Sulastri et al. 2023).

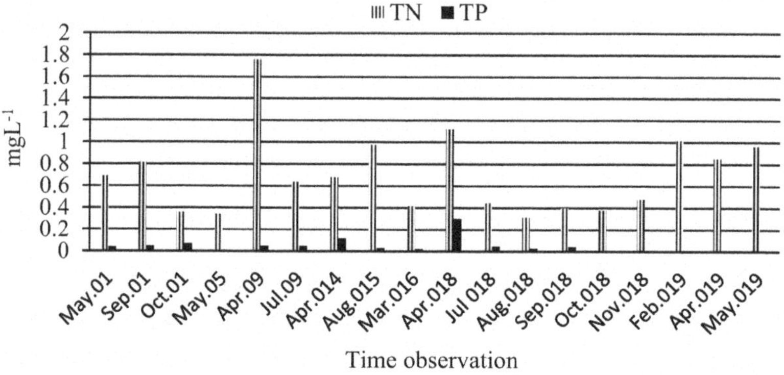

Fig. 8.5 Concentration total nitrogen (TN) and total phosphorus (TP) in Lake Maninjau, Indonesia (Sulastri et al. 2023)

8.3.2 Climate Change

The effects of climate change on lake ecosystems are substantial. Climate change has changed air temperature and precipitation pattern, which could aggravate eutrophication, particularly cyanobacteria blooms through the increase of nutrient load, both external and internal (Sinha et al. 2017; Orihel et al. 2017).

Cyanobacteria blooms thrive in stagnant and warm water such as lakes in reservoirs. As a consequence, it is anticipated that the frequency and extent of cyanobacteria blooms will intensify in association with an increase in water temperatures caused by climate change. At higher temperatures, cyanobacteria proliferate at a faster rate than other algae, thus increasing the occurrence of cyanobacterium blooms (Paerl 2014; Burford et al. 2019). Compared to other algae, cyanobacteria also demonstrate enhanced mobility both vertically and laterally within the water column. They can obtain nutrients from the colder, darker, bottom layers and utilize them to their growth in the more light, upper, warmer layers (Burford et al. 2019; Paerl et al. 2016).

Climate change has also caused an increase in precipitation patterns, variability, and intensity. Increasing phenomenon of tropical storms such as cyclone has occurred frequently. This condition increases significant changes in hydrologic variability in the water body seasonally during the wet and dry periods (Paerl 2014). Significant hydrologic alterations and nutrient concentrations have resulted in cyanobacteria blooms in large lakes such as Laguna Lake, Philippines, due to the reported increase in the frequency of tropical cyclones (Caballero and Navarro 2021). During periods of intense precipitation, nutrient inputs from runoff into water bodies are increased. Conversely, prolonged periods of drought during dry periods lead to increased residence times and nutrient concentrations in bodies of water. This environmental condition is favorable for cyanobacteria blooms (Paerl 2014). Prolonged drought causes water level reductions and increases phosphorus concentration, which also leads to perennial cyanobacterial dominance in shallow tropical lakes (Vanderly et al. 2021).

The potential impact of climate change on cyanobacteria bloom mechanisms might vary depending on the type of body of water and the species that produce the blooms (Ibelings et al. 2021). An increase in water temperature due to climate change supports stable water columns, which allows buoyant cyanobacteria such as Microcystis to form dense green scums at the water's surface (Barruffa et al. 2021). As recorded in Lake Maninjau, Indonesia *Microcystis* blooms occurred and formed dense green scums at surface water under conditions of high-water column stability (Sulastri et al. 2019). Elevated wind speed and cloud cover weaken thermal stratification, resulting in water column instability. This, in turn, diminishes the competitive advantage of buoyant cyanobacteria, which rely on regulating their position to obtain advantageous light access and shading over deeper non-buoyant phytoplankton (Huisman et al. 2018). Warmer climates may also have an impact on non-buoyant phytoplankton, such as *Planktothrix agardhii*, particularly in hypertrophic lakes where this species develops year-round blooming or extended blooming events (Ibelings et al. 2021).

An additional consequence of global warming is the escalation in atmospheric greenhouse gas concentrations, specifically carbon dioxide (CO_2). In nutrient-rich waters, atmospheric CO_2 concentrations probably also intensify cyanobacterial blooms (Paerl et al. 2011). Cyanobacteria can fix CO_2 for photosynthesis and growth. High rates of photosynthesis are observed in cyanobacteria blooms during bloom conditions, resulting in an increased demand for CO_2 concentration. Extremely low levels of CO_2 in the water result in cyanobacteria striving to fix for CO_2 from atmospheric fluxes. In addition, CO_2 depletion-induced cyanobacteria shift their fixation mechanism via CO_2-concentrating mechanisms (CCMs) (Huisman et al. 2018).

8.3.3 Physical Characteristics of the Water Body

Lake physical factors such as water exchange, body size, and depth influence the dominance of planktonic cyanobacteria (Huisman et al. 2018). In shallow water bodies with a depth < 2 m, cyanobacteria blooms may persist for many years because the euphotic depth and mixing depth ratio (Z_{eu}/Z_m) are higher than the critical value range of the Z_{eu}/Z_m ratio, which inhibits low-light-adapted cyanobacteria growth (Khanna et al. 2009). An example of a shallow lake in Sunter Indonesia, with a depth of 1–1.5 m and a Secchi depth of 0.3 m, or an euphotic depth of 0.81 m, shows that the ratio Zeu/Zm was 0.648 or greater than 0.2 (Sulastri et al. 2022). When wind exposes shallow lakes, water turbulence and mixing frequently occur throughout the water column. This condition usually causes cyanobacteria blooms, which are dominated by non-buoyancy-regulating species such as *Planktothrix agardhii* (Huisman et al. 2018). A persistent bloom of *Planktothrix agardhii* has been documented for years in Lake Sunter, Indonesia, which has a depth of only 1–1.5 m (Sulastri et al. 2023).

Water residence time is a critical aspect of the hydrology of a body of water in relation to cyanobacteria blooms (Ibelings et al. 2021). Long residence time lakes are typically more susceptible to cyanobacteria blooms. A longer period of residence time under less rainfall increased cyanobacteria biomass in Lake Kinneret, Israel (Gophen 2021), and similar case was observed in the upper parts of the Patos Lagoon, Brazil, and Lake Maninjau, Indonesia. *Microcystis* blooms have occurred more frequently in Lake Maninjau due to its long residence time of 25 years (Lukman et al. 2020; Sulastri et al. 2019). Increasing flushing with the conscience reduces water retention time, which is one of the strategies to minimize cyanobacteria blooms (Giani et al. 2020).

8.4 Threats: Cyanotoxin and the Threat to the Aquatic Ecosystem

Cyanotoxin produced by bloom-forming cyanobacteria is of particular concern. The toxin of cyanobacteria poses a threat in aquatic ecosystems and is harmful to organisms and humans (Zang et al. 2022). Potential human exposure to the toxins may occur via ingestion of cyanotoxin contaminated drinking water or direct contact with contaminated water (Wiley and McPherson 2024). There are approximately forty cyanobacteria species that have the capability to produce cyanotoxins, including those capable of both fixation and non-fixation of nitrogen (N_2) (Aklakur et al. 2023; Wang et al. 2021). Table 8.3 contains a list of toxic effects that target organ injury in animals induced by various cyanotoxins, as well as the genera of cyanobacteria responsible for producing these toxins.

The occurrence of cyanotoxin-producing cyanobacteria blooms raises significant concerns regarding the long-term viability of aquatic ecosystems on a global scale (Huisman et al. 2018; Namsaraev et al. 2018). The occurrence of cyanobacteria blooms and the prevalence of cyanotoxins in tropical lakes were also documented. The study indicated that *Microcystis*, which produced microcystin, was identified as the predominant bloom-forming cyanobacteria in the study (Mowe et al. 2015). The increasing frequency of cyanobacteria blooms in tropical pose significant threats to the sustainability of freshwater ecosystems, affecting not only human and environmental health but also fisheries production (Mishra et al. 2022). Cyanobacteria blooms decrease fisheries production, drinking water quality, recreational activity, and pose a threat to human health (Zhang et al. 2022). The further effect of cyanobacteria blooms will decrease the socioeconomic value of lakes in this region since most lakes are multi-use water resources such as for fisheries, recreational, aquaculture production, and domestic water supply (Hamilton et al. 2013).

Related to the water quality, the level of microcystin in some countries is below the WHO drinking water guideline (1 $\mu g.L^{-1}$), as recorded in some water bodies in Malaysia, Singapore, Vietnam, Thailand, and the Philippines (Table 8.4). Certain water bodies in Vietnam contained concentrations of microcystin that exceeded the WHO's threshold of 20 $\mu g.L^{-1}$, which is considered unsafe for recreational purposes. There are currently no available data regarding the concentration of microcystin in lakes such as Lake Maninjau in Indonesia. Nevertheless, the lake ecosystem has been significantly affected by cyanobacteria blooms, as evidenced by the mass mortality of fish and the deterioration of water quality (Makmur et al. 2020).

Table 8.3 Types of cyanotoxin, identified primary target organ damage in mammals and cyanobacteria genera that produce toxins

Cyanotoxin group and toxin type	Toxic effect	Genera of main producer	References
Hepatotoxin			
Microcystin	Liver, kidney, damage gastroenteritis, cancer abdominal pain	*Microcystis, Anabaena, Oscillatoria, Nostoc, Hapalosiphon, Anabaenopsis*	Huisman et al. (2018), Thawabteh et al. (2023)
Nodularin	Injury liver, kidney, gastroenteritis,	*Nodularia*	Huisman et al. (2018), Niamien-Ebrottie et al. (2015)
Cylindrospermopsin	Damage to multiple organs, gastroenteritis, hepatoenteritis, carcinogenic. Liver sirrhoosis and liver nicrosis	*Cylindrospermopsis, Anabaena, Raphidiopsis Aphanizomenon*	Niamien-Ebrottie et al. (2015), Thawabteh et al. (2023)
Neurotoxin			
Anatoxin-a	Nerve synapse	*Anabaena, Planktothrix (Oscillatoria) Aphanizomenon*	Chorus and Batram (1999)
Anatoxin-a(S)	Nerve synapse	*Anabaena*	Chorus and Batram (1999)
Saxitoxins	Paralysis and respiratory failure	*Aphanizomenon, Anabaena, Lyngbya, Cylindrospermopsis*	Huisman et al. (2018)
Demartotoxins			
Aplysiatoxins	Skin	*Lyngbya, Schizothrix, Planktothrix (Oscillatoria)*	Chorus and Batram (1999)
Lyngbyatoxin-a	Skin	*Lyngbya*	Chorus and Batram (1999)
Lipopolysaccharides (LPS)	Skin irritation Gastrointestinal irritants	*Microcystis, Oscillatoria*	Zanchett and Oliveira-Filho (2013)

8.5 Impact of Cyanobacteria Bloom on Aquatic Ecosystems

Eutrophication has been the most challenging worldwide threat to the water quality of freshwater resources over the past four decades. Excessive algal growth and subsequent harmful algal blooms (HABs) lead to reduced water clarity and dissolved

Table 8.4 Summary of dominant species and toxin recorded in some water bodies in South East Asia

Country	Name of water body	Dominant species/ genera	Microcystin (MC) concentration	References
Indonesia	Lake Maninjau Lake Sunter Lake Telaga Warna	*Microcystis aeruginosa* *Planktotrix agardhii* *Cylindrospermopsis riciborskii*	No data No data No data No data	Sulastri et al. (2019) Sulastri et al. (2022) Sulastri et al. (2022)
Malaysia	10 Urban lakes	*Microcystis* *Microcystis* spp. *Phormidium* spp. *Lyngbia* spp	1–10 μg.L^{-1}	Sinang et al. (2015)
Philiphine	Laguna Lake	*Microcystis* spp.	649–4049 μg.g^{-1} No data	Cuvin-Aralar et al. (2002) Caballero and Navarro (2021)
Singapore	Kranji Reservoir 17 Reservoirs	*Microcystis* spp. *Anabaena* spp. *Microcystis* spp., *Cylindrospermopsis*	2.66–2.80 μg.L^{-1} < 0.025–8.89 μg.L^{-1}	Te and Gin (2011) Lim et al. (2020)
Vietnam	Hoan Kiem Lake Nui Coc reservoir Dau Teng Reservoir TriAn-Reservoir, DauTieng Reservoir, AnGiang Reservoir	*Microcystis* *Microcystis* *Microcystis* *Microcystis* *Microcystis* *Planktothrix*	115.9–184.6 μg.L^{-1} 726.5–1116 μg.L^{-1} 2.13–2.50 μg.L^{-1} 2610 μg.L^{-1} 30 μg.L^{-1} 0.31 μg.L^{-1}	Duong et al. (2014) Pham et al. (2017) Trung et al. (2018)
Thailand	Huay Luang Reservoir Nam Un Reservoir Bueng Nong Khot Reservoir	*Microcystis Aaeruginosa* *Microcystis aeruginosa* *Microcystis*	No data No data 1.093–1.235 μg.L^{-1}	Fukushima et al. (2017a, b) Somdee et al. (2013)
Cambodia	Tonle Sap Lake	*Microcystis aeruginosa* *Microcystis aeruginosa*	No data 8.2–70,692 ng.L^{-1}	Fukushima et al. (2017a, b) Nakatani et al. (2022)

oxygen depletion (DO) in the water. Decayed cells originating from cyanobacteria blooms have the potential to generate a substantial quantity of organic substances and release soluble nutrients, which cause oxygen depletion in lakes and reservoirs. The depletion of oxygen is a common issue associated with eutrophication, which can result in prolonged hypoxia in the lake. Lake hypoxia releases substantial phosphate and other toxic substances such as hydrogen sulfide into the water column, which results from anaerobic microbial processes in the sediments (Henny and Nomosatryo 2012, 2016). The occurrence of low levels of dissolved oxygen has an impact on the organisms living in the water, such as the disappearance of invertebrates, reducing fish populations, reducing biodiversity, even resulting in the death of aquatic species that live below the surface of water bodies, and commonly, fish kill (Li 2021).

Hypereutrophic Lake Maninjau, Indonesia, has often experienced Lake hypoxia, especially in extreme weather conditions during the rainy season, which has led to massive fish kills (Henny 2015; Jasalesmana et al. 2023; Subehi et al. 2020; Fukushima et al. 2017a, b, 2021). *Microcystis aeruginosa* blooms in Lake Maninjau and *Planktothrix agardhii* blooms in Lake Sunter, Indonesia, have reduced the local fish population and phytoplankton diversity, even causing fish deaths (Makmur et al. 2020; Sulastri et al. 2019, 2022). Cyanobacteria blooms also depleted oxygen and has caused massive fish kills in Laguna Lake Philiphine (Cabelliro and Navarro 2021).

Cyanophyta blooms not only contribute to oxygen depletion but also diminish the transparency of the water in the lake (Zhang et al. 2022). Eutrophic lakes and reservoirs with cyanobacteria blooms typically display very low water clarity, and the water turns greenish with dark green scum. Lake Maninjau and shallow urban lakes, Indonesia, has transparency less < 1 m during algal blooms (Henny and Nomosatryo 2016; Henny and Meutia 2014). Secchi depth transparency of less than one meter in Lake Maninjau, Indonesia, were observed during *Microcystis* blooms in April 2018 and October 2011 (Fig. 8.6). The turbidity of water ranged from 26 to 50 NTU for the eutrophic Laguna Lake, Philippines, seasonally (Herrera and Nadaoka 2021). Low water clarity inhibits sunlight penetration and photosynthetic activity, hence suppressing the growth of aquatic macrophytes and leading to negative effects on invertebrate and fish habitat (Paerl 2014). Cyanobacteria blooms inhibit the sunlight penetration required by other aquatic plants, such as submerged plants, which play a pivotal role in controlling excess nutrients and maintaining lake water clarity (Li et al. 2009; Henny 2018; Kolada 2014). Cyanobacteria blooms also reduce species diversity of meta zooplankton and total bacteria in waters (Tian et al. 2023).

Cyanobacteria can produce a high volume of toxins, such as microcystin, which is highly toxic to aquatic life and humans. This may be harmful and even deadly to aquatic life. The consumption of aquatic organisms such as fish with certain levels of bioaccumulated toxins poses a great risk to the health of humans due to the movement of the toxin through the food chain from lower to higher trophic levels (Backović and Tokodi 2024; Zhang et al. 2022).

The toxins of cyanobacteria have an effect at all taxonomic levels on aquatic organisms through ingestion of cyanobacterial cells or direct contact with cyanotoxin (Mishra et al. 2022). Toxin release from cyanobacteria bloom causes severe damage to the antennal gland, gills, hepatopancreas, lymphoid organ, muscle, and dorsal cecum and kills post-larvae of *L. vannamei* shrimp, which would adversely affect other aquatic invertebrates (Morales-Covarruabias et al. 2016). Fish-induced cyanotoxin also showed various immunomodulatory, inflammatory, and endocrine responses (Flafushynska et al. 2023).

The effect of cyanotoxins would influence several trophic levels and lead to a decrease in the number of secondary producers. As a result of the reduced edibility of

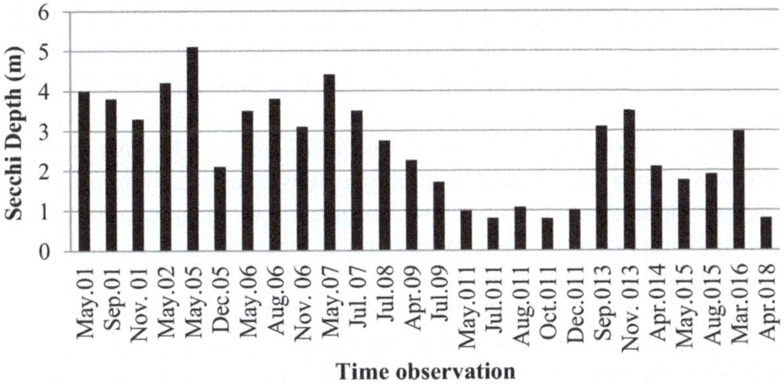

Fig. 8.6 Long-term monitoring Secchi Depth transparency in Lake Maninjau Indonesia (Sulastri et al. 2019)

cyanobacteria to mesozooplankton, the food chain from primary production to fish is extended resulting in increased carbon-based energy losses along the pathway to fish (Moustaka-Gouni et al. 2020). Significant functional alterations that occur in aquatic systems as a result of cyanobacteria blooms include a reduction in the efficient transfer of energy from the primary producer to higher trophic levels (Schwarzenberger 2022).

The excess cyanobacteria production of harmful algae poses a threat to aquaculture and fisheries activity. Combined oxygen depletion and toxins production caused by cyanobacteria blooms have caused fish kills in lakes and loss of aquaculture and fisheries (Makmur et al. 2020; Subehi et al. 2020). Massive fish kills almost occur every year in Lake Maninjau, Indonesia (Fig. 8.7). Fishing activity in this lake was also reduced because native fish species such as *Rasbora* sp. and *Hampala macrolepodota* were rare in this lake, and it was difficult to catch endemic species such as *Psylopsis sp.* (Makmur et al. 2020).

Cyanobacteria blooms in global water bodies will likely persist and pose a threat to the sustainability of the ecosystem, if preventative measures are not implemented.

Fig. 8.7 Excessive cyanobacteria growth and fish kill in in Lake Sunter Jakarta, Indonesia (**a**) (Private collection. Photo credit: BRIN/Siti Aisyah) and Fish kill (**b**) (Private collection. Photo credit: BRIN/Cynthia Henny)

Significant contributors to the proliferation and persistence of cyanobacteria blooms are global changes caused by human activities, including climate change, nutrient enrichment, and hydrological modifications of waterways (Burford et al. 2019; O'Neil et al. 2012).

The threat and current problem caused by cyanobacteria blooms to the aquatic ecosystem need immediate mitigation through extensive efforts monitoring the status cyanobacteria bloom formation and the occurrence of cyanotoxins. Locally and regionally, it is important to develop guidelines for cyanobacteria bloom, nutrient bloom criteria, and cyanotoxin threshold values to anticipate further effects on the aquatic ecosystem. To prevent cyanobacteria blooms, it also requires effective control of nutrient input into aquatic ecosystems. Reducing both nitrogen and phosphorus in the waterbody is the most effective method to prevent cyanobacteria bloom. Vertical stability, or thermal stratification, and long residence times of the water body favor cyanobacteria bloom. Therefore, flushing of the water and destratification of the water column are effective controls for cyanobacteria bloom, especially in shallow waters.

References

Aeriyanie AR, Sinang SC, Nayan N, Poh KB (2021) Role of internal and external nutrients loading in regulating in-lake nutrient concentrations in a eutrophic shallow lake. Environ Asia 14(1):63–68. https://doi.org/10.14456/ea.2021.7

Aklakur MD, Bakli S, Ashutosh D Deo AD, Singh DK, Pailan GH (2023) Cyanobacteria toxicity in aquaculture system and its impact on fish physiology. Aquac Mar Biol 12(1):28–33

Arianto T, Qayim I, Damar A (2019) The dynamic structure of phytoplankton community in Segara Anak Lake: a volcanic lake in Mount Rinjani National Park. In: IOP conference series: earth and environmental science, vol 299, p 012045. https://doi.org/10.1088/1755-1315/299/1/012045

Backović DD, Tokodi N (2024) Cyanotoxins in food: exposure assessment and health impact. Food Res Int 114271

Barathan BP, Chen W, Su Yuping, Wang X, Chen Y (2023) The effects of nutrient loading from diferent sources on eutrophication in a large shallow lake in Southeast China. Environ Geochem Health 45:7603–7620. https://doi.org/10.1007/s10653-023-01641-5

Barruffa AS, Victor Sposito V, Faggian R (2021) Climate change and cyanobacteria harmful algae blooms: adaptation practices for developing countries. Mar Freshw Res 72:1722–1734. https://doi.org/10.1071/MF21081

Burford MA, Carey CC, Hamilton DP, Huisman J, Paerl HW, Wood SA, Wulff A (2019) Perspective: advancing the research agenda for improving understanding of cyanobacteria in a future of global change. Harmful Alga. https://doi.org/10.1016/j.hal.2019.04.004

Caballero I, Navarro G (2021) Monitoring cyanobacteria blooms and water quality in Laguna Lake (Philippines) with Sentinel-2 satellites during the 2020 Pacific typhoon season. Sci Total Environ 788:147700. https://doi.org/10.1016/j.scitotenv.2021.147700

Chen N, Peng B, Hong H, Turyaheebwa N, Cui S, Mo X (2013) Nutrient enrichment and N: P ratio decline in a coastal bay river system in southeast China: the need for a dual nutrient (N and P) management strategy. Ocean Coast Manag 81:7–13. https://doi.org/10.1016/j.ocecoaman.2012.07.013

Chorus I, Welker M (2021) Toxic cyanobacteria in water: a guide to their public health consequences, monitoring and management, 2nd ed. CRC Press. https://doi.org/10.1201/9781003081449

Chorus I, Bartram J (1999) Toxin cyanobacteria in water, in guide in to their public health consequences, monitoring and management. CRC Press, London and New York

Cocquyt C, Plisnier PD, Mulimbwa N, Nshombo M (2021) Unusual massive phytoplankton bloom in the oligotrophic Lake Tanganyika. Plant Ecol Evol 154(3):351–361. https://doi.org/10.5091/plecevo.2021.1890

Crowe SA, Neill AHO, Katsev S, Hehanussa P, Haffner GD, Sundby B, Mucci A, Fowle DA (2008) The biogeochemistry of tropical lakes: a case study from Lake Matano Indonesia. Limnol Oceanogr 53(1):319–331. https://doi.org/10.4319/lo.2008.53.1.0319

Cuvin-Aralar ML, Fastner J, Focken U, Becker K, Aralar EV (2002) Microcystins in natural blooms and laboratory cultured Microcystis aeruginosa from Laguna de Bay. Philippines Syst Appl Microbiol 25:179–182

Ding S, Chen M, Gong M, Fan X, Qin B, Xu H, Gao S, Jin Z, Tsang DCW, Zhang C (2018) Internal phosphorus loading from sediments causes seasonal nitrogen limitation for harmful algal blooms. Sci Total Environ 625:872–884. https://doi.org/10.1016/j.scitotenv.2017.12.348

De Oliveira FHPC, da Silva JDB, Costa ANSF, Ramalho WP, Moreira CHP, Calazans TLS (2015) Cyanobacteria community in two tropical eutrophic reservoirs in northeastern Brazil. Acta Scientiarum 37(2):169–176. https://doi.org/10.4025/actascibiolsci.v37i2.26418

Dos Santos Silva RD, dos Santos SJ, de Oliveira DA, Camila Ferreira Mendes CF, Barbosa VV, Chia MA, de Lucena Barbosa JE (2019) Spatio-temporal variation of cyanobacteria and cyanotoxins in public supply reservoirs of the semi-arid region of Brazil. J Limnol 79(1):13–29. https://doi.org/10.4081/jlimnol.2019.1893

Dokulil MT, Teubner K (2000) Cyanobacterial dominance in lakes. Hydrobiologia 438:1–12. https://doi.org/10.1023/A:1004155810302

Duong TT, Ja¨hnichen S, Le TPQ, Ho CT, Hoang TK, Nguyen TK, Vu TN, Dang DK (2014) The occurrence of cyanobacteria and microcystins in the Hoan Kiem Lake and the Nui Coc reservoir (North Vietnam). Environ Earth Sci 71:2419–2427. https://doi.org/10.1007/s12665-013-2642-2

EPA (2019) Recommendations for cyanobacteria and cyanotoxin monitoring in recreational waters. Office of Water EPA 823-R-19-001. https://www.epa.gov/sites/default/files/2019-09/documents/recommend-cyano-rec-water-2019-update.pdf

Fong LS, Leng MJ, Taylor D (2020) A century of anthropogenic environmental change in tropical Asia: multi-proxy palaeolimnological evidence from Singapore's Central Catchment. Holocene 30(1):162–177

Fukushima T, Matsushita B, Subehi L, Setiawan F, Wibowo H (2017a) Will hypolimnetic waters become anoxic in all deep tropical lakes? Sci Rep 7:1–9. https://doi.org/10.1038/srep45320

Fukushima M, Tomioka N, Jutagate T, Hiroki M, Murata T, Preecha C, Avakul P, Phomikong P, Imai A (2017b) The dynamics of pico-sized and bloom forming cyanobacteria in large water bodies in the Mekong River Basin, River Basin. PLoS ONE 12(12):e0189609. https://doi.org/10.1371/journal.pone.018960

Fukushima T, Setiawan F, Subehi L, Fakhrudin M, Triwisesa E, Dianto A, Matsushita B (2021) Convection of waters in Lakes Maninjau and Singkarak, tropical oligomictic lakes. Limnology 23:375–383. https://doi.org/10.1007/s10201-021-00686-8

Falfushynska H, Kasianchuk N, Siemens E, Henao E, Rzymski P (2023) A review of common cyanotoxins and their effects on fish. Toxics 11:118. https://doi.org/10.3390/toxics11020118

Giani A, Taranub ZE, von Rückerta G, Gregory-Eaves I (2020) Comparing key drivers of cyanobacteria biomass in temperate and tropical systems. Harmful Algae 97:1018. https://doi.org/10.1016/j.hal.2020.10185

Gobler CJ (2020) Climate change and harmful algal blooms: insights and perspective. Harmful Algae 91:101731. https://doi.org/10.1016/j.hal.2019.101731

Gold AJ, Sims JT (2005) Eutrophication. In: Hillel D (ed) Encyclopedia of soils in the environment. Elsevier, pp 486–494. https://doi.org/10.1016/B0-12-348530-4/00093-X

Gophen M (2021) Climate change-enhanced cyanobacteria domination in lake Kinneret: a retrospective overview. Water 13:163. https://doi.org/10.3390/w13020163

Hamilton DP, Wood SA, Dietrich DR, Puddick J (2013) Costs of harmful blooms of freshwater cyanobacteria. Cyanobacteria 245–256. https://doi.org/10.1002/9781118402238.ch15

Haande S, Rohrlack T, Semyalob RP, Brettum P, Edvardsen B, Lyche-Solheim A, Sørensen K, Larsson P (2011) Phytoplankton dynamics and cyanobacterial dominance in Murchison Bay of Lake Victoria (Uganda) in relation to environmental conditions. Limnologica 41(1):20–29. https://doi.org/10.1016/j.limno.2010.04.001

Haryani GS (2022) Penanganan ikan invasif berbasis pendekatan bio-ekologi dan model kompetisi multispesies sebagai dasar optimasi pengelolaan di Danau Batur. Laporan Akhir Pendanaan Ekspedisi dan atau Eksplorasi. Deputi Bidang Fasilitasi Riset dan Inovasi, Badan Riset dan Inovasi Nasional. Jakarta. 34 hal

Hasim RT, Sunani MRFH (2023) Phytoplankton and chlorophyll-a: distribution, abundance, and trophic state index (TSI) in the transitional season in Limboto Lake. Egypt J Aquat Biol Fish 27(2):407–422. https://doi.org/10.21608/EJABF.2023.293103

Henny C, Nomosatryo S (2016) Changes in water quality and trophic status associated with cage aquaculture in Lake Maninjau, Indonesia. IOP Conf Ser Earth Environ Sci 31(1). https://doi.org/10.1088/1755-1315/31/1/012027

Henny C, Nomosatryo S (2012) Dinamika sulfida di Danau Maninjau: Implikasi terhadap pelepasan fosfat di lapisan hipolimnion. Limnotek 19(2):102–112

Henny C (2023) Pengelolaan Kualitas Air Melalui Pendekatan Biogeokimia. Orasi Pengukuhan Professor Riset. https://doi.org/10.55981/brin.875

Henny C (2018) Danau bekas tambang timah di Bangka Belitung: sumber daya air baru dan pengelolaan kualitas airnya. Prosiding Masyarakat Limnologi Indonesia 2018:30–37

Henny C, Meutia AA (2014) Urban LAKES in Megacity Jakarta: risk and management plan for future sustainability. Procedia Environ Sci 20:737–746. https://doi.org/10.1016/j.proenv.2014.03.088

Henny C (2015) Ecotechnology and ecosystem-based management tools as alternative solutions to reduce risks of ecosystem disaster and fish kills in Lake Maninjau. Proc Nat Sem Limnology VII:231–240

Herrera EC, Nadaoka K (2021) Temporal dynamics and drivers of lake ecosystem metabolism using high resolution observations in a shallow, tropical, eutrophic lake (Laguna Lake, Philippines). J Great Lakes Res 47(4):997–1020. https://doi.org/10.1016/j.jglr.2021.03.017

Ho JC, Michalak AM (2015) Challenges in tracking harmful algal blooms: a synthesis of evidence from Lake Erie. J Great Lakes Res. https://doi.org/10.1016/j.jglr.2015.01.001

Huang J, Xu Q, Wang X, Ji H, Quigley EJ, Sharbatmaleki M, Li S, Xi B, Sun B, Li C (2021) Effects of hydrological and climatic variables on cyanobacterial blooms in four large shallow lakes fed by the Yangtze River. Environ Sci Ecotechnology 5:100069

Huisman J, Codd GA, Paerl HW, Ibelings BW, Verspage JMH, Visser P (2018) Cyanobacteria bloom. Nat Rev Microbiol 16(8):471–483. https://doi.org/10.1038/s41579-018-0040-1

Ibelings BW, Kurmayer R, Azevedo SMFO, Wood SA, Chorus I, Welker M, (2021) Understanding the occurrence of cyanobacteria and cyanotoxin. In: Book toxic cyanobacteria in water, 2nd ed. CRC Press, p 82. ISBN 9781003081449

Jasalesman T, Putri MR, Abdurrachman M, Henny C, Nomosatryo S, Akhdiana I (2023) Water Quality assessment of Lake Maninjau after the mass fish kill event. IOP Conf Ser Earth Environ Sci 1245:012013. https://doi.org/10.1088/1755-1315/1245/1/012013

Jilbert T, Jokinen S, Saarinen T, Mattus-Kumpunen U, Simojoki A, Saarni S, Salminen S, Niemisto J, Horppila J (2020) Impacts of a deep reactive layer on sedimentary phosphorus dynamics in a boreal lake recovering from eutrophication. Hydrobiologia 847:4401–4423. https://doi.org/10.1007/s10750-020-04289-9

Junaidi, Syandri H, Azrita (2014) Loading and distribution of organic materials in Maninjau Lake west sumatra province-Indonesia. J Aquac Res Dev 5(7). https://doi.org/10.4172/2155-9546.1000278

Kadim MK, Sahami FM, Hasiru DJ (2020) Spatial and temporal distribution of phytoplankton with emphasis on the harmful and toxic algae in the Limboto Lake. Tomini J Aquat Sci 1(2):37–44

Kinsman R, Ibelings BW, Walsby A (1991) Gas vesicle collapse by turgor pressure and its role in buoyancy regulation by Anabaena flos-aquae. Microbiology 137:1171–1178

Khanna DR, Bhutiani R, Chandra KS (2009) Effect of the euphotic depth and mixing depth on phytoplanktonic growth mechanism. Int J Environ Res 3(2):223–228

Kolada (2014) The effect of lake morphology on aquatic vegetation development and changes under the influence of eutrophication. Ecol Ind 38:282–293. https://doi.org/10.1016/j.ecolind.2013.11.015

Komárek J, Komárková J, Kling H (2003) Filamentous cyanobacteria, in freshwater algae of North America. Weir JD, Sheath RG (eds), pp 117–196

Lee JK, Oh JM (2018) A study on the characteristics of organic matter and nutrients released from sediments into agricultural reservoirs. Water (Switzerland) 10(8):980. https://doi.org/10.3390/w10080980

Lehmusluoto P, Machbub B, Terangna N, Rusmiputro S, Achmad F, Boer L, Brahmana SS, Priadi B, Setiadji B, Sayuman O, Margana A (1997) National inventory of the Major Lakes. In: Expedition indodanau technical report. https://www.academia.edu/1025268/National_inventory_of_the_major_lakes_and_reservoirs_in_Indonesia

Lehtoranta J, Ekhol P, Pitkänen H (2009) Coastal eutrophication thresholds: a matter of sediment microbial processes. Ambio 38(6):303–308. https://doi.org/10.1579/09-A-656.1

Lewis WM (2010) Biogeochemistry of tropical lakes. SIL Proceedings, 1922–2010, Internationale Vereinigung für Theoretische und Angewandte Limnologie: Verhandlungen 30(10):1595–1603. https://doi.org/10.1080/03680770.2009.11902383

Li D, Li G, Chen W, Liu Y (2009) Interactions between a cyanobacterial bloom (Microcystis) and the submerged aquatic plant *Ceratophyllum oryzetorum*. Kom Chin J Oceanol Limn 27:38–42

Li D, (2021) Ecological threshold for prevention and control of cyanobacterial blooms in Baiyangdian Lake. Chin Res Acad Environ Sci

Lim MH, Tay HSM, Denise A, Devotta DA, Mowe MAD, Mitrovic SM (2020) Risk management of cyanotoxins in Singapore. J Water Resour Prot 12:512–525. https://doi.org/10.4236/jwarp.2020.126031

Lukman, Syawal MS, Maghfiroh, M (2020) Sumatran major lakes: limnological overviews. IOP Conf Series Earth Environ Sci 535:012064. https://doi.org/10.1088/1755-1315/535/1/012064

Lukman, Setyobudiandi I, Muchsin I, Hariyadi S (2015) Impact of cage aquaculture on water quality condition in Lake Maninjau, West Sumatera Indonesia. Int J Sci Basic Appl Res (IJSBAR) 23(1):120–137. https://www.academia.edu/103750791/Impact_of_Cage_Aquaculture_on_Water_Quality_Condition_in_Lake_Maninjau_West_Sumatera_Indonesia?uc-sb-sw=8369184

Lukman HH, Subehi L, Dina R, Mayasari N, Melati I, Sudariani Y (2021) Loads of pollution to Lake Toba and their impacts. Int J Adv Eng Inf Technol 11(3):930–936. https://doi.org/10.18517/ijaseit.11.3.12733

Luthfiani F, Kasmara H, Sunardi (2020) The dynamic of blue-green algae (cyanobacteria) in eutrophic tropical waters, the Cirata Reservoir. Indones J Limnol 1(1):1–6. https://doi.org/10.51264/inajl.v1i1.4

Makmur S, Muthmainnah D, Subagdja (2020) Fishery activities and environmental condition of Maninjau Lake, West Sumatra. IOP Conf Ser Earth Environ Sci 564:01202. https://doi.org/10.1088/1755-1315/564/1/012025

Marganof (2007) Model Pengendalaian Pencemaran di Danau Maninjau, Sumatera Barat. Disertasi Doktor Program Studi Pengelolaan Sumberdaya Alam dan Lingkungan. Institut Pertanian Bogor. https://repository.ipb.ac.id/handle/123456789/40787

Moraes MAB, Rodrigues RAM, Schlüter L, Podduturi R, Jørgensen NOG, Calijuri MC (2021) Influence of environmental factors on occurrence of cyanobacteria and abundance of saxitoxin-producing cyanobacteria in a subtropical drinking water reservoir in Brazil. Water 13:1716. https://doi.org/10.3390/w13121716

Moustaka-Gouni M, Sommer U (2020) Effects of harmful blooms of large-sized and colonial cyanobacteria on aquatic food webs. Water 12:1587. https://doi.org/10.3390/w12061587

Mowe MAD, Mitrovic SM, Lim RP, Furey A, Yeo DCJ (2015) Tropical cyanobacterial blooms: a review of prevalence, problem taxa, toxins and influencing environmental factors. J Limnol 74(2):205–224. https://doi.org/10.4081/jlimnol.2014.1005
Mishra S, Neha Kumari N, Häder D, Rajeshwar P, Sinha RP (2022) Cyanobacterial blooms and their implications in the changing environment. Adv Environ Eng Res 3(1). https://doi.org/10.21926/aeer.2201011
Morales-Covarrubias MS, Alonso-Rodríguez R, Velázquez-Garay JA, Flores-Chavarría AMGA, Federico Alva-Martín AF (2016) Effect of microcystis aeruginosa on organs and tissues of white shrimp (*Litopenaeus vannamei*) postlarvae in low-salinity water. Revista Científica FCV-LUZ 26(3):181–187
Moura AN, Dantas EW, Oliveira HSB, Bittencourt-Oliveira MC (2011) Vertical and temporal dynamics of cyanobacteria in the Carpina potable water reservoir in northeastern Brazil. Braz J Biol 71(2):451–459
Nakatani K, Nasukawa K, Kikuchi T, Fujii M, Miyanaga K, Tanji Y, Aan V (2022) Microcystin production and oxidative stress. In: Yoshimura C, Khanal R, Uk S (eds) Water and life at Tonle Sap Lake edition: 1 chapter: 28. Springer, Singapore, pp 285–295. https://doi.org/10.1007/978-981-16-6632-2_28
Namsaraev Z, Melnikova A, Ivanov V, Komova A, Teslyuk A (2018) Cyanobacterial bloom in the world largest freshwater lake Baikal. Conf Series Earth Environ Sci 121:032039. https://doi.org/10.1088/1755-1315/121/3/032039
Niamien-Ebrottie JE, Bhattacharyya S, Deep PR, Nayak B (2015) Cyanobacteria and cyanotoxins in the world: review. Int J Appl Res 1(8):563–569
Orihel DM, Baulch HM, Casson NJ, North RL, Parsons CT, Seckar DCM, Venkiteswaran JJ (2017) Internal phosphorus loading in Canadian fresh waters: a critical review and data analysis. Can J Fish Aquat Sci 74:2005–2029. https://doi.org/10.1139/cjfas-2016-0500
O'Neil JM, Davis TW, Burford MA, Gobler CJ (2012) The rise of harmful cyanobacteria blooms (CHABs): role of eutrophication and climate change in freshwater, estuarine and marine ecosystems. Harmful Algae 14:313–334.
Paerl HW (2014) Mitigating harmful cyanobacterial blooms in a human and climatically-impacted world. Life, 988–1012. https://doi.org/10.3390/life4040988
Paerl HW, Hall NS, Calandrino ES (2011) Controlling harmful cyanobacterial blooms in a world experiencing anthropogenic and climatic-induced change. Sci Total Environ 409(10):1739–1745
Paerl HW, Gardner WS, Havens KE, Joyner AR, McCarthy MJ, Newell SE, Scott JT (2016) Mitigating cyanobacterial harmful algal blooms in aquatic ecosystems impacted by climate change and anthropogenic nutrients. Harmful Algae 54:213–222. https://doi.org/10.1016/j.hal.2015.09.00
Perkins S (2019) Going beneath the surface: the ecology of metalimnetic cyanobacteria layers New Hampshire Lakes. Master's Theses. University of New Hampshire, Durham, 70 pp
Pham TL, Dao TS, Tran ND, Nimptsch J, Wiegand C, Motoo U (2017) Influence of environmental factors on cyanobacterial biomass and microcystin concentration in the Dau Tieng Reservoir, a tropical eutrophic water body in Vietnam. Ann Limnol Intl J Lim 53:89–100. https://doi.org/10.1051/limn/2016038
Piccin-Santos V, Bittencourt-Oliveira MC (2012) Toxic cyanobacteria in four Brazilian water supply reservoirs. J Environ Prot 3:68–73. https://doi.org/10.4236/jep.2012.31009. Published Online Jan 2012. http://www.SciRP.org/journal/jep
Piranti AS, Rahayu DR, Waluyo G (2017) Input of nutrient (Nitrogen and Phosphorus) from the catchment area into Rawapening Lake of Central Java. In: UNEJ e-proceeding, pp 50–51. https://jurnal.unej.ac.id/index.php/prosiding/article/view/4133/3826
Plisnier PD, Cocquyt C, Cornet Y, Poncelet N, Nshombo M, Ntakimazi G, Nahimana D, Makasa L, MacIntyre S (2023) Phytoplankton blooms and fish kills in Lake Tanganyika related to upwelling and the limnological cycle. J Great Lakes Res 49(6):102247

Qin B, Zhou J, Elser JJ, Gardner WS, Deng J, Brooks JD (2020) Water depth underpins the relative roles and fates of nitrogen and phosphorus in lakes. Env Sci Technol 54:3191–3198. https://doi.org/10.1021/acs.est.9b05858

Reynolds CS, Huszar V, Kruk C, Naselli-Flores L, Melo S (2002) Review, towards a functional classification of the freshwater phytoplankton. J Plankton Res 24(5):417–428

Schindler DW (2012) The dilemma of controlling cultural eutrophication of lakes. Proc R Soc B 279:4322–4333. https://doi.org/10.1098/rspb.2012.1032

Schindler DW, Carpenter SR, Chapra SC, Hecky RE, Orihel DM (2016) Reducing phosphorus to curb lake eutrophication is a success. Environ Sci Technol 50:8923–8929. https://doi.org/10.1021/acs.est.6b02204

Schwarzenberger A (2022) Negative effects of cyanotoxins and adaptative responses of Daphnia. Toxins 14:770. https://doi.org/10.3390/toxins14110770

Sinang SC, Poh KB, Shamsudin S, Sinden A (2015) Preliminary assessment of cyanobacteria diversity and toxic potential in ten freshwater lakes in Selangor, Malaysia. Bull Environ Contam Toxicol. https://doi.org/10.1007/s00128-015-1620-7

Sinha E, Michalak AM, Balaji V (2017) Eutrophication will increase during the 21st century as a result of precipitation changes. Science 357:405–408

Sirunda J, Oberholster P, Wolfaardt G, Botes M, Truter C (2021) The assessment of phytoplankton dynamics in two reservoirs in Southern Africa with special reference to water abstraction for inter-basin transfers and potable water production. Water 13:3045. https://doi.org/10.3390/w13213045

Sivonen K (1996) Cyanobacterial toxins and toxin production. Phycologia 35(l):12–24

Soares MCS, Huszar VLM, Miranda MN, Mello MM, Roland F, Lurling M (2013) Cyanobacterial dominance in Brazil: distribution and environmental preferences. Hydrobiologia 717:1–12. https://doi.org/10.1007/s10750-013-1562-1

Somdee T, Kaewkhiaw K, Somdee A (2013) Detection of toxic cyanobacteria and quantifi cation of microcystins in four recreational water reservoirs in Khon Kaen, Thailand. KKU Res J 18(1):1–8

Søndergaard (2007) Blackwell publishing ltd lake restoration: successes, failures and long-term effects. J Appl Ecol 44:1095–1105. https://doi.org/10.1111/j.1365-2664.2007.01363.x

Søndergaard M, Bjerring R, Jeppesen E (2013) Persistent internal phosphorus loading during summer in shallow eutrophic lakes. Hydrobiologia 710(1):95–107. https://doi.org/10.1007/s10750-012-1091-3

Sok T, Oeurng C, Kaing V, Sauvage S, Xixi L, Sanches JM (2022) Nutrient transport and exchange between the Mekong River and Tonle Sap Lake in Cambodia. Ecol Eng 176. https://doi.org/10.1016/j.ecoleng.2021.106527

Subehi L, Ridwansyah I, Fukushima T (2020) Dissolved oxygen profiles and its problems at Lake Maninjau, West Sumatra—Indonesia. Indones J Limnol 1(1):7–11. https://doi.org/10.51264/inajl.v1i1.3

Sulastri, Sulawesty F, Nomosatryo S (2015) Longterm monitoring of water quality and phytoplankton changes in Lake Maninjau, West Sumatera, Indonesia. Oseanologi dan Limnologi di Indonesia 41:339–353. https://jurnal-oldi.or.id/index.php/oldi

Sulastri, Henny C, Handoko U (2016a) Environmental condition and trophic status of Lake Rawa Pening in Central Java Kondisi Lingkungan dan Status Trofik Danau Rawa Pening di Jawa Tengah. Oseanologi dan Limnologi di Indonesia 1(3):23–38. https://scholar.google.com/scholar?cluster=2158951029872487246&hl=en&oi=scholarr

Sulastri, Nomosatryo S, Sulawesty F (2016b) Keterkaitan unsur hara dan biomasa fitoplankton (klorofil-a) di Danau Maninjau, Sumatera Barat. Prosiding Pertemuan Ilmiah MLI tahun 2015: Tantangan Terkini Perairan Darat Di Wilayah Regional Tropis Menyongsong World Lake Conference. Masyarakat Limnologi Indinesia (MLI), Cibinong, pp 129–141. http://lipi.go.id/publikasi/keterkaitan-unsur-hara-dan-biomasa-fitoplankton-klorofil-a-di-danau--maninjau-sumatera-barat/20972

Sulastri, Henny C, Arianto AB (2019) phytoplankton composition and the occurrence of cyanobacterial bloom in Lake Maninjau, Indonesia. IOP Conf Ser Earth Environ Sci 380:012020

Sulastri, Aisyah S, Dina R (2022) Trophic status and phytoplankton community structure of four small lakes in ciliwung watershed, indonesia. IOP Conf Ser Earth Environ Sci 950:012037. https://doi.org/10.1088/1755-1315/950/1/012037

Sulastri, Henny C, Nomosatryo S, Susanti E, Sulawesty F (2023) Monitoring planktonic cyanobacteria in Lake Maninjau, West Sumatra, Indonesia. IOP Conf Ser Earth Environ Sci 1260:012018. https://doi.org/10.1088/1755-1315/1260/1/012018

Sulawesty F (2007) Distribusi Vertikal Fitoplankton di Danau Singkarak. Limnotek Perairan Darat Indonesia XIV(1):37–46. https://karya.brin.go.id/id/eprint/16677/1/Jurnal_Fachmijany%20Sulawesty_Puslit%20Limnologi_37-46_2007.pdf

Sulawesty F (2018) Struktur komunitas fitoplankton Danau Hanjalutung Kalimantan Tengah. Prosiding Pertemuan Ilmiah Masyarakat Limnologi Indonesia tahun 2017. Masyarakat Limnologi Indonesia. Bogor. Hal. 221–228. https://masyarakatlimnologi.org/2018/05/30/prosiding-mli-2017-telah-terbit/

Sulawesty F, Sunanisari S, Mulyana E, Syawal MS, Nomosatryo S, Fauzi H (2002) Evaluasi kondisi limnologi Danau Singkarak. Prosiding Seminar Nasional Limnologi. Hal.: 95–108

Sulawesty F, Sumarni (2004) Komunitas fitoplankton di Situ Pondok, Kabupaten Tangerang. Limnotek Perairan Darat Indonesia XI(2):36–44

Sulawesty F, Awalina, Chrismadha T (2012) The changes of phytoplankton community structure of Situ Rawa Kalong, a shallow polluted tropical lake in West of Java. In: Proceeding the 10th international symposium on Southeast Asian water environment. The University of Tokyo, pp 335–341

Sulawesty F, Awalina (2013) Phytoplankton community structure and other related eutrophication symptoms parameter in Lake Batur, Bali—Indonesia. Oseanologi dan Limnologi di Indonesia 39(2):181–199

Sulawesty F, Awalina, Chrismadha T (2015) Struktur Komunitas Fitoplankton Situ Rawa Kalong Pada Musim Peralihan. Prosiding Seminar Nasional Limnologi VII Tahun 2014. Pusat Penelitian Limnologi LIPI. Hal. 475–483

Sulawesty F, Shohihah H dan Widiyanto T (2018) Biodiversitas fitoplankton dan kualitas air di Danau Aneuk Laot – Sabang, Nangroe Aceh Darussakam. Prosiding Pertemuan Ilmiah Masyarakat Limnologi Indonesia tahun 2017. Masyarakat Limnologi Indonesia. Bogor. Hal. 145–151. https://masyarakatlimnologi.org/2018/05/30/prosiding-mli-2017-telah-terbit/

Sulawesty F, Yustiawati, Aisyah S (2020) Komunitas Fitoplankton di Daerah Litoral Danau Maninjau dan Sungai Ranggeh, Kabupaten Agam; Kaitannya dengan Kandungan Nutrien di Perairan. Oseanologi dan Limnologi di Indonesia 5(1):47–59. https://doi.org/10.14203/oldi.2020.v5i1.289

Sulawesty F, Yustiawati Y (2021) Spatial variation of phytoplankton and some physico-chemical variables in Lake Tempe, South Sulawesi. IOP Conf Ser Earth Environ Sci 744:012084. https://doi.org/10.1088/1755-1315/744/1/012084

Sulawesty F, Yoga GP, Subehi L, Rosidah (2021) Phytoplankton community structure in Menjer Lake, Central Java. IOP Conf Ser Earth Environ Sci 869:012037. https://doi.org/10.1088/1755-1315/869/1/012037

Suwanpakdee S, Gutierrez R, Pithakpol S, Jampeetong A, Pathomaree W, Nomura N, Itayama T and Whangchai N (2016) Earthy-musty odour and off-flavour taints in Phayao Lake, Thailand. Chiang Mai J Sci 43(1):22–31. https://www.thaiscience.info/journals/Article/CMJS/10980093.pdf

Te SH, Gin KYH (2011) The dynamics of cyanobacteria and microcystin production in a tropical reservoir of Singapore. Harmful Algae 10:319–329

Thawabteh AM, Hani A Naseef HA, Donia Karaman D, Bufo SA, Scrano L, Karaman R (2023) Understanding the risks of diffusion of cyanobacteria toxins in rivers, lakes, and potable water. Toxins 15:582. https://doi.org/10.3390/toxins15090582

Tian X, Yuan Y, Zou Y, Qin L, Zhu X, Zhu Y, Zhao Y, Jiang M, Jiang M (2023) Cyanobacterial blooms increase functional diversity of Metazooplankton in a shallow eutrophic lake. Water 15:953. https://doi.org/10.3390/w15050953

Tingson KN, Tamayo-Zafaralla M (2018) Water quality and seasonal dynamics of Phytoplankton and Zooplankton in the West Bay of Laguna De Bay, Philippines. Ecosyst Dev J 8(1):41–52. https://ovcre.uplb.edu.ph/journals-uplb/index.php/EDJ/article/view/238/222

Trung B, Dao T, Faassen E, Lürling M (2018) Cyanobacterial blooms and microcystins in southern Vietnam. Toxins 10:471. https://doi.org/10.3390/toxins10110471

Vanderley RF, Ger KA, Becker V, Gabriela M, Bezerra TA, Panosso R (2021) Abiotic factors driving cyanobacterial biomass and composition under perennial bloom conditions in tropical latitudes. Hydrobiologia 848:943–960. https://doi.org/10.1007/s10750-020-04504-7

Wang Z, Akbar S, Sun Y, Gu L, Zhang L, Lyu K, Huang Y, Yang Z (2021) Cyanobacterial dominance and succession: factors, mechanisms, predictions and managements. J Environ Manage 297:113281

Whitton BA, Potts M (2000) Chapter 1 Introduction to cyanobacteria. In: Whitton B, Potts M (eds) The ecology of cyanobacteria. Kluwer Academic Publishers, The Netherlands, pp 1–11

Wiley DY, Mc Pherson RA (2024) The role of climate change in the proliferation of freshwater harmful algal blooms in inland water bodies of the United States. Earth Interact 28:1–16

Zanchett G, Oliveira-Filho EC (2013) Cyanobacteria and cyanotoxins: from impacts on aquatic ecosystems and human health to anticarcinogenic effects. Toxins 5:1896–1917. https://doi.org/10.3390/toxins5101896

Zhang Y, Song C, Ji L, Liu Y, Xiao J, Cao X, Zhou Y (2018) Cause and effect of N/P ratio decline with eutrophication aggravation in shallow lakes. Sci Total Environ 627:1294–1302. https://doi.org/10.1016/j.scitotenv.2018.01.327

Zhang W, Liu J, Xiao Y, Zhang Y, Yu Y, Zheng Z, Liu Y, Li Q (2022) The impact of cyanobacteria bloomson the aquatic environment and human health. Toxins 14:658. https://doi.org/10.3390/toxins14100658

Zimmer V, Bendoricchio G (2001) Nutrient and suspended solid loads in the Laguna de Bay, Philippines. Water Sci Technol 44(7):77–86. https://pubmed.ncbi.nlm.nih.gov/11724498/

Chapter 9
Biotic Communities: Invasive Macrophytes

Syarif Prasetyo, Luki Subehi, and Siti Norasikin Ismail

Macrophytes have an important role in ecological functions, especially in determining the structure and function of freshwater ecosystems. Macrophytes are classified into submerged macrophytes, emerging macrophytes, and free-floating macrophytes. Macrophytes provide complex habitat and food for semi-aquatic biota ranging from macroinvertebrates to mammals, and they participate in nutrient cycles. Global lake ecosystems have experienced significant changes in macrophyte abundance and composition due to multiple types of interactions, including human activity and climate change. Eutrophication due to anthropogenic activities causes some macrophytes to grow out of control in lakes and other locations across the world. The macrophyte known as water hyacinth (*Eichhornia crassipes* (Mart.) Solms) grows uncontrolled in freshwater environments that are high in nutrients, particularly phosphorus and nitrogen. A detrimental effect of blooming water hyacinth has been observed in many regions of the world, especially the lake ecosystems and river networks including the Lake Rawapening in Indonesia. The most dangerous environmentally destructive weed in the world, water hyacinth is almost impossible to eradicate because it is able to grow quickly both generatively and vegetatively and forms very dense masses that cover the surface of the water. The water hyacinth explosion has damaged the ecosystem, causing significant water loss in Lake Rawapening through evapotranspiration. The water hyacinth invasion also causes a decrease in water quality, depletion of oxygen levels, a decrease in fish production, as well as a decrease in natural beauty and public health. There are three ways to control water hyacinth in Lake Rawapening, namely physical, chemical, and biological. All control

---

S. Prasetyo (✉) · L. Subehi
Research Center for Limnology and Water Resources, National Research and Innovation Agency (BRIN), Jakarta, Indonesia
e-mail: syar026@brin.go.id

S. N. Ismail
Institute of Tropical Biodiversity and Sustainable Development, Universiti Malaysia Trengganu, Kuala Terengganu, Malaysia

© The Author(s), under exclusive license to Springer Nature Singapore Pte Ltd. 2025
A. C. Santos-Borja et al. (eds.), *Progress on Ecosystem Restoration of Tropical Inland Waters*, https://doi.org/10.1007/978-981-96-2284-9_9

methods have their respective advantages and disadvantages, and to achieve the best control, integrating two or more control methods may be an option. This article aims to provide deeper insight into the biological aspects of water hyacinth and the problems it can cause, how to control it, and how to use it to increase the economic value of this weed. The various benefits that can result from water hyacinth management include animal feed, raw material for making fish food, compost production, bioenergy and biogas production, as well as raw material for making handicrafts and other applications. The application of water hyacinth control methods by integrating the use of these plants as commodities with economic value is very important and must be integrated on a large scale. The existence of water hyacinth in Lake Rawapening has not completely disappeared because it has ecological and economic functions for society. The encouragement of socioeconomic development toward the presence of water hyacinth needs to be increased because it has the potential to create employment opportunities for the community around Lake Rawapening. In addition, this strategy is part of efforts to achieve global sustainable development goals (SDGs) through an integrated relationship between the availability of water, food, energy, and the environment.

9.1 Water Hyacinth Problems

Water hyacinth (*E. crassipes* (Mart.) Solms) is an invasive freshwater weed that causes various problems for the environment, social development, human health, the economy, and access to clean water in various parts of the world (Sharma et al. 2016; Dersseh et al. 2019). This species was accidentally discovered by Karl von Mortius in 1824 while on an expedition in the Brazilian Amazon (Setyanto and Warniningsih 2011). Subsequently, it was introduced as an ornamental plant to North America in the late 1800s (Brown et al. 2020). Water hyacinth appeared in Africa in the early 1900s and in Europe in the 1930s (Yan and Guo 2017). Water hyacinth has now spread throughout the world except for the Antarctic continent (Jiménez and Balandra 2007; Gaikwad and Gavande 2017). In Indonesia, water hyacinth was first introduced to the Bogor Botanical Gardens in 1894. Water hyacinth, which was originally brought and used as a shade plant for ponds, eventually filled ponds and was discharged into the river at the Bogor Botanical Gardens. From there, this plant spread to fill rivers, swamps, and reservoirs in various regions throughout Indonesia (Coniwanti et al. 2009).

Water hyacinth can grow optimally at a pH range of 6–8 and a temperature of 25–27.5 °C, but it can still grow at a temperature range of 1–40 °C (Heard and Winterton 2000). This worst and most dangerous water weed in the world can reproduce quickly and grow to a density of 60 kg/m (Ruan et al. 2016). Water hyacinths take 1–2 weeks to reproduce (Proctor 2016). A study in India showed that one water hyacinth plant can reproduce within 14 days (Mengist and Moges 2019). Prasetyo et al. (2021) found that the relative growth rate of water hyacinth at Lake Rawapening ranged from 6.40 to 7.26%/day, the ability to double time was 9.6–10.8 days, and one clump of water hyacinth weighing 160 g could cover 1 m^2 of mesocosm in 21 days. As for one hectare of water hyacinth, according to Sharma and Chauhan (2017), it might contain 2 million plants weighing between 270 and 400 tons. The explosion of water hyacinth in an ecosystem is a very serious threat to aquatic systems and inhibits human use of surface water (Chander et al. 2018). The explosion of water hyacinth is caused by the enrichment of nutrients in the water, especially nitrogen and phosphorus (Verma and Sivappa 2017; Guignard et al. 2017). The rapid and uncontrolled growth of water hyacinth causes problems such as clogging of waterways and flooding (Kriticos and Brunel 2016). Water hyacinth that covers the surface of the water consumes dissolved oxygen (Hasibuan et al. 2020), reducing water quality and biodiversity (Guerra et al. 2015). Due to its adverse ecological, social, and economic impacts, water hyacinth is included in the 100 most dangerous invasive species in the world (Enyew et al. 2020).

The magnitude of the water hyacinth problem is increasing exponentially in many lakes in Indonesia. This tropical country has 5807 lakes spread throughout its archipelago, with categories of natural lakes (1022), artificial lakes (1314), and 3471 lakes that have not been identified in this category (Luthfi 2021). Of the many lakes, some have been invaded by water hyacinth, which has resulted in severe environmental damage. In the 2014–2019 National Medium-Term Development Plan (RPJMN), the Government of Indonesia has determined the 15 lakes national priority. They are Lake Rawapening and Lake Rawa Danau (Java island) Lake Batur (Bali island), Lake Toba, Lake Singkarak, Lake Maninjau, and Lake Kerinci (Sumatera island), Lake Poso, Lake Tondano, Lake Matano, Lake Tempe and Lake Limboto (Sulawesi island), Lake Semayang Melintang-Jempang and Lake Sentarum (Kalimantan island), and Lake Sentani (Papua island).

Various programs to restore the condition of priority lakes from environmental damage, including water hyacinth control until 2020, have not been completed. There are still many lakes that have not been handled optimally. Furthermore, the government, through Presidential Regulation Number 18 of 2020, focuses more on maintaining, restoring, and conserving water resources and ecosystems through revitalizing the national priority of five lakes that are still in a state of severe environmental damage, namely Lake Maninjau, Lake Sentarum, Lake Limboto, Lake Sentani, and Lake Rawapening.

Water hyacinth has a negative impact on lake ecosystems and river networks in Indonesia, one of which is Lake Rawapening. The explosion of water hyacinth is currently very dominant and has resulted in damage to the Lake Rawapening ecosystem (Madian et al. 2019; Prasetyo et al. 2021a). Uncontrolled growth poses a serious threat to biodiversity (Plaza et al. 2010) and has a major impact on the daily lives of the people who depend on Lake Rawapening. The weed blast also disrupts hydroelectric power systems, blocks turbines (Enyew et al. 2020), and causes rapid loss of water through evapotranspiration (Kamau et al. 2015; Goshu and Aynalem 2017; Honlah et al. 2019). Health problems arise because the water hyacinth is a habitat for vectors of diseases and pests (Ndimele et al. 2011; Waithaka 2013; Gupta and Yadav 2020), including malaria and filariasis (McBeath et al. 2014; Van Oijstaeijen et al. 2020; Nega et al. 2021). Fishermen find it increasingly difficult to catch fish in Lake Rawapening due to decreased numbers and diversity (Stave et al. 2017). Another problem is the blockage of irrigation canals. Farmers who use irrigation water must spend time and energy clearing blocks in order to irrigate their fields (Enyew et al. 2020).

Various methods of dealing with water hyacinth, both biologically and chemically, are not economically profitable and lead to environmental damage (Dechassa and Abate 2020). It is necessary to look for technological breakthroughs that can increase the economic value of this invasive weed so that its eradication can generate returns. In this way, water hyacinth control plan could provide environmental benefits while still providing sustainable income (Harun et al. 2021).

There have been many studies related to the potential use of water hyacinth weed from Lake Rawapening as a material of economic value after being harvested manually or mechanically. Utilization of water hyacinth can be in the form of handicraft and furniture products (Salas-Ruiz et al. 2019), organic fertilizers, both liquid and compost (Prasetyo et al. 2022), composite and biopolymer industrial raw materials (Saratale et al. 2020), basic ingredients for fish feed (Prasetyo et al. 2021b), unconventional protein sources for animal feed (de Vasconcelos et al. 2016), the role of heavy metal phytoremediators and efficient adsorbents for textile waste (Priya and Selvan 2017; Premalatha et al. 2018; Shen et al. 2018), raw materials for bioenergy production (charcoal briquettes, biogas, and bioethanol), and others (Setyaningsih et al. 2017; Laitinen et al. 2017; Carnaje et al. 2018; Madian et al. 2019; Gaurav et al. 2020). Many applications of water hyacinth utilization have been carried out but do not automatically solve the problem of the existence of these weeds in the water. The basic problems that must be faced in efforts to control water hyacinth are related to a lack of public awareness, not yet integrated institutions, and the absence of appropriate and comprehensive policies.

The article mainly discusses about macrophytes in Lake Rawapening, Indonesia, since we have been conducting research for a significant period. Through the use of water hyacinth with economic value, it will help sustainability in efforts to control the rate of its invasion so that the Lake Rawapening ecosystem will improve ecologically. Additionally, this article hopefully can increase the awareness of people regionally and internationally. It could also have implications for further research that produces a legal framework for lake management policies in the future. Thus, Indonesia's participation in implementing the globally agreed sustainable development goals (SDGs), especially related to water, energy, and agriculture, can be achieved.

9.2 Lake Rawapening: Geographical Features

The area of the study is Lake Rawapening, which is located in Semarang Regency, Central Java Province, the Unitary State of the Republic of Indonesia (Fig. 9.1). Geographically, Lake Rawapening is located at coordinates 7° 04′–7° 30′ South Latitude and 110° 24′ 46″–110° 49′ 06″ East Longitude and has a height of 460 m above sea level (MASL). Lake Rawapening is a semi-natural lake type with an area of 2667 ha in the rainy season and 1650 ha in the dry season (Izzati et al. 2022; Prasetyo et al. 2022). This lake, which is located in the tropics, is surrounded by mountains and hills with a slope of between 8 and 30%. The mountains surrounding Lake Rawapening include Mount Telomoyo (1895 MASL), Mount Butak (1000 MASL), Mount Balak (700 MASL), Mount Payung, and Mount Rong (600 MASL) (Mardiatno et al. 2023). The water that floods Lake Rawapening comes from rainfall, groundwater, and 20 rivers that empty into the lake. The water comes out only through one outlet (Sungai Tuntang).

Lake Rawapening is very vulnerable to environmental changes because it has a high utilization rate. Various forms of utilization of Lake Rawapening include agriculture, fisheries, sources of electrical energy, raw water, tourism, religion, and culture, as well as flood control (Prasetyo et al. 2021b; Soeprobowati, 2017). Various activities to utilize water bodies and water catchment areas in the Lake Rawapening area have resulted in a decrease in environmental quality, namely sedimentation, pollution, and a decrease in water quality and eutrophication. Eutrophication that occurred in Lake Rawapening can be seen from the increased primary productivity and led to the uncontrolled dominance of aquatic plants (water hyacinth weed) (Grasset et al. 2016; Sutadian 2017; Sudarmadji and Pudjiastuti 2018; Chander et al. 2018).

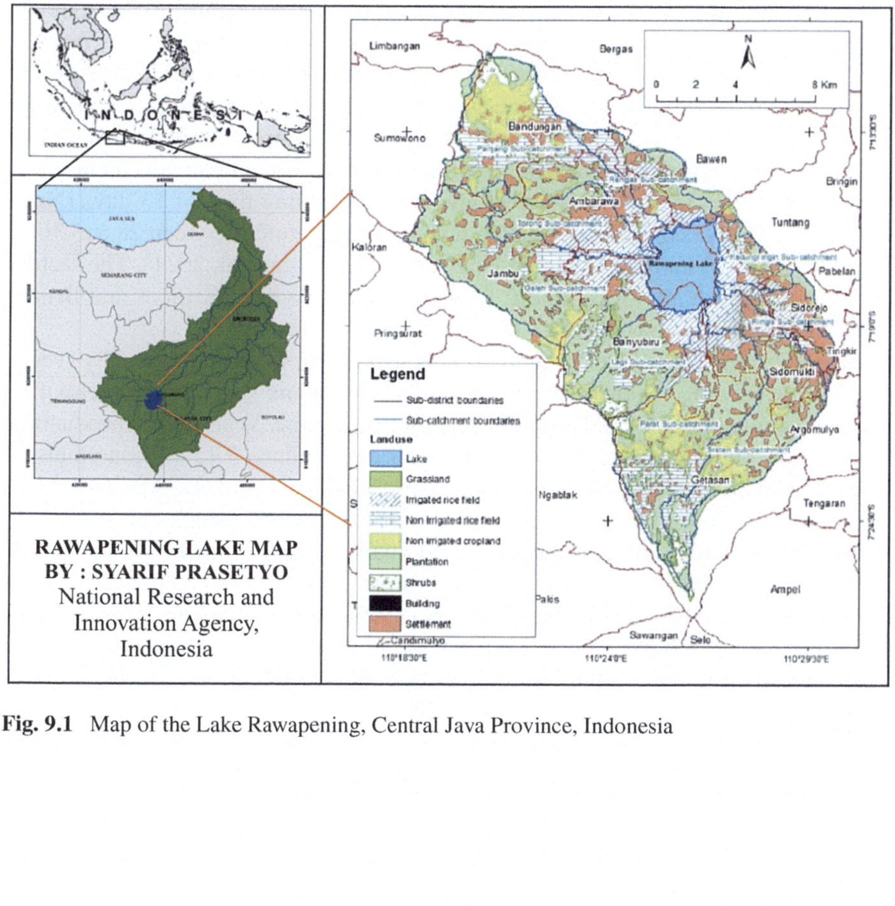

Fig. 9.1 Map of the Lake Rawapening, Central Java Province, Indonesia

9.3 The Biology of Water Hyacinth

Water hyacinth *E. crassipes* (Mart.) Solms is an invasive aquatic weed belonging to the Pontederiaceae family (Roopnarain et al. 2019). The water hyacinth has a structure with an air cavity that functions to float in the water. It can reach 1 m above the water level but generally ranges from 20 to 30 cm. Water hyacinth has a long stem, round, and supple. This exotic weed has flowers in bouquets of tubular petals, light purple on the corolla. Water hyacinth produces green fruit with a three-chamber box shape; the seeds are round with black color. Each flower can produce 400 seeds, and each plant can produce 5000 seeds (Gaikwad and Gavande 2017). The leaves of the water hyacinth are green with an oval shape and taper at the ends and base (Gaikwad and Gavande 2017; Bakrim et al. 2022). At the base of the weed stalk has a smooth bubble shape on its surface. Chlorophyll, a material that plays a role in the process of photosynthesis, is located in the epidermal cells. On the upper surface of the leaf, there are leaf mouths (stomata) that fill the surface. All parts of the water hyacinth organ have air cavities, which function to float on the surface of the water and also to store oxygen produced during the process of photosynthesis.

Water hyacinth has very plastic, black, fibrous roots with a single main root that has many sides, forming a very large root system (Bakrim et al. 2022). Stolons form a dense mass 20–60 cm long or even extend up to 300 cm. Periodically, axillary shoots develop as stolons, growing horizontally 10–50 cm before forming tillers (Yan and Guo 2017). There is a red sac at the tip of the root that functions to collect sediment or particles dissolved in water. Water hyacinth is able to exploit nutrients in water bodies that are low in nutrients, resulting in longer and denser lateral roots with low phosphorus levels (Bakrim et al. 2022). With this ability, water hyacinth can increase its biomass weight by up to 4.8% per day, with a doubling time of 13 days (Nega et al. 2021). This weed grows optimally at water temperatures between 28 and 36 C (Harun et al. 2021) and a pH ranging from 4 to 8 (Ajithram et al. 2021). The factors that limit its growth are low humidity in the range of 15–40% (Kriticos and Brunel 2016) and salinity in coastal areas (De Groote et al. 2003).

Water hyacinth reproduces sexually (generatively) by seeds and asexually (vegetatively) by budding. Both of these reproductive systems have great potential to increase the number of water hyacinths in a very short time. Vegetative reproduction through new shoots growing from the leaf axils can produce 3.42 million plants in about 200 days and 43 leaf tillers in just 50 days (Gaikwad and Gavande 2017). The sexual reproduction of water hyacinth is by means of seeds, which can germinate in a few days or remain dormant for 15–20 years in silt and mud (Thomas et al. 1990). Water hyacinth seeds that sink to the bottom of the water remain inactive until they experience environmental stress (drought). After experiencing re-inundation by water, the water hyacinth seeds will germinate and grow as new individuals (Sharma et al. 2016; Mathur and Mathur 2018). The resilience of water hyacinths in adapting to this environment has made it a serious problem for ecology and the economy in the tropical and subtropical hemispheres this century (Harun et al. 2021). The invasion of water hyacinth in Lake Rawapening is presented in Fig. 9.2.

Fig. 9.2 Water hyacinth invasion in Lake Rawapening (Private collection. *Photo credit* BRIN/ Syarif Prasetyo)

9.4 The Chemistry of Water Hyacinth

Water hyacinth has various characteristics and chemical compositions according to the environmental conditions in which it lives. Several studies have shown that wet hyacinth contains 95% water, 0.04% nitrogen, 1% ash, 0.2% K_2O, 0.06% P_2O_5, and as much as 3.5% organic matter. The chemical composition of water hyacinth based on dry weight is 75.8% organic matter, 1.5% nitrogen, and 24.2% ash (Sivasankari and David Ravindran 2016; Sukarni et al. 2019; Ayanda et al. 2020). The composition of primary metabolites, chemical elements, and results of proximate analysis are presented in Fig. 9.3.

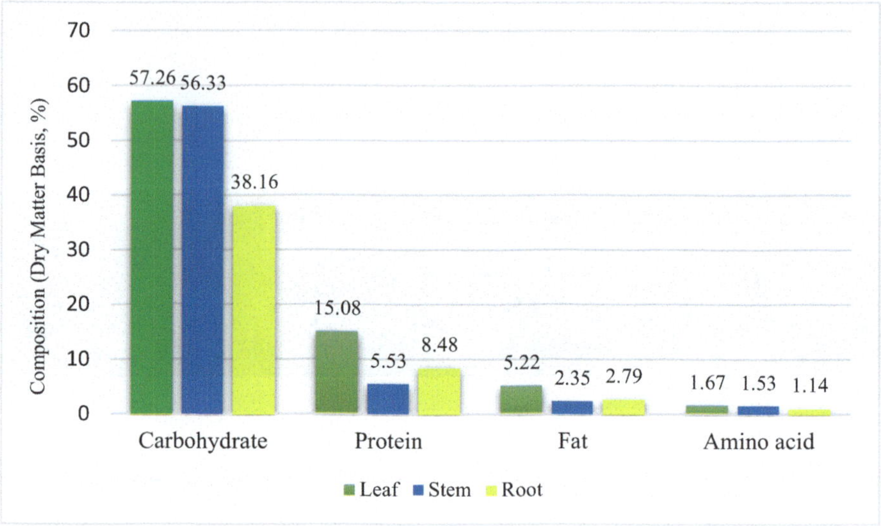

Fig. 9.3 Primary metabolite content is present in each part of the water hyacinth (Adapted from Ogamba et al. 2015)

9.5 Environmental, Social, and Economic Impacts

The invasion of water hyacinth causes serious environmental problems and has a negative impact on the socioeconomic conditions of people who depend on water bodies for their livelihood (Xu et al. 2017; Segbefia et al. 2019; Honlah et al. 2019). The subsystems disrupted by the explosion of water hyacinth in the waters include fisheries, agriculture, boat navigation, and human health (Ndimele et al. 2011; Odame Appiah et al. 2019; Harun et al. 2021).

The water hyacinth explosion in Lake Rawapening has disturbed the ecological balance and changed the biodiversity in it. Asmare (2017) explained that the rate of water loss caused by evapotranspiration can reach 1.8 times the respiration rate of surface water free from plants. This increase in evapotranspiration results in disruption of the water balance in the hydrological system and has a major impact on water-scarce areas (Dersseh et al. 2019).

The expanse of water hyacinth on the surface of the lake inhibits the penetration of sunlight and reduces temperature variability and other water problems (Tobias et al. 2019). The transfer of oxygen from the air to the surface of the water or the production of oxygen by other plants and algae is disrupted so that oxygen levels in the water decrease (Harun et al. 2021). Aquatic habitats are becoming less suitable and threatening the survival of lake biota because they have to compete for food, space, and sunlight with water hyacinth weeds that grow uncontrollably (Harun et al. 2021). The lower oxygen levels in the waters also encourage the release of phosphorus from the sediment into the water, which triggers the acceleration of eutrophication. In this way, the growth of water hyacinth is increasing rapidly because it is supported by nutritional intake (Dersseh et al. 2019).

The invasion of the water hyacinth in Lake Rawapening has affected the community's life and social sustainability. Water hyacinth is a nuisance to agricultural activities because it clogs the irrigation systems used to irrigate agricultural land (Enyew et al. 2020; Endgaw 2021). Blockage of water hyacinth in the river will reduce the water supply and trigger sedimentation (Dersseh et al. 2019). Rivers clogged with water hyacinth can cause flooding and have a negative impact on riparian communities, which are weak in flood disaster mitigation (Ilo et al. 2020).

The water hyacinth in the lake provides a wider habitat for the breeding of disease vectors such as the malaria-causing Anopheles mosquito (Segbefia et al. 2019; Van Oijstaeijen et al. 2020). Water hyacinth is an ideal habitat for snails, which are vectors for Bilharzia parasites and Mansoniodes mosquitoes, which endanger the health of people around the waters (Enyew et al. 2020; Nega et al. 2021). Water hyacinth provides protection for reptiles, including venomous snakes, which often attack humans (Patel 2012).

Water hyacinth that floats on the surface of the water will always move with the wind, and when it is in large numbers, it will cause problems for fishermen in Rawapening Lake. The presence of these weeds disrupts boat navigation and fishing schedules and damages fishing nets (Aloo et al. 2013; Waithaka 2013). Water hyacinth expanses will also limit the access of migratory fish to breeding grounds and foraging for several species of fish with high economic value (Harun et al. 2021). Many fish species die due to a lack of oxygen, so their populations decrease and fishermen's catches decrease (Witte et al. 2012; Genissa et al. 2018; Degaga 2019). Water hyacinth also causes damage to hydroelectric generators in dams that function as power plants (Lu et al. 2007; Tewabe 2015); this also occurs in Lake Rawapening, which supplies water for the Jelog hydropower plant and the Timo hydropower plant. Mass mortality of water hyacinth in large quantities is also a problem because, after decomposing, it can reduce the quality and quantity of water and have an impact on increasing the operational costs of raw water treatment (Villamagna and Murphy 2010; Patel 2012). When the dry season arrives and the water level drops drastically, the unpleasant odor from the rotting remains of the water hyacinth is very disturbing to the residents around Lake Rawapening. Handling water hyacinth from Lake Rawapening also requires high costs and burdens the government budget, which should be used for more productive programs.

9.6 Water Hyacinth Control Mechanisms and Challenges

Aquatic life in Lake Rawapening can suffer if water hyacinth weeds are not managed properly. Clearing water hyacinth weeds from Lake Rawapening can be difficult because these plants can reproduce and spread quickly. However, any control should aim to remove the plant faster than it can grow. One important component that must be considered in determining water hyacinth control is the uniqueness of the infested area (utilization of water bodies, water area, spatial planning, and weather) (Harun et al. 2021). Because the invasive nature of weeds varies in various environments, there may be no universal weed control method; no single method is suitable for all areas infested with weeds, and each method has its own limitations and challenges (Ajithram et al. 2021). Physical control (manual and mechanical removal), chemical and biological control, as well as an integrated approach (using two or three control mechanisms simultaneously) can be used to eradicate the water hyacinth weed invasion (Ayanda et al. 2020).

9.6.1 Physical Method

Physical control (manual and mechanical removal) (Fig. 9.4) in preventing the spread of water hyacinth invasion refers to the physical weeds (Bordoloi et al. 2018). This method involves taking or cutting weeds directly from their habitat (Villamagna and Murphy 2010). Physical control of water hyacinth in Lake Rawapening is carried out by the community traditionally by pulling it directly from the water by hand, cutting it with a sickle, and localizing the distribution of water hyacinth by installing bamboo barriers (Prasetyo et al. 2021a). Periodically, the people around Lake Rawapening collectively raise water hyacinth to the mainland in an effort to reduce the density of these weeds in water bodies.

The Pemali Juwana Watershed Agency (BBWSPJ), which is a government agency of Central Java Province, has formed an Operational Rawapening Control Unit IV (OP IV) to clear water hyacinth weeds from bodies of water. OP IV operates various modern machines and equipment to accelerate the cleaning of Lake Rawapening. The equipment includes Berky Harvesters (11 units), Amphibious Weed Harvesters (3 units), Excavators (3 units), and Dump Trucks (5 units) to transport water hyacinth. The use of machines has indeed significantly reduced the area of water hyacinth covering water bodies by up to 200 ha each year (Prasetyo et al. 2021c). However, this control method may be expensive given the nature of the equipment used, which requires fuel and maintenance (Villamagna and Murphy 2010; Ayanda et al. 2020).

Fig. 9.4 Manual and mechanical water hyacinth control in Lake Rawapening (Private collection. *Photo credit* BRIN/Syarif Prasetyo)

Handling water hyacinth mechanically in Lake Rawapening is considered ineffective and tends to fail. This happened because the water hyacinth cleaning operation did not consider the most basic aspects related to water conditions. It is important to understand and control the eutrophication that occurs in Lake Rawapening, but it has not been used as the basis for weed control. Eutrophication has occurred in Lake Rawapening as a result of excessive phosphate and nitrogen entering water bodies from the watershed area (Piranti et al. 2018).

As much as 10.32 mg/s of phosphorus enters water bodies, and 64.9% of it precipitates in sediments. The eutrophication that occurred in Lake Rawapening has succeeded in triggering the rapid growth of water hyacinth. The rapid growth of water hyacinth in Lake Rawapening is able to fill 1 m^2 of water within 22 days (Soeprobowati 2017). This means that when mechanical equipment is unable to keep up with the reproductive capacity and growth rate of water hyacinth, it will be difficult to control the weeds. Mechanical water hyacinth control methods are only effective in small water conditions and are therefore not efficient in lakes or large bodies of water. Applying mechanical methods to large lakes such as Lake Rawapening will require a large amount of time, energy, and cost (Hidayati et al. 2018).

9.6.2 Chemical Method

Chemical control of water hyacinth can be carried out using chemical herbicides such as Imazapyr, Glyphosate, Bispyribac, Triclopyr, Imazamox, Florpyrauxifen benzyl, and Diquad (Murugesan 2014; Carvalho and Cerveira Junior 2019; Garlich et al. 2019). The herbicide glyphosate is used efficiently to properly control water hyacinth and does not significantly change water quality parameters such as dissolved oxygen and pH (Souza et al. 2020). Spraying glyphosate herbicide at a dose of 450 g/ha was able to eradicate water hyacinth with a success rate of 92.13% (Al-Wagga and Al-Gburi 2018). The herbicides that are sprayed will be absorbed by the green parts of the plants and enter the cells, thereby disrupting the photosynthesis process. This chemical method is fast, efficient, and inexpensive when compared to mechanical or manual methods. Despite having a number of advantages, this method requires certain skills to be efficient (Ray and Hill 2013).

The use of herbicides to control water hyacinth is very limited because of the high risk of polluting the environment (Carvalho and Cerveira Junior 2019). Stakeholders on issues related to the environment call for more care and caution in the application of chemical methods because they will have an impact on non-target organisms (El-Shahawy 2015). Chemical herbicides are considered less selective against countermeasures when compared to mechanical and manual methods. Chemical elements pose a dangerous risk for non-target microorganisms and plants, which are habitats for other organisms. Another bad impact is affecting fish production in the waters (Gupta and Yadav 2020). Carcinogenic chemical elements also pose a risk to human health if water contaminated with herbicides is used for drinking and washing raw materials (Chander et al. 2018). Chemical water hyacinth control has proven successful for small water areas (Koutika and Rainey 2015), but is not suitable for large areas such as Lake Rawapening. Besides that, Lake Rawapening has a high level of complexity in its utilization for drinking water raw materials, aquaculture, tourism, agriculture, and so on. For this reason, chemical water hyacinth control has never been applied to Lake Rawapening. Because in general, the use of herbicides is feared to damage the diversity of biota and the balance of the Lake Rawapening ecosystem (Prasetyo et al. 2022).

9.6.3 Biological Method

The use of biological agents, including the exploitation of hosts as natural enemies of weeds, can reduce the size of water hyacinth populations in waters (Tipping et al. 2017). Until now, there have been several natural enemies that are generally used to control aquatic weeds, including moths (*Bellura* sp., *Xubida* sp., and *Niphograpta* sp.); flies (*Thrypticus* sp., *Megamelus scutellaris*); beetles (*Neochetina* sp.); grasshoppers (*Cornops aquaticum*); and Mites (*Orthogaluma terebrantis*) (Admas et al. 2017). A number of biological agents have been applied to Lake Rawapening as a way to control water hyacinth weeds. These biological agents include koan fish (*Ctenopharyngodon idella*), gouramy fish (*Osphronemus goramy*), and beetles *Neochetina bruchi* and *Neochetina eichhorniae*, which are natural predators of water hyacinth (Archambault et al. 2015).

Koan fish stocked in Lake Rawapening (Fig. 9.5) is not a special agent for water hyacinth control (Silva et al. 2014), but is one of the most important cultivated species in the world which is very effective for eliminating water weeds in many countries (Bala Singla et al. 2016). As a fish species known to be greedy, koan fish are used for biological control in the Egyptian Nile River (Sharma et al. 2016) and in Lake Kerinci Indonesia (Samuel et al. 2015). The Central Java government through the Maritime Affairs and Fisheries Service (DKP) is spreading koan fish as an effort to control water hyacinth in Lake Rawapening. Five million fish have been targeted to clear weeds that cover 70% of the surface of Lake Rawapening (Rahardjo et al. 2019). In detail, the koan stocked was 100,000 in 2012 and 2013, then 105,000 in 2014 and 148,300 in 2015. The following year the stocking of koan was stopped due to the enactment of Law Number 23 of 2014 concerning Regional Government (Hidayati et al. 2018).

Koan fish are able to eat water hyacinth up to 1.5 times their body weight (Prasetyo et al. 2022), so they have the potential to control these weeds in the waters. However, efforts to control water hyacinth with small amounts of koan in Lake Rawapening are very ineffective and tend to fail. In addition to the amount that is not proportional to the expanse of water hyacinth, which has a high growth rate, the fish that are stocked are also caught by fishermen. The success of cleaning water hyacinth in Lake Kerinci, which only released 2000 koans for 10 years (Samuel et al. 2015), turned out to be impractical in Lake Rawapening because the program was stopped before reaching the target of releasing five million fish. The use of koan fish in controlling water

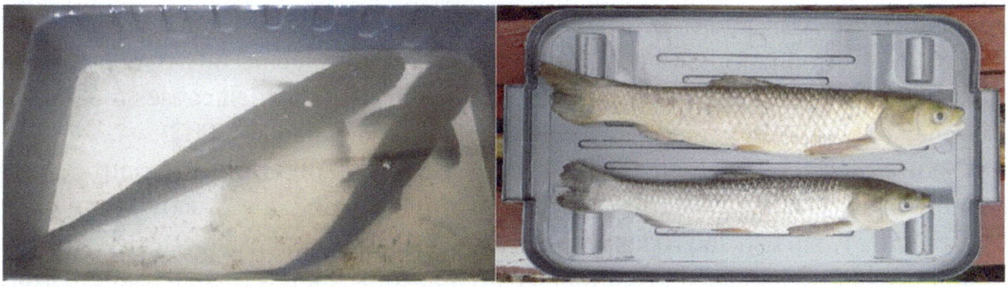

Fig. 9.5 Koan fish species (*Ctenopharyngodon idella* Valenciennes) used to control water hyacinth in Rawapening Lake (Private collection. *Photo credit* BRIN/Syarif Prasetyo)

hyacinth is an ecologically safe way, although it takes a relatively long time and requires supervision to maintain the population as predators in the waters.

Water hyacinth weed control in Lake Rawapening has also utilized *N. bruchi* and *N. eichhorniae* beetles as natural predators (Pratiwi et al. 2018). *N. bruchi* and *N. eichhorniae* are natural predators of water hyacinth (Fig. 9.6), which have specifically succeeded in controlling these weeds (Gichuki et al. 2012; Onyango and Ondeng 2015; Firehun et al. 2015). The beetle was introduced to the states of Louisiana, Texas, and Florida, where it invaded thousands of hectares of water hyacinth weeds in 1980, and the results were able to reduce the weed bed by 33% (Archambault et al. 2015).

The beetles *N. bruchi* and *N. eichhorniae* were chosen as biological agents because they have a short cycle of 90 days (Archambault et al. 2015). *N. bruchi* and *N. eichhorniae* eat different parts of the water hyacinth, so they can coexist without having to compete. Pratiwi et al. (2018) reported that *N. eichhorniae* was found in Lake Rawapening while *N. bruchi* was not found. The same thing was also reported by Buchori et al. (2006): *N. eichhorniae* was widespread in West Java while *N. bruchi* was not found. The rate of damage to water hyacinth leaves due to *N. eichhorniae* attacks was 0.43% per leaf (Pratiwi et al. 2018). The beetles used to control water hyacinth in Lake Rawapening are able to limit the growth of the weeds so that they

Fig. 9.6 *Neochetina bruchi*: biological agents controlling water hyacinth in Lake Rawapening (Private collection. *Photo credit* BRIN/Syarif Prasetyo)

are small in size. In addition, *N. eichhorniae* interferes with vegetative propagation and seed production and becomes a vector for pathogenic microorganisms in water hyacinth. This beetle also eats water hyacinth tissue, which causes the plant's buoyancy to be lost and causes it to sink (Jiménez and Balandra 2007).

The use of several biological agents at the same time in efforts to control water hyacinth really helps the process of physically cleaning the weeds from Lake Rawapening. Koan fish and rodent beetles can be an appropriate and effective collaboration because it has been scientifically proven to be more efficient and environmentally friendly (Hidayati et al. 2018). The existence of biological agents in Lake Rawapening is important to improve planning and proper calculations. Koan fish and beetles that are released are better spayed so that they can minimize the negative impact of their presence on the cleaning water hyacinth in Rawapening Lake.

9.7 Opportunity to Utilize the High Economic Value of Water Hyacinth

Recently, there has been a lot of concern about the harm that water hyacinth poses to a variety of aquatic biota. The explosion of water hyacinth has a negative impact on the health and safety of the aquatic environment, thereby triggering significant potential environmental and ecological risks (Su et al. 2018). Even though it has a negative impact on aquatic ecosystems, water hyacinth can be used for various needs of high economic value (Hossain et al. 2015). Some forms of water hyacinth utilization that can be cultivated by the community include the following:

9.7.1 Animal Feeds

Water hyacinth fiber content has the potential to be an alternative raw material for animal feed for cattle, sheep, rabbits, pigs, chickens, and so on. Water hyacinth is a viable source of animal feed because it contains protein and minerals that can be digested by ruminants (Sharma et al. 2016; Adeyemi and Osubor 2016). Water hyacinth leaves contain valuable nutrients, including 56.38% protein, 33% carbohydrates, and as many as 17 amino acids (Adeyemi and Osubor 2016). High water and mineral content make water hyacinth a useful source of animal feed and fiber (Mahmood et al. 2018; Su et al. 2018). The content of important phytochemicals, including alkaloids and saponins, in water hyacinth is considered to be a good animal feed supplement (Mujere 2016).

The water hyacinth protein content has high digestibility. It has been reported that including water hyacinth as a combination in animal feed can increase cow's milk production by up to 20% (Akankali and Elenwo 2019). Despite its potential as animal feed, water hyacinth contains sharp, needle-shaped oxalic acid and can damage the digestive tract of animals if it is not dissolved by digestive acids (Sharma and Chauhan 2017) Making water hyacinth a raw material for animal feed requires preprocessing to produce feed that is safe for livestock health (de Vasconcelos et al. 2016). Water hyacinth has a great opportunity to meet the demand for raw materials for animal feed, especially in areas experiencing scarcity (Wimalarathne and Perera 2019). In addition, its use as animal feed can be an alternative method for controlling these weeds in Lake Rawapening.

9.7.2 Fish Feed

There is a wide range of eating habits among fish, and some species, such as tilapia and carp, can eat certain aquatic plants. Water hyacinth weed can be used as a raw material for making fish feed (Prasetyo et al. 2021b). The leaves and roots of water hyacinth have a high protein content, which can increase the crude protein content of the feed and possibly help digest the feed better. Feeding goldfish fry with water hyacinth raw material at 40% can replace fish meal. After 70 days of feeding the water hyacinth, the golden fish fry showed a significant increase in weight (Mohapatra and Patra 2013). Incorporating 15% water hyacinth flour into the feed is the best for nourishing fish. The addition of water hyacinth flour to fish feed can change fish nutrition, and besides that, it can reduce the cost of making feed, which has implications for increasing farmer profit margins (Sarker and Aziz 2017).

Koan fish fed with a water hyacinth powder formulation produced higher weight, higher crude protein content, and better digestibility. In addition, this feeding has no impact on histological changes in the liver and kidneys of fish (Mahmood et al. 2018). This condition shows that the selection of water hyacinth for fish feed formulations is very suitable to be considered as one of the efforts to contribute to controlling this weed in Lake Rawapening.

9.7.3 Compost

The most efficient method for producing biological fertilizers made from water hyacinth is through composting (Prasetyo et al. 2021c). The water hyacinth root system, which contains inorganic elements such as nitrogen (N) and phosphorus (P), makes it the right raw material for producing fertilizer. Producing compost made from water hyacinth can help reduce the use of chemical fertilizers so that agricultural production becomes more environmentally friendly. The compost formulation based on 10,000 kg of water hyacinth mixed with poultry manure was reported to have a quality above the quality standard (Montoya et al. 2013). Various nutrient elements in water hyacinth make quality compost and are able to meet the nutritional needs of cultivated plants. Research by Yunindanova et al. (2020) states that water hyacinth compost has a C/N ratio of 10.65, a pH value of 8.08, organic C (18.93%), total N (1.78%), P Total (1.10%), and K Total (1.26%). In addition, water hyacinth compost has a black color and an earthy aroma. Based on its nutrient content, water hyacinth organic fertilizer can be used as a basic fertilizer and planting medium for the cultivation of various plants. The use of organic fertilizers contributes to sustainable agricultural activities (Sönmez et al. 2013; John 2016).

Water hyacinth organic material is useful in the process of improving soil conditions, including its structure and nutrients (Sittadewi 2007). The soil becomes fertile, and the nutrients needed for plant growth and development can be available. Compost made from water hyacinth from Rawapening (Fig. 9.7) can be applied to agricultural areas that have experienced nutrient degradation in the top soil and also damage to soil structure due to saturation during agricultural activities. Water hyacinth compost has also been applied to seed nurseries for horticultural and plantation crops. One of its applications is as a seedling medium for oil palm seeds. In trials conducted by Sitindaon and Tyasmoro (2018), compost made from water hyacinth was proven to have a real effect on the growth process of oil palm seedlings, especially on the length and diameter of the stems and roots. Oil palm seedlings grow better than those sown in medium without water hyacinth compost. Utilization of water hyacinth compost for tomato plants as much as 20 tons/ha can increase growth and yields (Akbar and Amir 2018), making it more financially profitable and more friendly to the environment because fertilizers come from organic materials. The application of water hyacinth compost with a ratio of 80% to lettuce plants has also been shown to have an effect on increasing the density of these vegetables (Prasetyo et al. 2021c).

The nutrient content of water hyacinth-based compost from Lake Rawapening has met the standards required for plant growth. The appearance of black compost with a soft structure and the smell of wet soil make it suitable for packaging and trading. Good packaging and labeling, as well as adding detailed information on nutrient content, will increase the attractiveness of the water hyacinth compost product. The higher the production of compost and the expansion of its marketing reach, the more it will help reduce the density of these weeds in Lake Rawapening. So that the process

Fig. 9.7 Compost is made from water hyacinth (Private collection. *Photo credit* BRIN/Syarif Prasetyo)

of controlling the growth rate and closing the water surface of Lake Rawapening by water hyacinth weeds carried out by the government can be assisted by creative economic activity in composting.

9.7.4 Biofuels

Water hyacinth contains low levels of lignin, so cellulose and hemicellulose can be easily converted into sugar. These sugars can be fermented to produce a lot of biomass for use in the biofuel industry (Wang et al. 2019). Under hydrolysis conditions, processing water hyacinth with alkali can increase the cellulose conversion rate up to 100% (Ruan et al. 2016). Manivannan et al. (2012) reported that the conversion of water hyacinth to bioethanol was carried out in two stages, namely acid hydrolysis and then fermentation with *Candida intermedia*. The process will produce a maximum of 0.21 g/l of bioethanol with a production capacity of 0.01 g/l/h. Bhattacharya et al. (2015) have found a way to produce xylose from water hyacinth and then convert it into xylitol, a poly-alcohol produced from hemicellulose. In addition, the extraction of potassium hydroxide from water hyacinth will produce promising biodiesel on the market because it burns cleanly (Efeovbokhan 2013).

Utilization of water hyacinth for biofuel production is in line with physical control of the weed and is a potential reaction to green advancement techniques (Wang et al. 2019). In addition, the conversion of water hyacinth into biofuel is a strategic step to control water hyacinth from Lake Rawapening while at the same time generating added value economically.

9.7.5 Biogas

Compared to other organic components, water hyacinth has a high concentration of hemicellulose, which will result in the production of biogas that contains carbon dioxide and methane, according to Bote et al. (2020). The basic idea behind producing biogas is to employ the anaerobic (aerobic means without oxygen) decomposition of organic materials to create methane (CH_4) gas. Numerous microorganisms, particularly bacteria that produce methane, aid in anaerobic decomposition. Without the need for oxygen, bacteria break down biodegradable materials in anaerobic digestion (Yonathan et al. 2012; Nugrha et al. 2020). Preparing tools and materials is the process of manufacturing. A biogas burner, a gas generator, a gallon or drum, galvanized pipe, and hose are the required tools and materials. Water, water hyacinth, and cow manure make up the ingredients. Cow dung is served as a biostarter, water is served as a nutrient, and water hyacinth is served as a substrate.

Water hyacinth was first cut into small pieces with a length of less than 1 cm, and then water and cow dung were added based on the dilution variable to produce biogas.

Enter the combined solution into the biodigester after selecting the appropriate dilution variable. The next step is to let it sit for a week so that the gas decomposes. Water hyacinth was first cut into small pieces with a length of less than 1 cm, then water and cow dung were added based on the dilution variable to produce biogas. Enter the combined solution into the biodigester after selecting the appropriate dilution variable. The next step is to let it sit for a week so that the gas decomposes. The resulting gas is then stored in drums or gallons and used to power electric generators and stoves by passing it through a hose. The fire and electricity can keep burning as long as the water hyacinth is decomposing properly. The fire and electricity can keep burning as long as the water hyacinth is decomposing properly.

Yonathan et al. (2012) conducted a study on the production of biogas from water hyacinth using the following ratios: water + cow dung: water hyacinth (2:1, 2:1.5, 2:2, and 2:2.5). The substrate weights for each successive variable are 326.5, 420, 490, and 543.2 g. Variable 2:2.5 has been used to improve the volume produced from variable 2:1. The maximum biogas output for a 2:1 mixture was 1162.97 mL, according to the results for the composition variable, while the minimum was 12.85 mL for a 2:1 ratio. The amount of biogas produced increases with the substrate's composition. The methane content generated was calculated from the findings of the gas chromatography analysis and was found to be 0.03 mol of methane per 100 g of water hyacinth. The analysis's findings suggest that biogas from water hyacinths can be exploited as a renewable energy source because it has been established that the gas contains methane. However, based on the findings of this study, additional research is still required to determine whether or not the methane gas content can still increase with a longer fermentation time and changes in other variables.

Biogas produced from water hyacinth is very environmentally friendly and low cost. Utilizing water hyacinth that grows uncontrollably in Lake Rawapening as biogas material can reduce the density of the weed population. Biogas production made from water hyacinth at low cost is very effective in helping people who have difficulty obtaining gas and electricity to meet domestic needs.

9.7.6 Handicraft

Water hyacinth may be used to create a variety of handicrafts with marketable value (Bruneckienė and Sinkienė 2014; Onyango and Ondeng 2015; Guna et al. 2017; Yan and Guo 2017; Sianturi et al. 2019; Mitan 2019). Water hyacinth leaf stalks are used by the locals in the area of Lake Rawapening to create a variety of inventive goods that are both useful and ecologically friendly (Sudana and Mohamad 2020). Furniture, including chairs and tables, as well as purses, sandals, baskets, and wallets, are among the handcrafted goods made by local businesses and communities near Lake Rawapening.

The drying of wet water hyacinth plants is the first step in the process of producing crafts from water hyacinth raw materials. It may be used in a variety of ways, such as needlework, in addition to being attractive and valuable economically (Putri and

Prasetyaningtyas 2019). The tools required to create this craft are a sickle to harvest the raw materials, a press to level the water hyacinth, a knife or pair of scissors, and a sewing machine to shape and join the supporting materials (cloth or leather). Then, in order to make a lot of sales, these water hyacinth art objects are disseminated either directly or online (using social media).

9.8 Water Hyacinth in Southeast Asian Countries

The water hyacinth (*E. crassipes*) is an invasive aquatic plant that has spread to many parts of the world, including Southeast Asia. The water hyacinth is reported to have arrived in Indonesia in 1894. The plant grew so quickly there that cartloads of it were thrown into the Tji Liwung River, which runs through the Bogor Botanical Garden, in less than a year. From there, it was expected to be transported to Japan in the Meiji era around 1900. It was also reportedly transported from Java to Bangkok in 1901, where it was claimed to have been brought by King Chulalongkorn's consort, who was enthralled with its beauty and filled the ponds in the royal garden (Molle and Keawkulaya 1998; Van Beek 1995). From there, it moved across the Mekong River, the Chao Phraya Delta, and on to Laos, Burma (now known as Myanmar), Cambodia, and Vietnam. Nonetheless, it was already there. It is unclear how it got there in the first place, although worries about canal congestion and mass development were documented as early as 1902 (Pieterse 1997; Teygeler 2000).

Water hyacinth was first reported in Thailand in the 1960s and has since spread to many parts of the country, particularly in the Chao Phraya River Basin and other major waterways. It has been causing significant economic and environmental impacts, including clogging irrigation canals, reducing water flow, and disrupting aquatic ecosystems. After being discovered for the first time in Vietnam in the 1980s, water hyacinth has since spread throughout the nation, especially in the Mekong Delta, which includes Dong Thap Province, Can Tho City, and Ho Chi Minh City. It has been posing issues for public health, water quality, local agriculture, and fisheries. It was also first reported in Cambodia in the 1990s and has since spread to many parts of the country, particularly in the Tonle Sap Lake, the Mekong River, and other major waterways. On the other hand, water hyacinth has expanded across the country since its discovery in Laos in the 2000s, reaching as far as the Nam Ngum Reservoir and the Mekong River. Aquatic habitats, local agriculture, fishing, and public health have all been adversely affected by the plant as well as the water quality. In the Philippines, water hyacinth first appeared in 1912 and has been reported to have spread in several areas, including Laguna Lake, Taal Lake, and the Pasig River (Francisco 2020). Currently, water hyacinth has filled most of the Philippine waters due to its fast-breeding rate (Philippine News Agency 2019). In Laguna Lake alone, approximately 800 ha are covered with water hyacinth (DENR 2021). This has caused problems for local agriculture, fisheries, and public health.

Located in the center of Myanmar there is a natural lake called Lake Inle. It is mainly recognized for its floating gardens and the way men row their boats while standing in front of it with one leg. The indigenous species that zoologists would include on the list are at least nine kinds of fish and over twenty species of snails are unique to this area. However, when it comes to water hyacinths, gardening is the most important thing. Like so many other areas, Lake Inle saw a mass development of them when they were initially introduced. There were concerns that the entire lake would be covered due to choked shorelines, shoreside canals, and river mouths entering the lake. For a while, there was even a municipal regulation that said that any boat with extra room should load up with water hyacinths and discard them on land. On the other hand, the Inthas discovered a more beneficial application for the plant. They either walled off a few-meter-wide section of the lake with bamboo poles and filled it with cut-out squares of grasses and soil from the surrounding marshes, or they rooted macrophytes from the lake's lower depths to build their typical floating gardens. As an alternative, they inserted bamboo poles through it to secure floating plant material that floods tore loose from the inlets and washed out into the lake. After that, they continued to add plant material and occasionally bottom sludge until the mixture of decomposing organic matter floating on the water's surface was one meter thick, with about a quarter of it protruding above the surface. After that, they planted flowers and tomatoes, among other vegetables. The gardens float, rising and falling in conjunction with the water level; they never flood, and they never dry out. Water hyacinths proved to be a great complement to the other plants that made up the gardens. These days, it frequently makes up most of the building supplies (Jernelöv 2017).

This aquatic plant has been found in several areas, including reservoirs, rivers, and canals. It is particularly common in Malaysia (Selangor, Pahang, Perak, Johor, and Kelantan), Singapore (Singapore River and Kallang River), Brunei (Brunei River), and Indonesia (Lake Toba, Lake Matano, Citarum River, and other major waterways).

In Southeast Asian nations, water hyacinth can have a major effect on human health, economics, and ecosystems. Among the effects are clogged irrigation canals and waterways, decreased water flow and increased flooding, disturbance of aquatic ecosystems and altered biodiversity, decreased water quality, and elevated risk of waterborne illnesses, negative effects on local agriculture and fisheries, and financial losses for tourism and recreation-related industries. To control the spread of water hyacinth in Southeast Asian countries, various measures have been implemented or proposed. Some of these measures include biological control methods, such as introducing natural predators or competitors; chemical control methods, such as herbicides or algaecides; and physical control methods, such as hand-picking or mechanical removal. Preventing its spread through ballast water management and inspecting boat hulls. Raising awareness among local communities about the risks associated with water hyacinth. It is worth noting that a combination of these control measures may be more effective than relying on a single approach. Additionally, early detection and rapid response are critical to preventing the spread of water hyacinth and minimizing its impacts on ecosystems and human communities.

9.9 Future Studies, Prospects, and Remarks

The ability of water hyacinth weeds to invade all types of water bodies in various geographic areas is a serious cause for concern. It is very important to carry out molecular analysis to determine the genetic similarity of water hyacinths that grow in various aquatic habitats and different geographical zones. Toxicological studies need to be examined in more depth, considering the use of plants as animal feed, to ensure that these plants are safe to use as animal feed.

The invasive ability of water hyacinth is a threat to the aquatic environment, but this weed also has prospects of high economic value, so its existence must still be managed. It would be fairer not to generalize water hyacinth as a threatening weed but to look at the peculiarities of each situation and condition where water hyacinth grows. In economically and ecologically favorable situations, the presence of water hyacinth can optimize its growth rate. If in the opposite (adverse) situation, then the application of the water hyacinth control method that is most effective, efficient, cost-effective, and environmentally friendly should be considered.

References

Adeyemi O, Osubor CC (2016) Assessment of nutritional quality of water hyacinth leaf protein concentrate. Egypt J Aquat Res 42(3):269–272. https://doi.org/10.1016/j.ejar.2016.08.002

Admas A, Samuael S, Belete E, Akliu A, Alebachew M (2017) Controlling water hyacinth in lake tana using biological method at green house and pond level. Eur J Exp Biol 07(01):1–5. https://doi.org/10.21767/2248-9215.100001

Ajithram A, Jappes JTW, Brintha NC (2021) Water hyacinth (*Eichhornia crassipes*) natural composite extraction methods and properties—a review. Mater Today Proc 45:1626–1632. https://doi.org/10.1016/J.MATPR.2020.08.472

Akankali J, Elenwo EI (2019) Use of water hyacinth as feed stuff for animals in Niger Delta, Nigeria. ASJ Int J Adv Sci Res Rev 4(01):91–97

Akbar Y, Amir Y (2018) Application of several doses of water hyacinth compost and its effect on growth and yield of tomato (*Solanum lycopersicum* L.). J Pertanian UMSB 2(1):67–73

Aloo PA, Ojwang WO, Omondi R, Njiru JM, Oyugi DO (2013) A review of the impacts of invasive aquatic weeds on the biodiversity of some tropical water bodies with special reference to Lake Victoria (Kenya)

Al-Wagga A, Al-Gburi O (2018) Using different methods of adding herbicides in controlling water hyacinth and reduce in water environment pollution

Archambault JM, Bergeron CM, Cope WG, Richardson RJ, Heilman MA, Corey JE, Netherland MD, Heise RJ (2015) Sensitivity of freshwater molluscs to hydrilla-targeting herbicides: providing context for invasive aquatic weed control in diverse ecosystems. J Freshw Ecol 30(3):335–348. https://doi.org/10.1080/02705060.2014.945104

Asmare E (2017) Current trend of water hyacinth expansion and its consequence on the fisheries around North Eastern Part of Lake Tana, Ethiopia. J Biodivers Endangered Species 05(02). https://doi.org/10.4172/2332-2543.1000189

Ayanda OI, Ajayi T, Asuwaju FP (2020) *Eichhornia crassipes* (Mart.) Solms: uses, challenges, threats, and prospects. Sci World J

Bakrim WB, Ezzariai A, Karouach F, Sobeh M, Kibret M, Hafidi M, Kouisni L, Yasri A (2022) *Eichhornia crassipes* (Mart.) Solms: a comprehensive review of its chemical composition, traditional use, and value-added products. Front Pharmacol 13

Bala Singla D, Renu Bala Singla C, Bala Singla R (2016) Morphometric alterations in exotic freshwater fish, *Ctenopharyngodon idella* (Cuv. and Val.) upon exposure to endosulfan. Int J Fish Aquatic Stud 4(3):590–593

Bhattacharya A, Haldar S, Chatterjee PK (2015) Geographical distribution and physiology of water hyacinth (*Eichhornia crassipses*)—the invasive hydrophyte and a biomass for producing xylitol. Int J Chemtech Res 7:1849–1861

Bordoloi S, Garg A, Sreedeep S, Lin P, Mei G (2018) Investigation of cracking and water availability of soil-biochar composite synthesized from invasive weed water hyacinth. Bioresour Technol 263:665–677. https://doi.org/10.1016/j.biortech.2018.05.011

Bote MA, Naik VR, Jagdeeshgouda KB (2020) Production of biogas with aquatic weed water hyacinth and development of briquette making machine. Mater Sci Energy Technol 3:64–71. https://doi.org/10.1016/J.MSET.2019.09.001

Brown AE, Adams JMM, Grasham OR, Camargo-Valero MA, Ross AB (2020) An assessment of different integration strategies of hydrothermal carbonisation and anaerobic digestion of water hyacinth. Energies (Basel) 13(22). https://doi.org/10.3390/en13225983

Bruneckienė J, Sinkienė J (2014) Critical analysis of approaches to smart economy

Buchori D, Kartosuwondo U, Tjitrosemito Dan Sahari SB, Kamper J, Darmaga K (2006) Persebaran Agens Hayati *Neochetina* spp. (Coleoptera: Curculionidae) Di Jawa Barat Dan DKI Jakarta

Carnaje NP, Talagon RB, Peralta JP, Shah K, Paz-Ferreiro J (2018) Development and characterisation of charcoal briquettes from water hyacinth (*Eichhornia crassipes*)-molasses blend. PLoS One 13(11). https://doi.org/10.1371/journal.pone.0207135

Carvalho LB de, Cerveira Junior WR (2019) Control of water hyacinth: a short review. Commun Plant Sci 9(1). https://doi.org/10.26814/cps2019021

Chander S, Pompapathi V, Gujrati A, Singh RP, Chaplot N, Patel UD (2018) Growth of invasive aquatic macrophytes over Tapi river. In: International archives of the photogrammetry, remote sensing and spatial information sciences—ISPRS archives. International Society for Photogrammetry and Remote Sensing, pp 829–833

Coniwanti P, Novalina S, Putri K (2009) Effect of ethanol solution concentration, temperature, and cooking time on each heater pulp through the organosolv process

Dechassa N, Abate B (2020) Current status of water hyacinth (*Eichhornia crassipes*) in Ethiopia: achievements, challenges and prospects: a review. J Environ Earth Sci 10(12):36–47. https://doi.org/10.7176/jees/10-12-03

Degaga AH (2019) Water hyacinth (*Eichhornia crassipes*) biology and its impacts on ecosystem, biodiversity, economy and human well-being. J Nat Sci Res 9(12):24–30. https://doi.org/10.7176/JNSR

De Groote H, Ajuonu O, Attignon S, Djessou R, Neuenschwander P (2003) Economic impact of biological control of water hyacinth in Southern Benin. Ecol Econ 45:105–117

Department of Environment and Natural Resources (DENR) (2021) Commits anew to partner with fisherfolk, clear out water hyacinth in Laguna Lake. https://denr.gov.ph/news-events/denr-commits-anew-to-partner-with-fisherfolk-clear-out-water-hyacinth-in-laguna-lake/

Dersseh MG, Melesse AM, Tilahun SA, Abate M, Dagnew DC (2019) Water hyacinth: review of its impacts on hydrology and ecosystem services—lessons for management of Lake Tana. In: Extreme hydrology and climate variability: monitoring, modelling, adaptation and mitigation. Elsevier, Amsterdam, pp 237–251

de Vasconcelos GA, Véras RML, de Lima SJ, Cardoso DB, de Castro SP, de Morais NNG, Souza AC (2016) Effect of water hyacinth (*Eichhornia crassipes*) hay inclusion in the diets of sheep. Trop Anim Health Prod 48(3):539–544. https://doi.org/10.1007/s11250-015-0988-z

Efeovbokhan VE (2013) Investigating bio-diesel production using potash from agricultural wastes. CORE Metadata, citation and similar papers at core.ac.uk provided by Covenant University Repository

El-Shahawy TAEG (2015) Chemicals with a natural reference for controlling water hyacinth, *Eichhornia crassipes* (Mart.) Solms. J Plant Prot Res 55(3):294–300. https://doi.org/10.1515/jppr-2015-0041

Endgaw F (2021) Socio-ecological impacts of water hyacinth, *Eichhornia crassipes* (MART.) in Lake Tana, Gulf of Gorgora, Ethiopia. J Appl Sci Environ Manag 24(12):2017–2025. https://doi.org/10.4314/jasem.v24i12.2

Enyew BG, Assefa WW, Gezie A (2020) Socioeconomic effects of water hyacinth (*Echhornia crassipes*) in Lake Tana, North Western Ethiopia. PLoS One 15(9). https://doi.org/10.1371/journal.pone.0237668

Firehun Y, Struik PC, Lantinga EA, Taye T (2015) Adaptability of two weevils (*Neochetina bruchi* and *Neochetina eichhorniae*) with potential to control water hyacinth in the Rift Valley of Ethiopia. Crop Prot 76:75–82. https://doi.org/10.1016/J.CROPRO.2015.06.015

Francisco M (2020) Why is the water *Hyacinth* a notorious aquatic pest? https://www.flipscience.ph/plants-and-animals/water-hyacinth/. Accessed June 2023

Gaikwad RP, Gavande S (2017) Major factors contributing growth of water hyacinth in natural water bodies. Int J Eng Res 6:304

Garlich N, Guarnieri CCO, Freitas RLG, Cervoni JHC, Cruz C, Ferreira MC (2019) Efficacy of imazamox with centrifugal energy spray nozzle on eichhornia crassipes and economic analysis of control viability. Planta Daninha 37. https://doi.org/10.1590/S0100-83582019370100102

Gaurav GK, Mehmood T, Cheng L, Klemeš JJ, Shrivastava DK (2020) Water hyacinth as a biomass: a review. J Clean Prod 277. https://doi.org/10.1016/j.jclepro.2020.122214

Genissa N, Sidik Boedoyo M, Yoesgiantoro D (2018) Analysis of the costs and benefits of water hyacinth wood pellets as alternative energy to increase energy security

Gichuki J, Omondi R, Boera P, Okorut T, Matano AS, Jembe T, Ofulla A (2012) Water hyacinth *Eichhornia crassipes* (Mart.) Solms-Laubach dynamics and succession in the Nyanza Gulf of Lake Victoria (East Africa): implications for water quality and biodiversity conservation. Sci World J. https://doi.org/10.1100/2012/106429

Goshu G, Aynalem S (2017) Problem overview of the Lake Tana Basin. In: Stave K, Goshu G, Aynalem S (eds) Social and ecological system dynamics: characteristics, trends, and integration in the Lake Tana Basin, Ethiopia. Springer International Publishing, Cham, pp 9–23

Grasset C, Abril G, Guillard L, Delolme C, Bornette G (2016) Carbon emission along a eutrophication gradient in temperate riverine wetlands: effect of primary productivity and plant community composition. Freshw Biol 61(9):1405–1420. https://doi.org/10.1111/fwb.12780

Guerra L, Gomes J, Puna J, Rodrigues J (2015) Preliminary study of synthesis gas production from water electrolysis, using the ELECTROFUEL® concept. Energy 89:1050–1056. https://doi.org/10.1016/j.energy.2015.06.048

Guignard MS, Leitch AR, Acquisti C, Eizaguirre C, Elser JJ, Hessen DO, Jeyasingh PD, Neiman M, Richardson AE, Soltis PS, Soltis DE, Stevens CJ, Trimmer M, Weider LJ, Woodward G, Leitch IJ (2017) Impacts of nitrogen and phosphorus: from genomes to natural ecosystems and agriculture. Front Ecol Evol 5

Guna V, Ilangovan M, Prasad M, Reddy N (2017) Water hyacinth: a unique source for sustainable materials and products. ACS Sustain Chem Eng 5:4478–4490

Gupta AK, Yadav D (2020) Biological control of water hyacinth. Environ Contam Rev 3(1):37–39. https://doi.org/10.26480/ecr.01.2020.37.39

Harun I, Pushiri H, Amirul-Aiman AJ, Zulkeflee Z (2021) Invasive water hyacinth: ecology, impacts and prospects for the rural economy. Plants 10

Hasibuan AA, Yuniati R, Wardhana W (2020) The growth rate and chlorophyll content of water hyacinth under different type of water sources. IOP Conf Ser Mater Sci Eng (IOP Publishing Ltd.)

Heard TA, Winterton SL (2000) Interactions between nutrient status and weevil herbivory in the biological control of water hyacinth. J Appl Ecol 37(1):117–127. https://doi.org/10.1046/j.1365-2664.2000.00480.x

Hidayati N, Soeprobowati TR, Helmi M (2018) The evaluation of water hyacinth (*Eichhornia crassiper*) control program in Rawapening Lake, Central Java Indonesia. IOP Conf Ser Earth Environ Sci (Institute of Physics Publishing)

Honlah E, Yao Segbefia A, Odame Appiah D, Mensah M, Atakora PO (2019) Effects of water hyacinth invasion on the health of the communities, and the education of children along River Tano and Abby-Tano Lagoon in Ghana. Cogent Soc Sci 5(1). https://doi.org/10.1080/23311886.2019.1619652

Hossain ME, Sikder H, Kabir MH, Sarma SM (2015) Nutritive value of water *Hyacinth* (*Eichhornia crassipes*). Online J Anim Feed Res 5(2):40–44

Ilo OP, Simatele MD, Nkomo SL, Mkhize NM, Prabhu NG (2020) The benefits of water hyacinth (*Eichhornia crassipes*) for Southern Africa: a review. Sustainability (Switz) 12:1–20

Izzati M, Soeprobowati TR, Prasetyo S (2022) Characterization of three selected macrophytes—an ecological engineering approach for effective rehabilitation of Rawapening Lake. J Ecol Eng 23(9):277–287. https://doi.org/10.12911/22998993/152047

Jernelöv A (2017) The long-term fate of invasive species: aliens forever or integrated immigrants with time? Springer. https://doi.org/10.1007/978-3-319-55396-2

Jiménez MMartínez, Balandra MAG (2007) Integrated control of *Eichhornia crassipes* by using insects and plant pathogens in Mexico. Crop Prot 26(8):1234–1238. https://doi.org/10.1016/j.cropro.2006.10.028

John M (2016) Production of organic compost from water hyacinth. Res Rev J Agric Allied Sci 5(2):50–57

Kamau AN, Njogu P, Kinyua R, Sessay M (2015) Sustainability challenges and opportunities of generating biogas from water. African Centre for Technology Studies (ACTS) Responsible Natural Resource Economy Series no 005/2015

Koutika LS, Rainey HJ (2015) A review of the invasive, biological and beneficial characteristics of aquatic species *Eichhornia crassipes* and *Salvinia molesta*. Appl Ecol Environ Res 13(1):263–275. https://doi.org/10.15666/aeer/1301_263275

Kriticos DJ, Brunel S (2016) Assessing and managing the current and future pest risk from water hyacinth, (*Eichhornia crassipes*), an invasive aquatic plant threatening the environment and water security. PLoS One 11(8). https://doi.org/10.1371/journal.pone.0120054

Laitinen J, Moliis K, Surakka M (2017) Resource efficient wastewater treatment in a developing area—climate change impacts and economic feasibility. Ecol Eng 103:217–225. https://doi.org/10.1016/J.ECOLENG.2017.04.017

Lu J, Wu J, Fu Z, Zhu L (2007) Water hyacinth in China: a sustainability science-based management framework. Environ Manag 40:823–830

Luthfi W (2021) Data Danau Diperbarui, Kini Ada 5.807 Danau yang Tersebar di Indonesia

Madian HR, Sidkey NM, Abo Elsoud MM, Hamouda HI, Elazzazy AM (2019) Bioethanol production from water hyacinth hydrolysate by *Candida tropicalis* Y-26. Arab J Sci Eng 44(1):33–41. https://doi.org/10.1007/s13369-018-3247-6

Mahmood S, Khan N, Iqbal KJ, Ashraf M, Khalique A (2018) Evaluation of water hyacinth (*Eichhornia crassipes*) supplemented diets on the growth, digestibility and histology of grass carp (*Ctenopharyngodon idella*) fingerlings. J Appl Anim Res 46(1):24–28. https://doi.org/10.1080/09712119.2016.1256291

Manivannan A, Jayarani PH, Narendhirakannan RT (2012) Enhanced acid hydrolysis for bioethanol production from water hyacinth (*Eichhornia crassipes*) using fermentating yeast *Candida intermedia* NRRL Y-981. J Sci Ind Res (India) 71:51–56

Mardiatno D, Faridah F, Listyaningrum N, Hastari NRF, Rhosadi I, da Costa ADS, Rahmadana ADW, Lisan ARK, Sunarno S, Setiawan MA (2023) A holistic review of lake Rawapening management practices, Indonesia: pillar-based and object-based management. Water (Switz) 15

Mathur P, Mathur SM (2018) Water hyacinth: a useful plant to improve rural economy. In: Singh VP, Yadav S, Yadava RN (eds) Energy and environment. Springer, Singapore, pp 31–38

McBeath AV, Smernik RJ, Krull ES, Lehmann J (2014) The influence of feedstock and production temperature on biochar carbon chemistry: a solid-state 13C NMR study. Biomass Bioenergy 60:121–129. https://doi.org/10.1016/J.BIOMBIOE.2013.11.002

Mengist Y, Moges Y (2019) Distribution, impacts and management option for water hyacinth (*Eichhnornia crassipes* [Mart.] Solms) in Ethiopia: a review. J Adv Agric 10:1764–1771. https://doi.org/10.24297/jaa.v10i0.8308

Mitan NMM (2019) Water hyacinth: potential and threat. Mater Today Proc 19:1408–1412. https://doi.org/10.1016/J.MATPR.2019.11.160

Mohapatra SB, Patra AK (2013) Utilization of water hyacinth (*Eichhornia crasipes*) meal as partial replacement for fish meal on the growth performance of *Cyprinus carpio* fry

Molle F, Keawkulaya J (1998) Water management and agricultural change: a case study in the Upper Chao Phraya Delta. Southeast Asian Stud 36(1):32–58

Montoya JE, Waliczek TM, Abbott ML (2013) Large scale composting as a means of managing water hyacinth (*Eichhornia crassipes*). Invasive Plant Sci Manag 6(2):243–249. https://doi.org/10.1614/ipsm-d-12-00013.1

Mujere N (2016) Water hyacinth: characteristics, problems, control options, and beneficial uses

Murugesan AG (2014) Conservation of fresh water bodies by biological control of the aquatic weed water hyacinth. Hydrol Curr Ress 05(04):157–157. https://doi.org/10.4172/2157-7587.s1.015

Ndimele PE, Kumolu-Johnson CA, Anetekhai M (2011) The invasive aquatic macrophyte, water hyacinth {*Eichhornia crassipes* (Mart.) Solm-Laubach: Pontedericeae}: problems and prospects. Res J Environ Sci 5(6):509–520

Nega DT, Ramayya AV, Manenti F, Amaral AF (2021) Turning curse into cure: potential of water hyacinth for bio-refining—a contextual investigation of Lake Tana. Environ Challenges 5. https://doi.org/10.1016/j.envc.2021.100387

Nugrha WD, Syafrudin, Pradita LL (2020) Biogas production from water hyacinth. In: Biogas production from water hyacinth, pp 1–17

Odame Appiah D, Yao Segbefia A, Honlah E (2019) Coping strategies to water hyacinth invasion among Riparian communities in Ghana

Ogamba EN, Izah SC, Oribu T (2015) Water quality and proximate analysis of *Eichhornia crassipes* from River Nun, Amassoma Axis, Nigeria. Res J Phytomed 1(1):43–48

Onyango JP, Ondeng MA (2015) The contribution of the multiple usage of water hyacinth on the economic development of Reparian Communities in Dunga and Kichinjio of Kisumu Central Sub County, Kenya

Patel S (2012) Threats, management and envisaged utilizations of aquatic weed *Eichhornia crassipes*: an overview. Rev Environ Sci Biotechnol 11:249–259

Philippine News Agency (2019) More water *Hyacinth* invasion looms over Laguna Lake. https://www.pna.gov.ph/articles/1074643

Pieterse AH (1997) Water *Hyacinth*: the world's most invasive aquatic plant. In: The ecology and management of aquatic-weeds. Oxford University Press, pp 27–32

Piranti AS, Rahayu DRUS, Waluyo G (2018) Nutrient limiting factor for enabling algae growth of Rawapening Lake, Indonesia. Biosaintifika J Biol Biol Educ 10(1):101–108. https://doi.org/10.15294/biosaintifika.v10i1.12500

Plaza MG, Pevida C, Martín CF, Fermoso J, Pis JJ, Rubiera F (2010) Developing almond shell-derived activated carbons as CO_2 adsorbents. Sep Purif Technol 71(1):102–106. https://doi.org/10.1016/j.seppur.2009.11.008

Prasetyo S, Anggoro S, Soeprobowati TR (2021a) The growth rate of water hyacinth (*Eichhornia crassipes* (Mart.) Solms) in Rawapening Lake, Central Java. J Ecol Eng 22(6). https://doi.org/10.12911/22998993/137678

Prasetyo S, Anggoro S, Soeprobowati TR (2021b) Potential of water hyacinth (*Eichhornia crassipes* (Mart.) Solms) in Rawapening lake as raw material for fish feed. J Phys Conf Ser

Prasetyo S, Anggoro S, Soeprobowati TR (2021c) Reducing the density of water hyacinth (*Eichhornia crassipes* (Mart.) Solm) in Rawapening lake by using it as a compost base material. Bioma Berkala Ilmiah Biologi 23(1):57–62. https://doi.org/10.14710/bioma.23.1.57-62

Prasetyo S, Anggoro S, Soeprobowati TR (2022) Water hyacinth *Eichhornia crassipes* (Mart.) Solms management in Rawapening Lake, Central Java. AACL Bioflux 15(1):532–543

Pratiwi FD, Zainuri M, Purnomo PW, Purwanti F (2018) Evaluation of weevils (*Neochetina* spp.) existence in Rawa Pening Lake (abundance and impact on target and non-target plant)

Premalatha R, Parameswari E, Malarvizhi P, Avudainayagam S, Davamani V (2018) Sequestration of hexavalent chromium from aqueous medium using biochar prepared from water hyacinth biomass. Chem Sci Int J 22(3):1–15. https://doi.org/10.9734/csji/2018/40547

Priya ES, Selvan PS (2017) Water hyacinth (*Eichhornia crassipes*)—an efficient and economic adsorbent for textile effluent treatment—a review. Arab J Chem 10:S3548–S3558. https://doi.org/10.1016/j.arabjc.2014.03.002

Proctor MD (2016) Weeds of North America. Richard Dickinson and France Royer, editors. 2014 The University of Chicago Press, Chicago, Illinois, USA. 797 pp. $35.00 paperback. ISBN: 978-0-226-07644-7. J Wildl Manag 80(5):949. https://doi.org/10.1002/jwmg.1061

Putri D, Prasetyaningtyas W (2019) Feasibility study of water hyacinth laptop bag with ribbon embroidery decoration 7(2):119–125

Rahardjo SAS, Faiqotul F, AndyS C (2019) Germadan Rawa Pening: collective action in managing common pool resources. J Penelitian Pengelolaan Daerah Aliran Sungai (J Watershed Manag Res) 3(1):1–12

Ray P, Hill MP (2013) Microbial agents for control of aquatic weeds and their role in integrated management. CAB Rev Perspect Agric Vet Sci Nutr Nat Resour 8

Roopnarain A, Nkuna R, Ndaba B, Adeleke R (2019) New insights into the metagenomic link between pre-treatment method, addition of an inoculum and biomethane yield during anaerobic digestion of water hyacinth (*Eichhornia crassipes*). J Chem Technol Biotechnol 94(10):3217–3226. https://doi.org/10.1002/jctb.6129

Ruan T, Zeng R, Yin X-Y, Zhang S-X, Yang Z-H (2016) Water hyacinth as biofuel feedstock

Salas-Ruiz A, Barbero-Barrera M del M, Ruiz-Téllez T (2019) Microstructural and thermo-physical characterization of a water hyacinth petiole for thermal insulation particle board manufacture. Materials 12(2). https://doi.org/10.3390/ma12040560

Samuel S, Suryati NK, Adiansyah V (2015) Limnological condition and estimation of potential fish production of Kerinci Lake Jambi, Sumatra. Indones Fish Res J 21(1):9. https://doi.org/10.15578/ifrj.21.1.2015.9-18

Saratale RG, Cho SK, Ghodake GS, Shin HS, Saratale GD, Park Y, Lee HS, Bharagava RN, Kim DS (2020) Utilization of noxious weed water hyacinth biomass as a potential feedstock for biopolymers production: a novel approach. Polymers (Basel) 12(8). https://doi.org/10.3390/POLYM12081704

Sarker MdA-A, Aziz I (2017) Incorporation of water hyacinth (*Eichhornia crassipes*) in feed for developing eco-friendly low-cost feed of mirror carp, *Cyprinus carpio* var. *specularis* (Linnaeus, 1758). J Agroecol Nat Resour Manag 4(1):5–9

Segbefia AY, Honlah E, Appiah DO (2019) Effects of water hyacinth invasion on sustainability of fishing livelihoods along the River Tano and Abby-Tano Lagoon, Ghana. Cogent Food Agric 5(1). https://doi.org/10.1080/23311932.2019.1654649

Setyaningsih LWN, Rizkiyaningrum UM, Andi R (2017) Pengaruh konsentrasi katalis dan reusability katalis pada sintesis triasetin dengan katalisator lewatit. Teknoin 23(1). https://doi.org/10.20885/teknoin.vol23.iss1.art7

Setyanto K, Warniningsih (2011) Utilization of water hyacinth to clean the water quality of the Gajahwong River, Yogyakarta. J Teknol Technosci 4(1):17–22

Sharma H, Chauhan MS (2017) Water hyacinth: control and utilization. Int J Civ Eng Technol (IJCIET) 8(5):765–772

Sharma A, Aggarwal NK, Saini A, Yadav A (2016) Beyond biocontrol: water hyacinth-opportunities and challenges. J Environ Sci Technol 9(1):26–48. https://doi.org/10.3923/jest.2016.26.48

Shen J, Huang G, An C, Song P, Xin X, Yao Y, Zheng R (2018) Biophysiological and factorial analyses in the treatment of rural domestic wastewater using multi-soil-layering systems. J Environ Manag 226:83–94. https://doi.org/10.1016/j.jenvman.2018.08.001

Sianturi OO, Tyas WP, Sunarti, Manullang OR, Manaf A (2019) The benefit of internet using to affect income for water hyacinth home-based entrepreneurs in Rawapening area-Indonesia. IOP Conf Ser Earth Environ Sci (Institute of Physics Publishing)

Silva AF, Cruz C, Pitelli RLCM,, Pitelli RA (2014) Use of grass carp (*Ctenopharyngodon idella*) as a biological control agent for submerged aquatic macrophytes a 1 Utilização da Carpa-capim (*Ctenopharyngodon idella*) como Agente de Controle Biológico de Macrófitas Aquáticas Submersas

Sitindaon PT, Tyasmoro SY (2018) The influence of compost water hyacinth on the growth of seeds on pre-nursery of oil palm. J Produksi Ternak 6(10):2713–2721

Sittadewi EH (2007) Processing of organic water hyacinth materials into growing media to support organic farming. J Teknik Lingkungan 8(3):229–234

Sivasankari B, David Ravindran RA (2016) A study on chemical analysis of water hyacinth (*Eichornia crassipes*), water lettuce (*Pistia stratiotes*). Int J Innov Res Sci Eng Technol 5(10):17566–17570. https://doi.org/10.15680/ijirset.2016.0510010

Soeprobowati TR (2017) Lake management: lesson learn from Rawapening lake. Adv Sci Lett 23(7):6495–6497. https://doi.org/10.1166/asl.2017.9664

Sönmez F, Çığ A, Gülser F, Başdoğan G (2013) The effects of some organic fertilizers on nutrient contents in hybrid gladiolus. Eurasian J Soil Sci 2:140–144

Souza ELC, Filho JT, Velini ED, Silva JRM, Tonello KC, Foloni LL, Barbosa AC, Freato TA (2020) Water hyacinth control by glyphosate herbicide and its impact on water quality. J Water Resour Prot 12(01):60–73. https://doi.org/10.4236/jwarp.2020.121004

Stave K, Goshu G, Aynalem S (2017) Social and ecological system dynamics characteristics, trends, and integration in the Lake Tana Basin, Ethiopia. Springer, Switzerland

Su W, Sun Q, Xia M, Wen Z, Yao Z (2018) The resource utilization of water hyacinth (*Eichhornia crassipes* [Mart.] Solms) and its challenges. Resources 7

Sudana IW, Mohamad I (2020) Characteristics of water hyacinth crafts Gorontalo. Dewa Ruci J Pengkajian dan Penciptaan Seni 15(1):38–47. https://doi.org/10.33153/dewaruci.v15i1.3171

Sudarmadji, Pudjiastuti H (2018) Impacts of agricultural practices and tourism activities on the sustainability of Telaga Warna and Telaga Pengilon lakes, Dieng Plateau, Central Java. In: E3S web of conferences. EDP Sciences

Sukarni S, Zakaria Y, Sumarli S, Wulandari R, Ayu Permanasari A, Suhermanto M (2019) Physical and chemical properties of water hyacinth (*Eichhornia crassipes*) as a sustainable biofuel feedstock. IOP Conf Ser Mater Sci Eng (Institute of Physics Publishing)

Sutadian AD (2017) Development of a cost-effective river water quality index: a case study of West Java province, Indonesia. VICTORIA

Tewabe D (2015) Preliminary survey of water hyacinth in lake. Glob J Allergy 013–018. https://doi.org/10.17352/2455-8141.000003

Teygeler R (2000) Water *Hyacinth* paper: contribution to a sustainable future. In: Paper and water: a guide for conservators. Gentenaar & Torley Publishers, pp 168–188

Thomas TH, Eden RD, Sayigh AAM (1990) Water hyacinth—a major neglected resource

Tipping PW, Gettys LA, Minteer CR, Foley JR, Sardes SN (2017) Herbivory by biological control agents improves herbicidal control of waterhyacinth (*Eichhornia crassipes*). Invasive Plant Sci Manag 10(3):271–276. https://doi.org/10.1017/inp.2017.30

Tobias VD, Conrad JL, Mahardja B, Khanna S (2019) Impacts of water hyacinth treatment on water quality in a tidal estuarine environment. Biol Invasions 21(12):3479–3490. https://doi.org/10.1007/s10530-019-02061-2

Van Beek S (1995) The Chao Phya: river in transition. Oxford University Press

Van Oijstaeijen W, Van Passel S, Cools J, Janssens de Bisthoven L, Hugé J, Berihun D, Ejigu N, Nyssen J (2020) Farmers' preferences towards water hyacinth control: a contingent valuation study. J Great Lakes Res 46(5):1459–1468. https://doi.org/10.1016/J.JGLR.2020.06.009

Verma R, Sivappa (2017) Calculating growth rate of water hyacinth pollution. Int J Eng Tech Res (IJETR) 7(7):85–90

Villamagna AM, Murphy BR (2010) Ecological and socio-economic impacts of invasive water hyacinth (*Eichhornia crassipes*): a review. Freshw Biol 55:282–298
Waithaka E (2013) Impacts of water Hyacinth (*Eichhornia crassipes*) on the fishing communities of Lake Naivasha, Kenya. J Biodivers Endangered Species 1(2):1–4
Wang Z, Zheng F, Xue S (2019) The economic feasibility of the valorization of water hyacinth for bioethanol production. Sustainability (Switz) 11(3). https://doi.org/10.3390/su11030905
Wimalarathne HDA, Perera PCD (2019) Potentials of water hyacinth as livestock feed in Sri Lanka. Indian J Weed Sci 51(2):101. https://doi.org/10.5958/0974-8164.2019.00024.8
Witte F, Silsbe GM, Hecky RE, Goudswaard PC, Guildford SJ, Kishe-Machumu MA, Wanink JH (2012) Did the loss of phytoplanktivorous fish contribute to algal blooms in the Mwanza Gulf of Lake Victoria? Hydrobiologia 679(1):283–296. https://doi.org/10.1007/s10750-011-0893-z
Xu D, Qiu X, Xu Z (2017) Effect of water hyacinth mulch on soil temperature, water content and maize yield (*Zea mays* L.) in Southeast China. Environ Eng Manag J (EEMJ) 16(1)
Yan SH, Guo Y (2017) Water Hyacinth environmental challenges, management and utilization
Yonathan A, Prasetya AR, Pramudono B (2012) Biogas production from water hyacinth (*Eichhornia crassipes*): consistency and pH study of the produced biogas. J Teknologi Kimia dan Industri 1(1):412–416
Yunindanova MB, Supriyono S, Hertanto BS (2020) Processing of invasive water hyacinth weeds into marketable organic fertilizers as a solution to the Lake Rawapening problem. PRIMA J Commun Empowering Serv 4(2):78. https://doi.org/10.20961/prima.v4i2.42053

Part III
Response, Approaches and Strategies

Chapter 10
Applied Technologies

Cynthia Henny, Irma Melati, Surono, Awalina Satya, Tjandra Chrismadha, and Miratul Maghfiroh

In the following section, special topics are chosen to describe the applied technologies that help to reduce environmental contamination and restore aquatic ecosystems. Constructed wetlands, natural materials, and bioremediation i.e., mycoremediation are the select topics discussed along with the research and/or studies, especially in tropical region.

10.1 Constructed Wetlands (CWs)

Constructed wetlands (CWs) for wastewater treatment are widely recognized as proven ecotechnologies that have become viable options for wastewater treatment and freshwater ecosystem protection and restoration (Henny et al. 2019a, b; Henny and Kurniawan 2020; Vymaza 2022). CWs resemble natural wetland systems in design and construction through the optimization of concurrent physical, chemical, and biological processes under integrated conditions. This enables the effective transformation and removal of pollutants from wastewaters and contaminated water bodies. CWs utilize the natural functions of aquatic plants, soil, and microorganisms to remove pollutants. As an alternative to conventional wastewater treatment systems, CWs have gained global acceptance due to their advantages of environmentally sustainable, low-cost and easy operation. By utilizing passive water flow and relying on the gravity system, CWs operate without consuming energy; therefore,

C. Henny (✉) · I. Melati · A. Satya · T. Chrismadha · M. Maghfiroh
Research Center for Limnology and Water Resources, National Research and Innovation Agency (BRIN), Bogor, Indonesia
e-mail: cynt001@brin.go.id

Surono
Research Center for Applied Microbiology, National Research and Innovation Agency (BRIN), Bogor, Indonesia

© The Author(s), under exclusive license to Springer Nature Singapore Pte Ltd. 2025
A. C. Santos-Borja et al. (eds.), *Progress on Ecosystem Restoration of Tropical Inland Waters*, https://doi.org/10.1007/978-981-96-2284-9_10

the site slope is a factor in the design of the system to reduce operational expenses. CWs serve many functions, encompassing the improvement of water quality in water resources, stormwater management, provision of habitat for biodiversity and microclimate, and offering aesthetic in urban landscapes (Zhang et al. 2014; Wu et al. 2018; Henny et al. 2019a; Vymazal 2019).

Diverse types of CWs have been widely studied and implemented on a global scale over the past six decades, in effort to treat variety of wastewaters and improve water quality (Vymazal 2011; Zhang et al. 2015; Vymaza 2022). A number of studies have thoroughly documented the application of CWs to treat different wastewaters in tropical regions. However, implementation is proceeding rather slowly, despite the fact that the year-round sunlight and high temperatures in tropical regions can increase plant productivity and microbial activity in the CW system, hence enhancing pollutant removal and treatment efficiency (Guino-o Ii et al. 2009; Nur Asmaliza et al. 2014; Zhang et al. 2015; Kurniadie et al. 2018).

Fundamentally, there are three types of constructed wetland systems based on the flow regime: free water surface (FWS) and subsurface flow (SSF). Generally, a free water surface flow (FWS) CW system resembles a shallow pond with soil to support rooted plants or without soil if free-floating aquatic plants are used. The primary design function is to facilitate wastewater contact with plants and associated microorganisms through the use of a slow surface inflow (Zhang et al. 2015; Vymaza 2022).

Subsurface flow (SSF) CWs are commonly designed to facilitate the passage of wastewater flow to pass through porous materials such as gravel, sand, soil, and other active materials planted with emergent plants. Subsurface flow systems can be grouped into two main categories based on the flow direction: horizontal (HSSF) and vertical (VSSF). VSSF The wastewater inflow slowly passes through the aerobic zone, the anoxic zone, and anaerobic zone. The aerobic zone encompasses the area of the plant roots and rhizomes where oxygen is released into the substrate. The anoxic and anaerobic zones are under the surface of the bed of filtration materials (Kadlec 2009; Vymazal 2009; Saeed and Sun 2012; Vymaza 2022).

An additional type of CW is constructed floating wetlands (CFWs), which is an innovative treatment approach that treats stormwater, contaminated and eutrophic lakes and rivers directly in the water bodies using rooted, emergent macrophytes growing on a floating mat. The benefit of CFWs is not only to improve water quality of water bodies, but also to enhance the habitat for biodiversity and the aesthetic of lake, river, and urban water landscape.

Applied CWs sometimes consist of several different systems in a series called hybrid systems to maximize pollutant removal efficiency for high-strength wastewater and to meet the permissible limits of water quality standards. Hybrid CWs may combine FWS with HSSF and VSSF systems, even with CFWs.

This section primarily focuses on the studies and application of CWs that have been developed to treat diverse wastewater and to restore water resources in the tropical regions, especially in South Asia. The aim is to provide pertinent details regarding the implementation of CWs, the treatment performance, and operational considerations, including the selection of the vegetation that has a significant contribution to pollutant removal in CW systems. An overview of lab and application scale studies on different CW systems and treatment efficiency established in South Asian countries is presented in Tables 10.1 (Indonesia), 2 (Vietnam), 3 (Malaysia), 4 (Thailand), and 5 (Philippines). Treatment efficacy varied across the studies on all types of CWs, which utilized a variety of vegetation types and operational designs.

10.1.1 Application of Constructed Wetlands (CWs)

CWs can be built in a more controlled environment compared to natural wetlands, in addition to having advantages of flexibility in site selection and sizing. Numerous studies of CWs have been documented in both controlled laboratory settings and open field conditions (Nilratnisakorn et al. 2009; Asmaliza et al. 2013; Van Thi Thanh Ho et al. 2020; Henny et al. 2022a, b). The majority of pilot and small-scale field studies were conducted on-site and in situ (Othman et al. 2020; Suwandhi et al. 2022; Komunikasi et al. 2023; Tran et al. 2023). There were few reported studies on applied scales (Shaharuddin et al. 2014; Napaldet and Buot 2019; Nguyen et al. 2020; Perdana et al. 2020; Agaton and Guila 2024).

The purpose of CWs is to prevent pollution from potential wastewater in the receiving waters or to remediate and restore contaminated or eutrophic water bodies. Various wastewaters have been tested to be treated by CWs, including common wastewaters such as domestic wastewater, greywater, stormwater, and even eutrophic lakes and rivers, and industrial wastewaters such as textile effluent, tapioca, food, and fish production industries.

The selection of types of CWs, in addition to the treatment goals, depends on the site-specific, available space, treatment components, and cost. Tropical regions characterized by year-round direct sunlight and warmer climates are ideal conditions for plant growth and increased microbial activity in both the biofilm of plant roots and the media used in CWs. These factors collectively enhance the transformation of pollutants and overall, the treatment efficiency of CWs. Nevertheless, the performance efficiency of CWs depends on the relevant operational factors and plant selection, which are highly variable among the systems.

Free water surface (FWS) CWs in applied scale are mostly reported in some part of Malaysia (Table 10.3), and subsurface flow (SSF) CWs vertical (VSSF) and typically horizontal subsurface flow (HSSF) are the popular type of CWs studied and implemented in Indonesia (Table 10.1), Vietnam (Table 10.2), and Thailand (Table 10.4). While. A hybrid system combining FWS and SSF CWs has also been implemented in Thailand (Table 10.4), the Philippines (Table 10.5), and Vietnam (Table 10.2) to achieve higher treatment efficiency, especially.

Table 10.1 A summary of constructed wetland system operation and treatment efficiency in Indonesia

Application scale (location)	Type of CWs	Type of wastewater (WW)	Plant species	HRT (d)*	Removal efficiency of pollutants (%)	References
Lab	Hybrid	Batik textile WW	*Canna Indica*	5	89.53 (COD); 98.74 (TSS)	Rahmadyanti and Audina (2020)
Pilot (University Campus, Bandung)	VSSF	Sewage	*Phargmites karka*		48–98 (COD); 74.19–97.95 (TN); 71.7 (TP); 99 (Coliforms)	Kurniadie (2011)
Lab	HSSF	Tofu wastewater	*Pennisetum purpureum*		27–85 (COD); 61–90 (TSS)	Suwandhi et al. (2022)
Lab	VSSF	Food industry	*Typha angustifolia*	5, 10, 15	85.83–94.17 (COD); 76.7–87.52 (TN); 73.78–84.71 (TSS)	Abdulgani et al. (2013)
Pilot (Community Residence—Serengseng District, Jakarta City)	SSF	Greywater	*Typha latifolia*		58–91 (COD); 62–89 (TN); 53–93 (TP)	Hendrawan et al. (2013)
Applied (Uluwatu Beach, Bali)	HSSF	Greywater	*Iris pseudacorus, Caladium bicolor, Rhoe discolor, Sansevieria trifasciata, Heliconia psittacorum*		80–90.66 (COD); 83.5 (TN); 89.9 (TP)	Perdana et al. (2020)
Pilot (Batik Industry, Surabaya)	VSSF	Batik WW	*Canna Indica*	21	91.82 (COD); 96.2 (TN); 91.25 (TSS)	Rahmadyanti and Wiyono (2020)
Lab	VSSF	Greywater	*Cyperus papyrus, Praghmites australis*		78–80 (COD); 55–75 (TN); 54–65 (TP); 50–59 (TSS)	Rahmadyanti and Wiyono (2017)
Pilot (Sumedang)	SSF	Tapioca WW	*Phragmites karka*	4	82–86.2 (COD); 25 (TSS)	Kurniadie et al. (2018)
Lab	HSSF	Textile WW	*Heliconia psittacorum*	2.7	80 (COD); (TN); (TSS)	Henny et al. (2022b)

(continued)

Table 10.1 (continued)

Application scale (location)	Type of CWs	Type of wastewater (WW)	Plant species	HRT (d)*	Removal efficiency of pollutants (%)	References
Pilot (Urban Lake, Megacity Jakarta)	CFW	Urban water	*Heliconia discolor, Vetiveria zizanioides*		12–68 (TN); 12–68 (TP); 19–46 (TSS)	Henny et al. (2019a)
Applied (Lake Maninjau)	CFW	Lake	*Echinodorus palaefolius*		50 (TN); 35 (TP); 50 (TSS)	Henny et al. (2020)
Pilot (Lake Maninjau)	CFW	Eutrophic lake	*Cyperus papyrus, Echinodorus sp., Canna variegata, Vetiveria zizanioides*		20–60 (TN); 30 (TP); 60 (TSS); 80 (Chlorophyll-a)	Henny et al. (2019b)

Lab: Laboratory; * HRT: Hydraulic Retention Time

10.1.2 Operational Design

The crucial operating designs of the CW system include hydraulic loading rate (HRL), hydraulic retention time (HRT), the type of vegetation, and the type of bed media for the SSF CW system, and other environmental factors, especially temperature. Employed HRL, or HRT determines the area of CW system. As for the size selection of the CW system, it is merely dependent on the loading rate of the wastewater and the targeted effluent. FWS CWs are typically engineered to treat stormwater and secondary municipal wastewater (WW) or tertiary treatment following conventional wastewater treatment facilities. In contrast, SSF CWs are specifically engineered to handle robust primary municipal WW, landfill leachate, and a variety of industrial WW. The size range in which FWS CWs are employed is a much larger size range than that of SSF CWs. Operating designs will almost certainly dictate the treatment performance of the CW system (Saeed and Sun 2012; Vymazal 2019; Vymaza 2022).

Hydraulic conditions determine the treatment performance of CWs by affecting microbial processes and the rate of pollutant transformation. HRT is the length of time necessary for pollutants in the wastewater to interact with the biofilm in the substrate and the plant rhizosphere, which promotes the transformation of pollutants. A high HRL or a low HRT reduces the contact time between the pollutants and biofilm, or the rhizosphere due to the wastewater rapidly flowing to the outlet. Conversely, low HRL or high HRT increases contact time between the pollutants, biofilm, and the rhizosphere, thereby contributing to enhanced removal of the pollutants. Different types of CWs have typical operational design regarding the HRL or HRT set-up. The setting up of HRL and HRT will determine the area of CWs installed. Kadlec (2009) reviewed data from 400 FWS and around 200 HSSF system treating BOD

Table 10.2 A summary of constructed wetland system operation and treatment efficiency in Vietnam

Application scale (location)	Type of CWs	Type of wastewater (WW)	Plant species	HRT (d)*	Removal efficiency of pollutants (%)	References
Pilot (Binh Chanh District, Ho Chi Minh City)	VSSF	Industrial WW	*Phragmites australis*, *Typha orientalis*	2, 4, 5.4, 9	47–55 (COD); 40 (TN); 67–86 (TSS)	Dan et al. (2020)
Lab	HSSF	Domestic WW	*Brachiaria mutica*	7.15	88.63 (COD); 63.43 (TN); 84.66 (TP); 90.4 (TSS); 99.47 (Coliforms)	Van Thanh Ho et al. (2020)
Lab	HSSF	Industrial WW	*Brachiaria mutica*	7.15	93.86 (COD); 97.72 (BOD); 97.46 (TN); 91.12 (TP); 83.07 (TSS)	Van Thanh Ho et al. (2020)
Lab	HSSF	Domestic WW	*Brachiaria mutica*	7.15	81.58 (COD); 93.3 (BOD); 58.6 (TN); 85.5 (TP); 67.3 (TSS); 90.4 (Coliform)	Van Thi Thanh Ho et al. (2020)
Pilot (Danang City)	SSF	Effluent wastewater treatment plant	*Canna generalis*, *Portulaca grandiflora*	2	40–50 (BOD/COD); 30–40 (TN); 40–60 (TP)	Tran et al. (2023)
Pilot (Bach Quang ward, Thai Nguyen City)	Hybrid-HSSF	Domestic WW	*Canna generalis*		81 (BOD), 85 (TSS), 93 (N-Ammonia) and 77 (P-phosphate)	Tran et al. (2019)
Applied (Hanoi City)	VSSF	Blackwater	*Typha orientalis*, *Phragmites communis*, *Dracaena fragrans*	2, 3	72.1–76 (COD); 70.4–78 (TSS)	Nguyen et al. (2007)

(continued)

Table 10.2 (continued)

Application scale (location)	Type of CWs	Type of wastewater (WW)	Plant species	HRT (d)*	Removal efficiency of pollutants (%)	References
Applied (Dormitory, Hue University -Quang Tri Campus)	Vertical filter (gravel and sand) + FWS	Sewage	*Colocasia esculenta, Dracaena sanderiana*		74 (BOD); 90 (N-Ammonia); 76 (TSS); 84 (Coliform)	Nguyen et al. (2020)
Pilot (Bach Quang ward, Thai Nguyen City)	HSSF	Domestic WW	*Cyperus alternifolius*	4.57	40 (BOD); 60 (TN); 45 (N-Ammonia); 30 (P-phosphate)	Thuy Anh et al. (2020)

Lab: Laboratory; * HRT: Hydraulic Retention Time

and nutrient. Overall, the median hydraulic loadings for FWS are approximately 3 cm.d^{-1}, while for HSSF they are 7 cm.d^{-1}, which is roughly double. Both FWS and HSSF operated for BOD treatment are typically at a higher HRL than for nutrient treatment. The FWS HRTs are 9.3 days compared to 2.9 days for HSSF, which is three times longer. In addition to operating within higher inlet concentration ranges for BOD and nutrients, SSF CWs remove nutrients more efficiently than FWS CWs.

Global previous studies reported the range of hydraulic loading rate (HLR) of the FWS system using different vegetation treated different types of wastewaters was 0.7–1800 m^3.d^{-1} and hydraulic retention time (HRT) ranging from < 1–10 days (Zhang et al. 2014). Very limited studies were reported on FWS system in Southeast Asia. Applied studies of FWS system to treat stormwater from around 0.7% catchment area were reported in Malaysia using mixed types of aquatic plants. The FWS with an area of 12,500 m^2, water depth 0.3–1 m treated stormwater with inflow rate of 900 m^3.d^{-1} and HRT of 3 days. The bed media with a depth of 0.6 m consisted of pea gravel and soil mixture, and was planted with *Hanguana malayana, Phragmites karka, Lepironia articulata and Typha latifolia, Eleocaris dulcis.* A similar study of FWS-treated stormwater was designed with an HRL of 300 m^3/d and an HRT of 1.2 days. The system removed up to 80% of BOD and nutrients (Asmaliza et al. 2013; Shaharuddin et al. 2014).

Meanwhile, more studies of SSF and some hybrid CW systems are widely reported. Dologuin et al. (2022) evaluated 3 HRTs (14, 21, and 48 days) in the VSSF CWs with substrate only (gravel, sand, and soil) and with two types of plants (Taro and Cattail) to treat landfill leachate containing 5000 mg/L of BOD_5 from the solid waste disposal facility of Iligan City, Philippines. VSSF with plants, especially Taro, and the highest HRT gave the highest BOD_5 removal efficiency (> 90%). A pilot study of VSSF for advanced treatment of industrial wastewater using *Phragmites australis* and *Typha orientalis* operated at different HRL and HRT of 2, 4, 5.4, and 9 days, respectively, removed COD effectively at HRT of 5.4 and 9 days

Table 10.3 A summary of constructed wetland system operation and treatment efficiency in Malaysia

Application scale (location)	Type of CWs	Type of wastewater (WW)	Plant species	HRT (d)*	Removal efficiency of pollutants (%)	References
Lab	FWS	Stormwater	*Eichornia crasssipes*	10	9.1–68.2 (BOD/COD); 31.4–70.3 (TN); 71–100 (TP)	Asmaliza et al. (2013)
Applied (University Sains Malaysia (USM) Campus)	FWS		*Typha aungustifolia, Phragmites karka, Lepironia articulata, Hanguana malayana, Eleocharis dulcis*		9.7–80 (COD/BOD); 24–46 (TP); 25.9–30 (TSS)	Asmaliza et al. (2013)
Applied (University Sains Malaysia (USM) Campus)	FWS	Stormwater	*Typha aungustifolia, Lepironia articulata, Hanguana malayana, Eleocharis dulcis*		27.7 (BOD); 46.43 (COD); 61.56 (N-Ammonia); 60.98 (TP); 67.86 (TSS)	Shaharuddin et al. (2014)
Pilot (Faringipani Resort Langkawi)	FWS	Stormwater	*Neptunia oleracea, Eichhornia crassipes, Ipomoea aquatica, Thalia Geniculata, Chrysopogon zizanioides, Nymphaea caerulea*		14.1–52.1 (BOD); 49.7–66.9 (COD); 4.3–80.4 (N-Ammonia); 4.3–80.6 (TP); 15.6–20.2 (TSS)	Akhir et al. (2017)
Pilot (University Pertahanan Malaysia Campus)	HSSF	Stormwater	*Typha angustifolia*	30	53.2 (COD); 64 (BOD); 74 (TSS)	Othman et al. (2020)
Pilot (University Pertahanan Malaysia Campus)	VSSF	Stormwater	*Typha angustifolia*	30	70.8 (COD); 67.7 (BOD); 83.5 (TSS)	Othman et al. (2020)

(continued)

Table 10.3 (continued)

Application scale (location)	Type of CWs	Type of wastewater (WW)	Plant species	HRT (d)*	Removal efficiency of pollutants (%)	References
Pilot (University Putra Malaysia Campus)	SSF	Domestic WW	*Lepironia articulata*	7	56–77 (COD); 20–88 (TP); 27–96 (N-Ammonia); 50–88 (TSS); 99 (coliforms)	Katayon et al. (2008)
Lab, Singapore	CFW	Reservoir	*Chrysopogon zizanioides, Typha angustifolia, Polygonum barbatum*		19.1–46 (TN); 7.8–40.8 (TP)	Chua et al. (2012)

Lab: Laboratory; * HRT: Hydraulic Retention Time

Table 10.4 A summary of constructed wetland system operation and treatment efficiency in Thailand

Application scale (location)	Type of CWs	Type of wastewater (WW)	Plant species	HRT (d)*	Removal efficiency of pollutants (%)	References
Pilot (Amphawa District, Samut Songkram Province)	SSF	Greywater	*Canna indica; Sterilitzia reginae; Pandandus tectorius; Philodendron*	5	81 (COD); 84 (BOD); 30 (TN); 40 (TP); 78 (TSS)	Benjawan et al. (2022)
Pilot (Asian Institute of Technology (AIT), Bangkok)	VSSF	Blackwater	*Typha augustifolia*	6	98 (COD); 94 (TN); 91 (N-Ammonia); 74 (TS)	Koottatep et al. (2002)
Lab	VSSF	Synthetic dyes	*Typha angustifolia*	15	60 (COD); 50 (Dye); 86 (TDS)	Nilratnisakorn et al. (2009)
Applied (Koh Phi Phi, Southern Thailand)	Hybrid	Greywater	*Canna, Heliconia, Papyrus*		91.5 (BOD); 38.9 (TN); 50 (TP); 90 (TSS); 92 (Coliform)	Brix et al. (2011)

Lab: Laboratory; * HRT: Hydraulic Retention Time

Table 10.5 A summary of constructed wetland system operation and treatment efficiency in Philippines

Application scale (location)	Type of CWs	Type of wastewater (WW)	Plant species	HRT (d)*	Removal efficiency of pollutants (%)	References
Applied (Bayawan City, Negros Oriental)	Hybrid	Domestic wastewater	Phragmites karka		99.33 (BOD); 62 (TP); 71 (Coliform)	Guino-o Ii et al. (2009)
Applied (Barangay Bonbonon, Iligan City)	VSSF	Landfill leachate	Typha latifolia, Colocasia esculenta	14, 21, 28	96–99 (BOD)	Dologuin et al. (2022)
Applied (Balili River, Benguet)	FWS + HSSF	Eutrophic river	Pennisetum purpureum, Eleusine indica, Amaranthus spinosus, Eichhornia crassipes	3, 4	87–95 (BOD); 85–99 (N-nitrate); 85–96 (P-phosphate); 96–99.99 (TSS); 99.99 (Colliform)	Napaldet and Buot (2019)

Lab: Laboratory; * HRT: Hydraulic Retention Time

at a loading rate of 130 $kg.h^{-1}.d^{-1}$ (Dan et al. 2020). Combining the novel VSSF using expanded clay and FWS for dormitory sewage purification and reuse operated at HRL ranges of 0.02–0.12 $m.d^{-1}$ gave effective removal of BOD, ammonia, and coliform (Nguyen et al. 2020).

Bed media for SSF CWs, such as gravel, sand, soils, or compost, is frequently employed (Nguyen et al. 2020; Henny et al. 2022b; Vymaza 2022). Nevertheless, to optimize the treatment efficacy, active media such as bentonite (Pat-Espadas et al. 2018), zeolite (Susanti et al. 2023), and rice husk (Rahmadyanti and Wiyono 2020) are incorporated.

10.1.3 Selection of Vegetation

Vegetation is a significant component that has several roles in relation to constructed wetland treatment processes. Constructed wetlands (CWs) combine physical, chemical, and biological processes in treating wastewater, where plants can synergize with microbes attached to the filtration media bed and associated with plant root systems. The most important role of plants in CWs is to remove pollutants from the contaminated water either by retaining the solids, nutrient uptake, pollutant transformation, or degradation via synergism with biofilm in the rhizomes. Plants with high biomass productivity provide the energy source for the microbes and other microorganisms in biofilm associated with the media and therefore increase CW treatment performance

(Brix et al. 2011; Vymazal 2019; Henny and Kurniawan 2020; Vásquez et al. 2022; Vymaza 2022).

Vegetation possesses distinct capabilities and capacities with respect to nutrient uptake, pollutant transformation, and toxic metal accumulation. Consequently, these variations impact the efficiency of CW treatment. Aquatic emergent plants, including *Cyperus papyrus, Typha angustifolia, Lepironia verticulta, Iris sp., and Echinodorus palaefolius* (Shaharuddin et al. 2014; Dan et al. 2020; Henny et al. 2020; Thuy Anh et al. 2020; Vymaza 2022), and several perennial plants and grasses such as common reeds like *Phragmites sp., Heliconia sp., Brachiaria mutica, Canna sp., Vetiveria zizanioides* (Nguyen et al. 2007; Henny and Kurniawan 2020; Van Thi Thanh Ho et al. 2020; Henny et al. 2022b; Rohaningsih et al. 2023; Tran et al. 2023) are mostly used in free surface flow (FWS), both horizontal and vertical flow subsurface types of CWs (HSSF and VSSF), and constructed floating wetlands (CFWs). The FWS system can also sometimes use floating plants such as *Pistisia, sp., Lemna sp.,* and *Eichornia crassipes,* floating leaves with roots in sediment like *Nymphea stellata,* and submerged plants, such as *Hydrilla verticillate* (Asmaliza et al. 2013; Henny and Kurniawan 2020).

Selection of vegetation in CW treatment system depends on the types of wastewater and target pollutants to be treated. Plants to be selected should be of a more robust type due to their high pollutant load tolerance. Robust types of plants used in CWs are still the global type of tropical plants. As in the CFW system, plants with long and dense root networks are important selection criteria to enhance system performance (Zhang et al. 2015; Henny et al. 2019a, b; Henny and Kurniawan 2020). The plant's root system and the microbial biofilm associated with the roots play an important role in nutrient or pollutant uptake, transformation, and degradation. When it comes to the utilization of vegetation, particularly in the CFW system, that is applied directly to bodies of water, it is crucial to avoid employing invasive species such as *Eichornia crassipes,* which is frequently observed covering the surface of major lakes. In tropical regions, this plant has potential to double its growth rapidly, which could cause lake siltation and a decline in the local fish population as a result of the lake's low dissolved oxygen level (Hidayati et al. 2018; Henny and Kurniawan 2020; Prasetyo et al. 2022; Vásquez et al. 2022).

CWs that are designed to treat stormwater and surface runoff in the form of a popular rain garden and CFWs implemented in the body of water may utilize exotic plants to attract butterflies and enhance the aesthetic of the city. Plants with radiant-colored flowers, such as *Iris sp., Canna sp.,* and *Heliconia sp.,* can be used for that purpose (Brix et al. 2011; Henny et al. 2019b, 2022b; Perdana et al. 2020; Tran et al. 2023). As in CFWs, both the plants and the roots provide the habitat for the lake's aquatic life and also certain types of terrestrial fauna, which serve either as feeding grounds or as places for foraging, reproduction, nursery, and sheltering (Henny et al. 2019b, 2020). As an eco-city concept evolves, CW application with a variety of vegetation becomes essential for improving the lake landscape, water quality, and urban environment, and good natural resources for the sustenance and maintenance of biodiversity (Brix et al. 2011; Shaharuddin et al. 2011; Chua et al. 2012; Henny et al. 2020; Vásquez et al. 2022; Vymaza 2022).

10.1.4 Treatment Performance

Different types of CWs work differently depending on the pollutant load in the influent, the loading rate, the hydraulic retention time, the type of vegetation, and the type of bed media for the subsurface flow. These all determine the removal efficiency (RE) depending on the targeted pollutant in various wastewater (Zhang et al. 2015; Vymazal 2019; Vymaza 2022).

It appeared that all varieties of CW systems eliminated BOD/COD, nutrients, and TSS effectively; HSSF and VSSF both demonstrated high removal of TP, but HSSF performed more effectively (Zhang et al. 2015; Dan et al. 2020; Henny et al. 2022b; Tran et al. 2023). Applied or pilot studies of SSF CW and Hybrid systems appeared to be more effective at eliminating BOD/COD (>90%), TN (>90%), TP (>80%), TSS (>90%), and coliform (99%) (Koottatep et al. 2002; Katayon et al. 2008; Hendrawan et al. 2013; Napaldet and Buot 2019; Perdana et al. 2020; Anh et al. 2022). Hybrid and both HSSF and VSSF demonstrated good removal of pollutants from blackwater, landfill leachate, industrial, and textile wastewater (Koottatep et al. 2002; Nguyen et al. 2007; Dan et al. 2020; Rahmadyanti and Audina 2020; Rahmadyanti and Wiyono 2020; Dologuin et al. 2022). Implemented FWS treating stormwater removed considerable amounts of BOD, TN, TP, and TSS with varied removal efficiency and significantly improved the water quality of the stormwater (Shaharuddin et al. 2011; Asmaliza et al. 2013; Noor et al. 2013). In comparison to other CWs, CFWs were not extensively researched in South Asian regions. Nevertheless, their utilization in the treatment of lakes and a reservoir revealed noteworthy effectiveness in the elimination of COD (80%), TN (50%), TP (40%), TSS (60%), and chlorophyll-a (80%) (Chua et al. 2012; Henny et al. 2019a, b, 2020, 2022a), slightly better than other global CFW studies in tropical regions (Zhang et al. 2015).

Despite the extensive research and global implementation of CWs as a sustainable environmentally friendly technology for the treatment of various wastewaters, as well as eutrophic and polluted lakes and reservoirs, their application in tropical regions, including Southeast Asian countries, remains rather limited. As previously stated, research conducted in South Asian regions has demonstrated a great potential of CWs in the treatment of a wide range of wastewaters. The findings have substantially increased understanding regarding the significance of constructed wetland technology. Consequently, their adoption for the purpose of improving the quality of water resources in tropical regions could grow in popularity.

Figure 10.1 represents several types of CWs application in Indonesia for different purposes, such as treating runoff from the lake catchment of an urban lake and paddy field runoff from the lake catchment of Lake Maninjau, treating laboratory wastewater, and using CFWs for water quality improvement in the water front area and floating cage fishery's area of Lake Maninjau. Lake Maninjau is a hyper-eutrophic lake and suffers from cyanobacteria blooms (Sulastri et al. 2023).

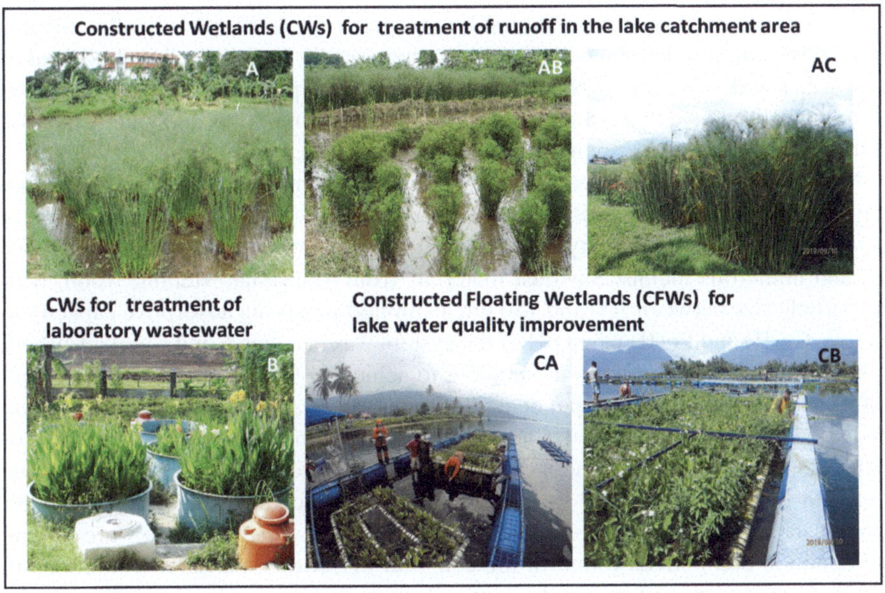

Fig. 10.1 Application of constructed wetlands (CWs) for different purposes: (A) Free water surface (FWS) CWs in an urban lake catchment with *Cyperus papyrus* (A) and *Cyperus isocladus* (AB) (Private collection. Photo credit: Ami. A. Meutia); FWS CWs in Lake Maninjau with *Cyperus papyrus* and *Heliconia sp.* (AC) (Private collection. Photo credit: Cynthia Henny); Subsurface flow (SSF) CWs with *Canna indica* (B) (Private collection. Photo credit: Gadis S. Haryani); Constructed floating wetlands (CFWs) in Lake Maninjau with Echinodorus palaefolius (CA, CB) (Private collection. Photo credit: BRIN/Cynthia Henny)

10.2 Natural Materials

10.2.1 Basics, Types, and Applications

Natural materials are substances that can be derived from plants, animals, or soil and originate from nature. Some of these materials may undergo minor or no treatment for use. Natural materials often offer a myriad of properties that are functional and beneficial to improve the quality of human life and the environment. One of the remarkable properties is the ability of natural materials to selectively remove pollutants via adsorption. The concept of adsorption underpins the separation of substances wherein the adsorbates adhere to the active surface of the adsorbents. Adsorption customarily involves several interactions e.g., chemisorption, physical adsorption, and ion exchange (Crini et al. 2019). In chemisorption, the bonding between the molecules of adsorbates and the active surface in adsorbents is bolstered by a typical covalent bonding. While physical adsorption underlines the importance of van der Waals forces in creating the attractive forces between adsorbates and adsorbents. Additionally, ion exchange is a reversible process that implies exchanging similarly charged ions between solution and solid (Gupta et al. 2009). Practically, adsorption using natural materials has long been scientifically investigated and utilized to treat water and wastewater. Natural materials as adsorptive substances can be the promising alternative that is versatile, economically competitive, and widely applied in the control and management of pollution.

Natural materials that have been known as adsorbents are, for example, wood, clay, natural zeolites, and chitosan (Gupta et al. 2009). These materials are usually capable of adsorbing certain types of unwanted substances such as salts, basic dyes, oil, and numerous metals. Sawdust materials from Malaysian Meranti wood, with the particle size between 100 and 150 μm, exhibited satisfying adsorptive capacity to remove Cu(II) and Pb(II) when pH of the system was between 6 and 7 and adsorbent concentration was set to more than 0.25 g to achieve > 90% removal (Ahmad et al. 2009). Another report (Praipipat et al. 2023) that used sawdust from Thai Andaman Redwood in the form of powder and bead had performed distinctive removal efficiencies of lead (Pb^{2+}) and reactive blue 4 (RB4) dye due to the difference in surface area. Sawdust bead used had three times greater surface area than that of powder form. It is recognized that cellulose, lignin, and some other phenolics are among the common compounds found in the cell wall of sawdust particles (Shukla et al. 2002). These compounds facilitate chemical structure and are functional for removal mechanisms. The mechanisms that could occur during adsorbate removal include electrostatic attractions, hydrogen bonding, n-π interactions (Praipipat et al. 2023), and possible ion exchange (Shukla et al. 2002).

Clay and natural zeolites are hydrated materials found abundantly in e.g., the soil, sediments, and volcanic deposits and have been known for pollutant removal since early times. They are different in their crystal structures whereas the similarities between them include an open structure that holds exchangeable ions and dihydrogen monoxide molecules. In addition, the substitutions of Al^{3+} for Si^{4+} in the tetrahedral layer facilitate negative structure charge that is concomitantly balanced with the cations (Bish 2013). As a result, these materials demonstrate selectivity of cations as well as remarkable Cation Exchange Capacities (CEC) which refers to the amount of exchangeable cations of the soil (Chapman 1965). The applications of clay and natural zeolites for pollutant removal varied. Thai kaolin and ballclay (black clay) could remove several metals i.e., Cd^{2+}, Cr^{3+}, Cu^{2+}, Ni^{2+}, Pb^{2+}, and Zn^{2+} in the prepared solutions. Cr^{3+} showed highest affinity onto Thai kaolin as well as ballclay (Chantawong et al. 2003).

Globally, natural zeolite production was supplied by top producers such as China, South Korea, the and United States, however, in Southeast Asian region, Indonesia and the Philippines were also among the global natural zeolite producers (Borsatto and Inglezakis 2012). Natural zeolites (Fig. 10.2) have helped environmental remediation by removing several types of pollutants such as heavy metals, or nutrients in ionic forms (Williams 2018). Indonesian's Cikalong natural zeolites that comprised of mordenite and clinoptilolite showed > 90% ammonium removal efficiency when > 1.5 g zeolite mass was used to adsorb 100 mg/L ammonium solution in a batch experiment (Kurniawan et al. 2022). The removal of heavy metals e.g., lead and cadmium had been investigated from Indonesian zeolitic tuff containing mainly mordenite

Fig. 10.2 Natural zeolites with differing particle sizes (Private collection. Photo credit: BRIN/ Miratul Maghfiroh)

and clinoptilolite. The efficiency ranged up to 76% and 51% for lead and cadmium removal, respectively (Budianta et al. 2020).

Chitosan is recognized as a derivative that can be found from natural materials such as shrimp and crab shells. As an alternative material for adsorption, chitosan harbors a notable adsorptive capacity toward toxic metal ions, for example, copper and lead (Zhang et al. 2021). High content of nitrogen enables chitosan to be an electron donor which elevates the adsorptive capacity (Lee et al. 2005). Numerous other examples of natural materials to remove unwanted substances have been demonstrated and recorded. Table 10.6 presents some studies on the application of natural materials from various solutions.

According to Table 10.6, natural materials include a vast array of materials from various resources available in our natural surroundings. With little or some treatments, these materials can act as adsorbents with significant adsorptive capacity to remove unwanted substances from solutions. This suggests that natural materials are promising alternatives as adsorbents.

Apart from being incredible adsorbents, certain natural materials also have demonstrated the ability as coagulants. Coagulation is an impurity removal process that includes destabilization and agglomeration of particles into larger aggregates (Ang and Mohammad 2020). Like any other natural materials, natural coagulants can be animal, microorganism, or plant-based compounds (Kurniawan et al. 2020). Plant-based coagulants from *Moringa oleifera* (Moringa), *Opuntia ficus-indica* (Cactus pear) and *Jatropha curcas* (Barbados nut) are commonly investigated for its remarkable capability to remove color, total suspended solids, and microorganisms e.g., coliform, fungi (Bahrodin et al. 2022). Plant-based coagulants made from Malaysian banana stems exhibited noticeable removal efficiency of COD (Chemical Oxygen Demand), turbidity and SS (Suspended Solids) of spent coolant wastewater up to 80.1%, 98.5%, and 88.6%, respectively (Alwi et al. 2013). Another natural coagulant produced from Malaysian bagasse (*Saccharum officinarum*) was experimented to remove suspended solids up to 95.9% (Bahrodin et al. 2022). The removal of COD concentration up to 75.64% was observed when experimenting the Indonesian rambutan (*Nephelium lappaceum* L.) seed as natural coagulant which suggested its

Table 10.6 Natural materials and their applications

Type of natural material	Material origin	Removed substances	Concentration of adsorbates (mg/L)	Removal capacity	References
Peat and activated carbon composite	Malaysia	Total Iron (Fe) from leachate	7.19	79.6%	Rosli et al. (2017)
Coal and peat soil	Indonesia	Hg(II) from aqueous solution	200	82%	Budiharjo et al. (2021)
Coffee husk	Vietnam	Pb^{2+} and Cd^{2+} from industrial wastewater	$Pb^{2+} = 118.4$ $Cd^{2+} = 107.5$	89.6 and 81.5%	Quyen et al. (2021)
Fruit peel	Thailand	Cu(II) from aqueous solution	100	52.3–97.0%	Phuengphai et al. (2021)
Fishscale and seashell	Myanmar	Anionic surfactants from aqueous solution	100	68.54 mg/g (fishscale) and 57.92 mg/g (seashell)	Than et al. (2020)
Marang fruit skin	Brunei Darussalam	Toxic dyes (methylene blue and methyl violet 2B) from aqueous solution	Up to 1000 mg/L	184.6 mg/g (methylene blue) and 137.3 mg/g (methyl violet)	Lim et al. (2015)
Chicken eggshell	Cambodia	As and Fe from acid mine drainage	1 mg/L (As) and 20 mg/L (Fe)	> 99% for both As and Fe	Heng et al. (2022)
Sugarcane bagasse	Nigeria	Pb and Ni from untreated refinery wastewater	0.8 mg/L (Pb) and 39.4 mg/L (Ni)	89.31% (Pb) and 96.33% (Ni)	Ezeonuegbu et al. (2021)
Water hyacinth, water lettuce, water lily	Nigeria	Nickel from refinery wastewater	0.54 mg/L	86–90%	Ugya et al. (2019)
Groundnut husk	Ethiopia	Cr from tannery effluent	12.25 mg/L	up to 94.41%	Tegegn et al. (2022)

(continued)

Table 10.6 (continued)

Type of natural material	Material origin	Removed substances	Concentration of adsorbates (mg/L)	Removal capacity	References
Coffee husk	Ethiopia	Methylene Blue from aqueous solution	20 mg/L	96.9%	Ayalew and Aragaw (2020)
Scales of tilapia and red drum fish	Ghana	Fluoride from aqueous solution	2–10 mg/L	up to 76% (tilapia) and 70% (red drum)	Nkansah et al. (2022)
Coffee pulp	Colombia	Pb(II) from aqueous solution	20–500 mg/L	86.45%	Gómez-Aguilar et al. (2021)

potential for dye wastewater treatment (Riyanto and Sari 2024). Although massive lab-scale experiments of natural coagulants have been conducted, their use for ecosystem interventions (real-life applications) is uncommon. Alternatively, the chemical coagulants e.g., aluminum sulfate (alum), Poly-Aluminum Chloride (PAC), and iron(III) chloride with the combination of other compounds (such as lanthanum-modified bentonites) have been utilized for in-lake measures/interventions (Lürling et al. 2020). The chemical coagulants with the combination of phosphorus sorbent promoted cyanobacterial removal, improved the water quality of the lake, and intercepted phosphorus released from the lake sediment (Waajen et al. 2016).

While many lab-scale experiments have been extensively conducted to discover more potential of natural materials, it is necessary to evaluate their performance in real-life or scale-up applications. The application of natural materials in the actual treatment processes is usually challenged by many factors such as performance consistency and cost-effectiveness (Ang and Mohammad 2020). These factors often affect the confidence of the stakeholders to move forward with scale-up applications; thus, its application is relatively limited.

10.2.2 Natural Materials and the Combination with Other Technologies

Each technology has strength and weakness; thereby, the combination of several technologies is encouraged to obtain optimum performance or efficiency. The use of natural materials may be coupled with other technologies to enhance the capacity of the system to remove pollutants. In the Constructed Wetlands (CWs), as discussed in the previous section, for example, the substrate can consist of different types of natural materials. It can be sand, soil, fly ash, or zeolite. Substrates support the growth of plants and facilitate surface area for sorption of pollutants (Li et al. 2014).

Biochar which is yielded from pyrolysis of biomass may be chosen as an option because of its remarkable surface area for efficient adsorptive capacity toward contaminants (Malyan et al. 2021). A mixture of biochar-soil as the substrate is also useful for odor absorption. In a Vertical Flow Constructed Wetland (VFCW) to treat formaldehyde-containing wastewater from a Glycol Company in Thailand, the use of biochar-soil substrate successfully enhanced the formaldehyde removal efficiency up to 94.9% (Wararam et al. 2015). Another experiment using rice husks as the substrates in VFCW found that the overall system could remove 81.8% of Cr and 96.2% ammonium from Batik (Indonesian traditional clothing) wastewater in Indonesia (Rahmadyanti and Wiyono 2020).

Likewise, the application of natural coagulant may be integrated with other preferred technologies to achieve maximum performance of the treatment system. The integration of natural coagulant can be carried out with adsorption system (activated carbon or sand), ozonation, disinfection, or membrane filtration and installed for pre- or post-treatment process of the overall system (Ang and Mohammad 2020).

10.3 Mycoremediation

Mycoremediation, a form of bioremediation, is a sustainable and natural technology for cleaning polluted environments using certain fungi to degrade, restore, and detoxify the contaminated area. The process of mycoremediation involves the complete breakdown of harmful contaminants into harmless compounds such as carbon dioxide (CO_2), water (H_2O), nitrogen (N_2), and hydrogen chloride (HCl), with the assistance of various fungi (Khatoon et al. 2020). Fungi, being eukaryotic organisms, are found everywhere in nature and thrive in multiple environmental conditions worldwide. Fungi are an excellent option for the remediation of a variety of pollutants due to their robust growth, extensive hyphal network, production of versatile extracellular ligninolytic enzymes, extensive metabolic capacity, high surface area to volume ratio, resistance to heavy metals, flexibility in responding to changes in pH and temperature, and presence of metal-binding proteins (Akhtar and Mannan 2020; Akpasi et al. 2023).

Mycoremediation is among the most effective, affordable, and environmentally responsible methods that have gained much attention lately. Numerous fungi employ various techniques to transform aquatic pollutants into less detrimental environmental substances. The rate at which an aquatic pollutant degrades depends on its chemical composition (Abuqamar et al. 2023; Akpasi et al. 2023). The aquatic pollutants consist of heavy metals, persistent organic pollutants like polycyclic aromatic hydrocarbons (PAHs), agrochemicals compounds (pesticides and herbicides), textile dyes, chemicals from the leather tanning industry, wastewater, petroleum fuels, pharmaceuticals, and personal care products (PPCs). The cleanup outcomes may also include valuable components like enzymes (such as laccase and tyrosinase) and biomass from nonedible, edible, or medicinal mushrooms, enhancing the profitability of the remediation procedure.

10.3.1 *Factors Influencing Mycoremediation*

Mycoremediation is closely linked to several significant factors. The nutrient level, physical (pH, temperature, salinity), chemical, biological (oxygen level) parameters, and genetic diversity all play a role in the remediation of hazardous materials.

10.3.1.1 Nutrient Levels

Nutrient levels, including substrate type, are crucial in mycoremediation as they influence fungal growth and colony development rate. High nutrient levels can lead to dense fungal colonies with reduced hyphal extension, affecting the overall efficiency of the remediation process. The availability of nutrients directly affects the metabolic activities of fungi involved in degrading contaminants (Singh 2006). Organic and inorganic fertilizers are frequently used to meet nutrient requirements (Akpasi et al. 2023). Compost can increase the mycoremediation efficiency of PAHs by up to 50% (Guerin 2000).

Further, nitrogen and phosphate may turn into limiting variables (Dickson et al. 2019). Nitrogen limitation did not significantly enhance the remediation efficiency of acetaminophen pharmaceuticals (APAP) (Esterhuizen-Londt et al. 2019). Elevated glucose levels can inhibit the effectiveness of remediation by lowering extracellular enzyme activity (Akpasi et al. 2023). Similarly, keeping sucrose and urea concentrations in the ideal 4–10 g/L and 5–7 g/L can decolorize over 97% of azo dyes in less than 24 h by *Candida tropicalis (C. tropicalis).* Decolorization may be inhibited when sucrose and urea concentrations are exceeded (Tan et al. 2013). Microorganisms prioritize utilizing the external carbon source over the dye when both concentration proportions exceed a certain level. Consequently, the efficiency of color removal decreases as the concentrations of sucrose and urea increase (Qu et al. 2012).

Glucose addition slightly improved the mineralization of lindane by *Phlebia* (*white rot fungi*). Around 10% of labeled lindane was converted to $^{14}CO_2$ after 60 days, with glucose further enhancing this process. Conversely, nitrogen levels in the fungal cultures did not affect lindane degradation, indicating that enzymes responsible for degradation are not influenced by nitrogen concentrations (Xiao and Kondo 2020). It is also acknowledged that the success of mycoremediation can be influenced by the type of substrates and the mobilization agents used (Akpasi et al. 2023; Melati et al. 2021a).

10.3.1.2 Physical, Chemical, and Biological Parameters

The efficiency of mycoremediation processes can be significantly affected by various parameters, including pH, temperature, oxygen levels, and other environmental chemicals. The concentration of hydrogen ions can control the bioavailability of harmful substances in the bio-ecosystem. The pre valence of various fungal species may vary depending on the pH level present. *P. oxalicum* showed the highest decolorization efficiency of AR 183, DB 15, and DR 75 at a pH range of 6.0–7.0 (Saroj et al. 2014). *Ganoderma* sp. was most effective at decolorizing remazol black at acidic pH (Sudiana et al. 2018), while white rot fungi (WRF) *Corilopsis* sp. degrades azo dyes best under acidic conditions (Cheng et al. 2016). *Aspergillus flavus* and *Penicillium canescens* achieve optimal direct blue decolorization at pH 4 and 5 (Hefnawy et al. 2017). Similarly, endophytic fungi were most efficient at decolorizing congo red, orange G, and rhodamine B at pH 6 (Agrawal et al. 2021). While the dark septate endophytes (DSE) strains (KSP, PP, and CPP) exhibited significant decolorization of Remazol Brilliant Blue R (74–99%) and Acid Blue 113 (79–99%) dyes over a wide pH range of 3–11 (Melati et al. 2023b).

Fungi can thrive in various temperature ranges, categorizing them as thermophilic (above 40 °C), mesophilic (5–35 °C), and psychrophilic (below 5 °C). As a result, their effectiveness as bio-remediating agents differs based on their classification (Gupta et al. 2019). The highest direct red decolorization of *A.niger* was observed at 25 °C (Verma et al. 2019). The highest phosphorus removal efficiency of 76.5% was achieved at 30 °C by the halotolerant fungus *Aureobasidium* sp. MSP8 (Zeng et al. 2021). *Ganoderma lucidum* showed that the optimal RBBR decolorization percentage (50.3%) was achieved at 35 °C, pH 4.0 (Qin et al. 2019). Temperature and pH significantly impact the availability of pollutants in wastewater as well as the development of fungus and enzyme activity.

The efficiency of mycoremediation can be significantly impacted by the chemicals or mediators applied. The addition of mobilizing agents such as Tween 20, Tween 80, and soybean oil stimulates the breakdown of polycyclic aromatic hydrocarbons by *Allescheriella* sp. and *Phlebia* sp. However, the fungal responses to these compounds vary (Giubilei et al. 2009). The presence of ethoxylated oleyl-cetyl alcohol changed the formation of organic acids and proteins, inhibited proteolytic activity, and improved the potential for mycoremediation in contaminated environments (Jakovljevic and Vrvic 2016). The addition of surfactant and rhamnolipids resulted in an increased degradation of PAHs in *Trametes hirsuta* (*T. hirsuta*) due to the enhanced production of enzymes, as compared to Triton x-100 and sodium dodecyl sulfate (SDS) (Rathankumar et al. 2022). *T. hirsuta* EDN084T's increased laccase enzyme activity resulted in a significant improvement (2–30-fold) in synthetic dye decolorization after the addition of 1 mM of violuric acid (VA) as a mediator (Yanto et al. 2019).

The effectiveness of remediation technologies, such as mycoremediation, heavily relies on the presence of oxygen. Fungi are both aerobic and anaerobic but grow well under aerobic conditions. Oxygen plays a vital role in bioremediation by facilitating the growth, activity, and metabolic function of microorganisms such as bacteria and fungi. Sufficient oxygen levels are essential for the breakdown of oil contaminants in water, as they stimulate the metabolic functions of remediation agents. Increased oxygen levels can improve bioremediation efficiency by supporting the respiratory activities of microorganisms, leading to faster decomposition of pollutants. Conversely, low oxygen levels can hinder the process by restricting microbial activity and slowing down pollutant degradation (Maxwell et al. 2023; Tan et al. 2013).

Higher oxygen concentrations positively correlate with decreased dichlorodiphenyltrichloroethane (DDT) residues by fungal enzyme laccase (Zhao and Yi 2010). Introducing oxygen improved the breakdown of phenoxy acid herbicides, resulting in shorter lag phases and higher degradation rate constants (Tuxen et al. 2006). *Candida tropicalis* and *Dipodascus ingens* demonstrated a relatively higher efficiency level in aerobically decolorizing six azo dyes. Nevertheless, under static conditions, this yeast could not grow and decolorize any azo dyes (Tan et al. 2013, 2014). *Galactomyces geotrichum* (GG) demonstrated the ability to aerobically decolorize 7 azo dyes, comprising four acid dyes and three direct dyes. Conversely, strain GG could not decolorize azo dyes under anaerobic conditions (Guo et al. 2019). Hadibarata et al. (2022) showed that *Pleurotus eryngii* F032 removed the highest amount (73%) of Benzo [a]pyrene (BaP) in aerobic condition (agitated culture) than in anaerobic condition (static culture) (59%) within 20 days.

10.3.1.3 Genetic Diversity

Genetic diversity is essential for improving the efficiency of mycoremediation processes. Fungal strains have a wide range of phenotypic traits due to their adaptation to different environments, which affects their ability to break down pollutants (Shourie and Vijayalakshmi 2022; Yu et al. 2023). Certain fungi have specific gene families, such as Class II peroxidase and DyP-type peroxidase, which are more prevalent in fast-decomposer genomes, contributing to their adequate dye-degradation capacity (Yu et al. 2023) . Additionally, genetic modification of fungi increases their tolerance to pollutants like arsenic, with genes like WaarsM, ACR2, and ACR3 providing arsenic tolerance and aiding in bioremediation efforts (Naseem et al. 2021). The genetic composition of fungal isolates impacts their ability to degrade pollutants. It provides insights into the mechanisms behind their remediation potential, emphasizing the importance of genetic diversity in optimizing mycoremediation strategies.

10.3.2 Mycoremediation Application in Southeast Asia

Mycoremediation is a promising way to restore these ecosystems in Southern Asia, where pollution from urbanization and industrial activities has caused widespread contamination of water bodies. Certain varieties of fungi can be introduced into contaminated rivers to aid in the breakdown of contaminants and encourage the development of aquatic life. Mycoremediation can also help clean up contaminated water sources and make them safe for use in agriculture and human consumption, which may assist with the region's water scarcity problem. Communities that depend on these water bodies for their livelihoods may profit significantly from this. However, there has been extensive research and global implementation of mycoremediation as a sustainable, environmentally friendly technology for treating various pollutants in aquatic environments; their use in Southeast Asian countries is still quite limited. This sub-chapter discusses mycoremediation applied technology to break down different kinds of dangerous and persistent pollutants in aquatic ecosystems, such as dyes, heavy metals (HMs), and another persistent organic pollutants such as pharmaceuticals, personal care products (PPCPs) and agrochemicals in Southeast Asian.

10.3.2.1 Mycoremediation of Synthetic Dyes

Synthetic dyes have the potential to harm aquatic life and contaminate aquatic environments. Moreover, synthetic dyes reduce the aesthetic value of water bodies by raising the demand for chemical and biological oxygen, which hinders photosynthesis, inhibits the growth of plants, gets into the food chain, causes recalcitrance and bioaccumulation, and may even be toxic, mutagenic, or carcinogenic (Al-Tohamy et al. 2022; Mudhoo et al. 2020; Patil et al. 2021). Being positioned at the high level of the food chain, humans possess a dye concentration that is 1000 times greater than water's (El Bouraie and El Din 2016; Melati et al. 2021a). The potential for dye pollution is expected to increase significantly with the increase in textile production in Indonesia. In 2022, economic growth in Indonesia will reach 5%, and one of the drivers of this economic progress is the textile industry sector (Indonesian Information Portal 2023). In addition, synthetic dyes are also extensively used in paper and pulp, pharmaceutical, tannery, and numerous other industries, so the potential for pollution by dyes will be more significant and more accurate.

The use of fungi in decolorizing and degrading dyes has been widely reported (Melati et al. 2021a; Vikrant et al. 2018). Various fungal mechanisms for removing dyes from aquatic environments, such as biosorption, bioaccumulation, and biodegradation, have also been widely studied. Following the processes of biosorption and bioaccumulation by fungal cells, the degradation of dyes takes place through the utilization of different intra and extracellular enzymes produced by fungi, such as laccases, peroxidases (manganese peroxidase, lignin peroxidase), tyrosinase, triphenylmethane reductase, 1, 2-dioxygenase, and 2, 3-dioxygenase enzymes (Chen et al. 2019; Melati et al. 2023b; Mostafa et al. 2019). Different types of fungi, such as *white rot fungi* (WRF), *brown rot fungi* (BRF), and *dark septate endophyte* (DSE) fungi, utilize their ligninolytic enzyme system to both decolorize and degrade synthetic dyes (Eichlerová and Baldrian 2020; Herath et al. 2024; Melati et al. 2021b, 2023b). However, the effectiveness of this process is influenced by the chemical structure of the dye, which determines the specific degrading enzyme involved and the removal rate (Abuqamar et al. 2023).

Various categories of fungi have been employed in the bioremediation of dyes, including WRF (*Trametes* sp., *Trichoderma* sp., *Cerrena*, and *Pleurotus* sp.) and filamentous fungi (*Aspergillus* sp., *Fusarium* sp., and *Penicillium* sp.), which are widely recognized for their dye bioremediation potential. Interestingly, new types of fungi from the DSE group are starting to be developed to remediate synthetic dyes.

Melati et al. (2023a, b) proved that DSE could degrade dyes from both the azo and anthraquinone groups up to 98% (Table 10.7).

Table 10.7 shows that filamentous fungi and WRF fungi are the most common worldwide and frequently utilized in Southeast Asia as biodecolorization agents for various synthetic dyes. The involvement of ligninolytic enzymes, particularly laccase, is evident in the degradation process based on the activity of the implicated enzymes.

10.3.2.2 Mycoremediation of Heavy Metals

The issue of heavy metals contamination in Southern Asia is a pressing environmental issue. It is primarily attributed to industrial activities and mining processes, prompting the need for fungi-based remediation. Heavy metals (HMs) are metallic elements that occur naturally and have an atomic number (Z) higher than 23. They also have a density greater than 5 gcm^{-3} (Saleh and Aglan 2018). Chromium, lead, cadmium, mercury, arsenic, manganese, nickel, copper, iron, and more are among the prevalent heavy metals in polluted environments. These metals significantly threaten ecosystems and human health due to their extreme toxicity. The issue of heavy metals pollution is increasingly becoming a matter of grave concern. Numerous industrial activities, including leather tanning, electroplating, paint production, and wastewater discharge from these sectors, contribute to this problem. The presence of these metals is essential in diminishing the biodegradability of organic pollutants, resulting in them being twice as detrimental as contaminants. The biodegradability of organic pollutants makes them twice as harmful as contaminants (Debnath et al. 2019). The accumulation of heavy metals in river sediments and their subsequent release into adjacent water bodies poses a significant risk to ecosystems and causes pollution. (Ali et al. 2019). Unfortunately, using such contaminated water for agricultural purposes has long-lasting detrimental impacts on the ecosystem through the food chain (Khatoon et al. 2020). It threatens human health (Balali-Mood et al. 2021). The accumulation of HMs in human cells and tissues due to prolonged exposure can have fatal consequences, leading to neurological, cardiovascular, hematologic, reproductive, and immunological disorders (Debnath et al. 2019; Tufail et al. 2022). HMs are typically toxic, even at deficient concentrations, and are non-biodegradable. Moreover, they often exhibit resistance to conventional elimination methods.

Fungi can mitigate heavy metals (HMs) in polluted water by directing them toward the fungal fruiting bodies and filaments. Moreover, the cell wall materials of fungi exhibit remarkable metal-binding properties, enabling them to selectively target specific HMs (Dugal and Gangawane 2012). Numerous fungal species can remediate HMs owing to the diverse components found in their cell wall, such as carboxylic, phosphate, amine, or sulfhydryl groups. These components contribute negatively charged characteristics, facilitating the binding of metals (Wang and Chen 2009). However, the interaction between metals and their compounds with fungi can vary depending on the specific metal, organism, and environment (Archana and Jaitly 2015). The ability of fungi to absorb and accumulate heavy metals presents a

Table 10.7 Mycoremediation of synthetic dyes in the Southeast Asian region

Fungi	Fungi origin	Dyes	Concentration	Decolorization efficiency (incubation time)	Enzyme	References
Dark Septate Endophytic (DSE)	Indonesia	Remazol Brilliant Blue R (RBBR) and Acid Blue113 (AB 113)	50 mg/L	89 and 98% (14 days)	Ligninolytic enzymes	Melati et al. (2023b)
Trametes Hirsuta (T. Hirsuta)	Indonesia	AB 129, AB 113, RBBR, AB 25, Reactive Black 4 (RB4), and RB5	100 mg/L	88, 73, 71, 72, 42, and 6%, (7 h)	Laccase	Yanto et al. (2022)
Gloeophyllum trabeum and Aspergillus orizae	Indonesia	Methylene Blue (MB)	95.24 mg/L	69.34% (7 days)	–	Purnomo et al. (2022)
Trichoderma yunnanense	Indonesia	Indigosol Golden Yellow	250 mg/L	96.3% (168 h)	–	Rohmawati and Sri Kasiamdari (2022)
T. polyzona	Indonesia	RBBR, AB129, and RB5	100 mg/L	95, 89 and 95% (96 h)	Laccase	Ramadhan et al. (2021)
T. Hirsuta D7 (Immobilized in LECA)	Indonesia	RB4, RBBR and AB129	100 mg/L	90, 95, and 96% (24 h)	Laccase	Alam et al. (2021)
DSE (CPP and KSP)	Indonesia	Congo Red (CR)	50 mg/L	97.00 and 85.00% (14 days)	Laccase	Melati et al. (2021b)
T. hirsuta EDN 082 (Immobilized in LECA)	Indonesia	RBBR	100 mg/L	99.29% (3 days)	Laccase	Anita et al. (2020)
Penicillium sp.	Indonesia	MB	40 mg/L	95.45% (8 days)		Permana et al. (2019)
Leiotrametes flavida Strain	Indonesia	RBBR	100 mg/L	62% (24 h)	Laccase	Falah et al. (2018)

(continued)

Table 10.7 (continued)

Fungi	Fungi origin	Dyes	Concentration	Decolorization efficiency (incubation time)	Enzyme	References
Strain KRUS G	Indonesia	RBBR	100 mg/L	> 79% (6 days)	Laccase, MnP	Sumandono et al. (2015)
Resinicium bicolor	Malaysia	CR	1.00%	96.79% (96 h)	Laccase	Shah et al. (2023)
Aspergillus iizukae	Malaysia	RBBR	10 mg/L	94.78% (9 days)	Laccase (Lac), manganese peroxidase (MnP), and lignin peroxidase (LiP)	Noman et al. (2020)
Aspergillus sp. *and Pleurotus* sp.	Malaysia	Reactive Red 120	10 mg/L	60–70% (21 days)	–	Ibrahim et al. (2018)
Cerrena sp. (WRF)	Malaysia	Reactive Black 5 (RB5) and MB	0.20%	86–88% (28 days)	Laccase	Hanapi et al. (2018)
Penicillium simplicissimum	Malaysia	Crystal violet (CV), methyl violet (MV), malachite green (MG)		98.7; 97.1 and 96.1% (2 h)	MnP, Tyrosinase, and triphenylmethane reductase	Chen et al. (2018)
Trichoderma asperellum	Malaysia	CV, Methyl Violet (MV), Malachite Green (MG)	100 mg/L	16.3%; 82.7%; 70.2; (14 days)	Laccase, MnP, LiP, ADH-DCIP reductase	Marcharchand and Ting (2017)
Pleurotus eryngii	Malaysia	Reactive Black 5 (RB5)	10 mg/L	93.56% (72 h)	Lac, MnP, and LiP	Hadibarata et al. (2013)
Aspergillus iizukae	Malaysia	RBBR, Methyl orange (MO)	10 mg/100 mL and 40 mg/100 mL	94.78% (9 days) and 84.9% (3 days)	Lac, MnP, and LiP	Noman et al. (2020)

(continued)

Table 10.7 (continued)

Fungi	Fungi origin	Dyes	Concentration	Decolorization efficiency (incubation time)	Enzyme	References
Curvularia clavata	Malaysia	Reactive Black 5 (RB5), Acid Orange 7 (AO7), and CR	100 mg/L	90.3, 46.3 and 96% (24 h)	Carboxymethylcellulose (CMCase), xylanase, laccase, and manganese peroxidase (MnP)	Neoh et al. (2015)
Resinicium bicolor (WRF)	Thailand	Congo red (CR)	1%	96.79% in	Laccase	Shah et al. (2023)
Trametes hirsuta	Thailand	Mixed textile dyes (Navy EC-R, Ruby S3B, and Super Black G)	4%	95.39% (2 days)	MnP and Laccase	Thampraphaphon et al. (2022)
TrametespolyzonaKU-RNW027	Thailand	RBBR and Reactive Blue 12	25 mg/L	100% (30 min)	Laccase	Lueangjaroenkit et al. (2019)
Trametes polyzona	Thailand	Bromophenol Blue (BB), RBBR,Methyl Orange	0.01%	> 80% (3 h)	Laccase	Chairin et al. (2013)
Lentinus polychrous	Thailand	Indigo Carmine	20 mg/L	93.1% (6 h)	Laccase	Sarnthima et al. (2009)
Aspergillus niger	Vietnam	CR	200 mg/L	90% (14 days)	–	Tran et al. (2020)
Fusarium oxysporum	Vietnam	Malachite green (MG)	1000 mg/L	88.1% (20 h)	Laccase	Thoa et al. (2022)
Polyporus sp.	Vietnam	Acid Red and RBBR	100 mg/L	60 and 83.00% (30 min)	Laccase	Hiền et al. (2016)

(continued)

Table 10.7 (continued)

Fungi	Fungi origin	Dyes	Concentration	Decolorization efficiency (incubation time)	Enzyme	References
Lentinus squarrosulus	Vietnam	Acid Blue 62; Porocion Brilliant blue HGR; Acid Blue 113	0.15 mg/mL	92; 73.5; 61% (48 h)	LiP	Giap et al. (2022)
Colletotrichum sp.	Philippines	CR	100 mg/L	92.82–99.51% (10 days)	Laccase, LiP, MnP	Obanan et al. (2022)
Penicillium citrinum		MB		55.45% (10 days)		
Mycelia sterilia		Malachite Green		85.19 (10 days)		
Aspergillus flavus		Rhodamine B		44.92%(10 days)		
Cladosporium sp., *Penicillium* sp.	Philippines	CR and Crystal Violet (CV)	0.01%	100% and 87–91% (4 weeks)		Torres et al. (2011)

promising opportunity for a cost-effective alternative approach to removing heavy metals from wastewater. Certain fungi originate from Southeast Asia, such as *Penicillium* sp. and *Tricoderma* sp. *Aspergillus* sp. and *Pleurotus* sp., even the newest type of fungi, DSE, were reported to be able to eliminate various kinds of heavy metals (Table 10.8).

It is interesting that filamentous fungi such as *Aspergillus* sp., *Penicillium* sp, and *Trichoderma* sp. are more commonly utilized than WRF in heavy metal remediation in Southeast Asia. One study (Ghosh et al. 2023) stated that filamentous fungi exhibit high removal efficiency and rapid elimination of diverse pollutant compounds, coupled with their ease of handling, rendering them valuable assets in the bioremediation of emerging contaminants.

10.3.2.3 Mycoremediation of Another Persistent Organic Pollutant (PPCPs, Agrochemicals Compound, PAHs, and Dioxin)

Using fungi for mycoremediation to address environmental pollutants has shown promise in Southeast Asia, particularly in dealing with persistent organic pollutants like agrochemical compounds, PPCPs, PAHs, and Dioxin. Agrochemicals affect aquatic life and lower the dissolved oxygen, which leads to eutrophication. This phenomenon also resulted in the build-up of harmful substances in various foods and vegetables, subsequently infiltrating the food chain and posing a threat to human well-being (Akhtar and Mannan 2020). Similarly, pharmaceuticals in aquatic environments are a major scientific challenge, with antibiotic resistance being a significant concern. The impact goes beyond just resistance, affecting the behavior and reproduction of aquatic organisms and potentially causing widespread ecosystem disruptions (Hassan et al. 2023). Innovative strategies like mycoremediation to mitigate their adverse effects on aquatic and terrestrial ecosystems in Southeast Asia are needed. Various studies in Southeast Asia have pointed out the effectiveness of fungi from the WRF group (*Pleurotus* sp., *Trametes* sp. *Fusarium* sp., and *Rigidoporus* sp.) or filamentous fungi (*Aspergillus* sp.) in degrading a wide range of pollutants from PPCs, agrochemicals (herbicides and insecticides) to PAH, phenol, and Dioxin (Table 10.9).

Limited studies on using fungi as mycoremediation agents for agrochemical compounds and PPCPs in Southeast Asia indicate that mycoremediation research holds great potential for advancement. Additionally, significant research in the usage of agrochemicals and PPCPs in Southeast Asia underscores the critical importance of conducting research in this area. Southeast Asian countries are the hotspot for pharmaceutical manufacturing and also have a growing market for pharmaceutical consumption (Sitharam 2020). Similarly, the usage of agrochemicals has been progressively rising in Southeast Asia due to the expansion of large-scale commercial farming and the goal of maximizing food production in rural subsistence economies (Hughes et al. 2021).

Table 10.8 Mycoremediation of Heavy metals in the Southeast Asian region

Fungus	Fungus origin	Removed HMs	Concentration of HM	Reduction percent (incubation time)	Mechanism	Reference
Penicillium sp.	Indonesia	Zn	Batik effluent	> 90%	Biosorption	Dewi and Hana (2021)
DSE Fungi (KSP, CPP, PP, and TKC)	Indonesia	Cr VI	10–50 mg/L	Up to 99%	Biosorption and enzymatic process	Melati et al. (2023a)
Penicillium sp. and *Aspergillus* sp.	Indonesia	Cr VI	< 0.12 to 0.101 mg/L	82.05 and 75.82% (5 days)	Adsorption	Dewi et al. (2020)
Aspergillus sp.	Indonesia	Cr VI	1.1 mg/L	89.09% (3 days)	–	Dewi et al. (2019a, b)
Strain XF02, XF09, and XF21	Indonesia	Pb	500–1000 mg/L	98.31–99.53 (10 days)	Bioaccumulation	Wijaya et al. (2019)
Yeast isolates ES9.3	Indonesia	Cu	10 mM	81% (30 h)	Biosorption	Irawati et al. (2017)
Trichoderma atroviride	Indonesia	Cr, Ni, and Co	10 mg/L	Up to 94, 68 and 60% (5 days)	–	Hernahadini et al. (2014)
Aspergillus flavus (A. flavus)	Indonesia	Hg^{2+}	10 mg/L	97.5% (7 days)	Biosorption or bio immobilization	Kurniati et al. (2014)
Pleurotus pulmonarius	Malaysia	Cu, Fe, Mn, and Zn	–	34.54; 76.82; 38.1; and 41.94% (144 h)	Biosorption	Muliyadi et al. (2022)
R. oryzae and *A. niger*	Malaysia	Hg^{2+}	10–100 mg/L	60–90%	Biosorption	Anuar et al. (2020)
Trichoderma atroviride	Malaysia	Cd	0.20 μM	Good growth	–	Jamali et al. (2020)
Trichoderma sp.	Malaysia	Cd and Ni	0.5 to 2.0 mg/L	91 and 75%; (72 h)	Biosorption	Ab Rahman et al. (2016)

(continued)

Table 10.8 (continued)

Fungus	Fungus origin	Removed HMs	Concentration of HM	Reduction percent (incubation time)	Mechanism	Reference
Pleurotus ostreatus	Malaysia	Fe;Cu; and Cr (VI)	20 mg/L	1.98; 56.56; 2.30% (10 h)	Biosorption	Arbanah et al. (2012)
A. restrictus and *Sterigmatomyces halophilus*	Thailand	Fe and Zn	1000 mg/L	85 and 84%	Biosorption	Bano et al. (2018)
A. terreus	Thailand	Pb	0.788 mg/L	95.94% (168 h)	Adsorption/ absorption	Neeratanaphan et al. (2017)
Penicillium sp.	Thailand	As	9.82 mg/L	20.69%(6 days)	Biosorption and biodegradation	Visoottiviseth and Panviroj (2001)
T. virens	Philippines	Pb, Cu and Cr	100 mg/L	91–96, 70, and 63% (7 days)	–	Tansengco et al. (2018)

Table 10.9 Mycoremediation of another persistent organic pollutant (Pharmaceuticals and Personal Care Products (PPCPs), agrochemicals compound, PAHs, and Dioxin) in Southeast Asian region

Fungi	Fungi origin	Removed pollutant	Concentration of pollutant (mg/L)	Removal percent (incubation time)	Mechanism	References
Geotrichum silvicola	Indonesia	Paraquat (herbicide)	1 mg/L	9.66% (192 h)	–	Bosha et al. (2022)
Pleurotus eryngii F032	Indonesia	Benzo [a]pyrene (BaP)	–	73% (20 days)	Biosortion and biotransformation	Hadibarata et al. (2022)
Aspergillus fumigatus	Indonesia	Chlorpyrifos (organophosphate insecticide)	10%	99.65% (9 days)	Enzymatic activity	Anggreini et al. (2020)
Trametes polyzona	Indonesia	PAHs (phenanthrene, fluoranthene and pyrene)	100 mg/L	> 80% (15 days)	Laccase, MnP	Wulandari (2020)
Aspergillus sp.	Indonesia	Phenol	0.33 mg/L	48.48%	–	Dewi et al. (2019a, b)
Trichoderma viride	Indonesia	Glyphosate (herbicide)	–	70%	–	Arfarita et al. (2016)
Tremetes versicolor	Indonesia	1,1,1-trichloro-2,2-bis (4-chlorophenyl) ethane (DDT)	–	75% (30 days)	Adsorption and ligninolytic system	Arum Sari et al. (2014)
Pleurotus ostreatus	Indonesia	Heptachlor	5 mM	89% (28 days)	Transformation	Purnomo et al. (2014)
Aspergillus tamarii (A. tamarii) and Cunninghamella sp.	Thailand	Paraquat (herbicide)	10 mg/L	80 and 68% (16 days)	Biosorption and biodegradation	Wongputtisin et al. (2021)
Trichoderma harzianum and Trametes versicolor	Thailand	Sulfamethoxazole (SMX) (antibiotic)	10 mg/L%	71 (7 days) and 90% (10 days)	Degradation	Piyaviriyakul et al. (2021)

(continued)

Table 10.9 (continued)

Fungi	Fungi origin	Removed pollutant	Concentration of pollutant (mg/L)	Removal percent (incubation time)	Mechanism	References
Trametes polyzona	Thailand	PPCPs (Tetracycline, doxycycline, and amoxicillin)	25 mg/L	100%, 100 and 25% (3 days)	Laccase	Lueangjaroenkit et al. (2019)
Trametes polyzona (laccase)	Thailand	Bisphenol	0.01%	100% (3 h)	Laccase	Chairin et al. (2013)
Ganoderma lucidum	Malaysia	PPCPs	–	< 10– > 90% (48 h)	–	Hanafiah et al. (2024a, b)
Pleurotus eryngii	Malaysia	Benzo9a)pyrene(BaP)/PAH	10 mg/L	73% (20 days)	–	Hadibarata et al. (2022)
Rigidoporus sp.; F. oxysporum T1 and V. dahliae T1-BH.2	Vietnam	Herbicides (2,4-Dichlorophenoxyacetic acid and 2,4,5-Trichlorophenoxyacetic acid)	200 and 100 mg/L	100% (6 days and 16 days)	Laccase	Nguyen et al. (2022)
Rigidoporus (WRF)	Vietnam	Tetrachlorodibenzo-p-dioxin (TCDD)	0.5 pg TEQ/μL T	73% (28 days)	Laccase and MnP	Dao et al. (2019)
Lasiodiplodia sp.	Vietnam	Tiamulin (antibiotic)	10 mg/L	> 85% (12 days)	–	Nguyen et al. (2017)

10.3.2.4 Mycoremediation of Real Wastewater

The primary goal of bioremediation research should always be the practical implementation of findings in real-life situations. However, studies have yet to be conducted to evaluate the efficacy of fungi in removing pollutants from actual wastewater in Southeast Asian region. Generally, in line with the mycoremediation of several pollutants described previously, the use of filamentous fungi and WRF is more widely reported for the remediation of real wastewater (Table 10.10).

According to Table 10.10, a variety of WRF have been utilized for the treatment of batik wastewater, such as *Trametes* sp., *Trichoderma* sp., *Pleurotus* sp., *Ganoderma* sp., and *Pleurotus* sp. These fungi appear to be quite robust, achieving removal rates of up to 90–95% for color and 23.74–53.74% for chemical oxygen demand (COD),

Table 10.10 Mycoremediation of real wastewater in the Southeast Asian region

Fungi	Fungi origin	Wastewater	Parameter	Reduction percentage (incubation times)	References
Trametes hirsuta EDN 082 (myco-LECA)	Indonesia	Printing batik dye wastewater	Color	90–95% (144 h)	Yanto et al. (2023)
Aspergillus tamarii (A. tamarii)	Indonesia	Batik wastewater	BOD, COD, TDS, and TSS	42.10, 37.13, 36.10, and 21.20% (5 days)	Ayu and Kasiamdari (2023)
A. sclerotiorum				37.44, 42, 42 and 31.3% (5 days)	
Trichoderma yunnanense	Indonesia	Batik wastewater	BOD, COD, TDS, Sulfide, TSS, pH, and oil -fat	32.68, 23, 0.74, 12.69, 56.63, 37.14, 18.21 and 52.78% (240 h)	Rohmawati and Sri Kasiamdari (2022)
Pleurotus ostreatus	Indonesia	Batik wastewater	COD	53.74% (24 h)	Dewi et al. (2019a)
Aspergillus sp.	Indonesia	Batik effluent	BOD, COD, TSS, TDS, and EC	88.34, 89.11, 75.77, 85.85 and 71.21% (3 days)	Dewi et al. (2018)
Ganoderma lucidum	Malaysia	Mining lake	Color	90% (48 h)	Hanafiah et al. (2024a, b)
Pleurotus pulmonarius	Malaysia	Textile wastewater	Color and BOD	59.45 and −3.72% (144 h)	Muliyadi et al. (2022)
Pleurotus florida	Philippines	Textile wastewater	Color	84.98% (22 h)	Montiague et al. (2020)
Trametes elegans PP17	Thailand	Pulp and paper mill effluent	Color	63.6% (5 days)	Ridtibud et al. (2023)

depending on the type of WRF and incubation time utilized. The biological oxygen demand (BOD) parameter shows variations in response to different WRF. *Trichoderma yunnanense* was found to decrease BOD levels in batik waste by 32.68% over 240 h (Rohmawati and Kasiamdari 2022), while *Pleurotus pulmonarius* was observed to increase BOD levels in textile wastewater by 3.72% (Muliyadi et al. 2022). Temperature is essential for the development and functioning of microorganisms, as well as the decomposition of organic substances, leading to increased availability of organic matter (Muliyadi et al. 2022).

Aspergillus, a type of filamentous fungus, appears to have a remarkable ability to degrade batik wastewater compared to WRF (Table 10.10). *Aspergillus tamarii* (*A. tamarii*) showed removal efficiencies of 42.10, 37.13, 36.10, and 21.20% for BOD, COD, total dissolved solids (TDS), and total suspended solids (TSS), respectively, while *A. sclerotiorum* showed high removal at 37.44, 42, 42, and 31.3%, with an incubation period of five days (Ayu and Kasiamdari 2023). *Aspergillus* sp. showed greater efficacy in reducing BOD, COD, TSS, TDS, and electrical conductivity (EC) in batik waste with percentages of 88.34%, 89.11%, 75.77%, 85.85%, and 71.21%, respectively, within three days of incubation (Dewi et al. 2018).

WRF has been documented to possess extensive capabilities in breaking down different types of waste. *Pleurotus pulmonarius* demonstrated a 59.45% decolorization rate of textile wastewater in 144 h, whereas *Pleurotus florida* exhibited an even higher efficiency of 84.98% within 22 h (Montiague et al. 2020; Muliyadi et al. 2022). *Ganoderma lucidum* can achieve a 90% decolorization rate of mining waste, while *Trametes elegans* PP17 has shown a 63.6% decolorization rate of pulp and paper mill effluent (Hanafiah et al. 2024a, b; Ridtibud et al. 2023).

10.4 Remarks

There are growing numbers of technologies which can be advantageous in helping to remove certain water pollutants, thereby, that also benefits to reduce the contamination and restores the water environment. Constructed wetland, natural materials along with bioremediation i.e., mycoremediation have shown such great potentials as alternative technologies for environmental restoration through extensive research and studies, however, the real-life or scale-up application is still relatively limited or underrated in tropical region, especially Southeast Asia. One of the influencing factors can be the consistency of the performance. More studies on pilot scale are, then, required to fill the knowledge gap prior to the wider/full-scale application.

References

Abdulgani H, Izzati M, Sudarmo (2013) Pengolahan LIimbah Cair Industri Kerupuk Dengan Sistem Subsurface Flow Constructed Wetland Menggunakan Tanaman *Typha Angustifolia*. Purifikasi 1(1):90–101

Ab Rahman NNN, Shahadat M, Omar FM, Chew AW, Kadir MOA (2016) Dry *Trichoderma* biomass: biosorption behavior for the treatment of toxic heavy metal ions. Desalin Water Treat 57:13106–13112. https://doi.org/10.1080/19443994.2015.1057767

Abuqamar SF, Abd El-Fattah HI, Nader MM, Zaghloul RA, Abd El-Mageed TA, Selim S, Omar BA, Mosa WF, Saad AM, El-Tarabily KA, El-Saadony MT (2023) Exploiting fungi in bioremediation for cleaning-up emerging pollutants in aquatic ecosystems. Mar Environ Res 190:106068. https://doi.org/10.1016/j.marenvres.2023.106068

Agaton CB, Guila PMC (2024) Success factors and challenges: implications of real options valuation of constructed wetlands as nature-based solutions for wastewater treatment. Resources 13(1). https://doi.org/10.3390/resources13010011

Agrawal PK, Upadhyay P, Shrivastava R, Sharma S, Garlapati VK (2021) Evaluation of the ability of endophytic fungi from *Cupressus torulosa* to decolorize synthetic textile dyes. J Hazardous Toxic Radioact Waste 25:06020005. https://doi.org/10.1061/(asce)hz.2153-5515.0000569

Ahmad A, Rafatullah M, Sulaiman O, Ibrahim MH, Chii YY, Siddique BM (2009) Removal of Cu (II) and Pb (II) ions from aqueous solutions by adsorption on sawdust of meranti wood. Desalination 247(1–3):636–646

Akhir MSAM, Amir AA, Bin Mokhtar M (2017) Nutrients and pollutants removal in small-scale constructed wetland in *Frangipani* resort Langkawi Malaysia. Nat Environ Pollut Technol 16(2):569–577

Akhtar N, Mannan MA (2020) Mycoremediation: Expunging environmental pollutants. Biotechnol Reports 26. https://doi.org/10.1016/j.btre.2020.e00452

Akpasi SO, Michael I, Anekwe S, Tetteh EK, Amune UO, Shoyiga HO, Mahlangu TP, Kiambi SL (2023) Mycoremediation as a potentially promising technology: current status and prospects—a review. https://doi.org/10.3390/app13084978

Al-Tohamy R, Ali SS, Li F, Okasha KM, Mahmoud YAG, Elsamahy T, Jiao H, Fu Y, Sun J (2022) A critical review on the treatment of dye-containing wastewater: ecotoxicological and health concerns of textile dyes and possible remediation approaches for environmental safety. Ecotoxicol Environ Saf 231:113160. https://doi.org/10.1016/j.ecoenv.2021.113160

Alam R, Ardiati FC, Solihat NN, Alam MB, Lee SH, Yanto DHY, Watanabe T, Kim S (2021) Biodegradation and metabolic pathway of anthraquinone dyes by *Trametes hirsuta* D7 immobilized in light expanded clay aggregate and cytotoxicity assessment. J Hazard Mater, 124176. https://doi.org/10.1016/j.jhazmat.2020.124176

Ali H, Khan E, Ilahi I (2019) Environmental chemistry and ecotoxicology of hazardous heavy metals: environmental persistence, toxicity, and bioaccumulation. J Chem. https://doi.org/10.1155/2019/6730305

Alwi H, Idris J, Musa M, Ku-Hamid KH (2013) A preliminary study of banana stem juice as a plant-based coagulant for treatment of spent coolant wastewater. J Chem 2013(1):165057

Ang WL, Mohammad AW (2020) State of the art and sustainability of natural coagulants in water and wastewater treatment. J Clean Prod 262:121267

Anggreini CD, Tazkiaturrizki T, Rinanti A (2020) Chlorpyrifos removal improvement in liquid media by *Aspergillus Fumigatus*. Int J Sci Technol Res 9(1)

Anh TT, Huyen DTT, Ha TD, Khai NM (2022) Dewatering of urban lake sediments using constructed wetlands: a case study in Hanoi, Vietnam. Int J GEOMATE 23(99):100–108. https://doi.org/10.21660/2022.99.3644

Anita SH, Ardiati FC, Oktaviani M, Sari FP, Nurhayat OD, Ramadhan KP, Yanto DHY (2020) Immobilization of laccase from *Trametes hirsuta* EDN 082 in light expanded clay aggregate for decolorization of Remazol Brilliant Blue R dye. Bioresour Technol Reports 12:100602. https://doi.org/10.1016/j.biteb.2020.100602

Anuar SNIS, Othman F, Chia Chay T, Atikah Husna Ahmad Nasir N (2020) Biosorption of mercury ion (Hg 2+) using live and dead cells of *Rhizopus oryzae* and *Aspergillus niger* : characterization, kinetic and isotherm studies. J Pure Appl Microbiol 14:1749–1760. https://doi.org/10.22207/JPAM.14.3.13

Arbanah M, Najwa MRM, Halim KHK (2012) Biosorption of Cr(III), Fe(II), Cu(II), Zn(II) ions from liquid laboratory chemical waste by *Pleurotus ostreatus*. Int J Biotechnol Wellness Ind 1(3):152

Archana A, Jaitly AK (2015) Mycoremediation: utilization of fungi for reclamation of heavy metals at their optimum remediation conditions. Biolife 3:77–106

Arfarita N, Djuhari, Prasetya B, Imai T (2016) The application of *Trichoderma viride* strain frp 3 for biodegradation of glyphosate herbicide in contaminated land. Agrivita 38:275–281. https://doi.org/10.17503/agrivita.v38i3.550

Arum Sari A, Kristiani A, Tachibana S, Sudiyani Y, Abimanyu H (2014) Mechanisms and optimization of oil palm empty fruit bunch as a pre-grown source for white-rot fungus to degrade DDT. J Environ Chem Eng 2:1410–1415. https://doi.org/10.1016/j.jece.2014.07.018

Asmaliza MNN, Lariyah MS, Rozi A, Aminuddin AG (2013) Performance of artificial wetland in removing contaminants from storm water under tropical climate. IAHS-AISH Publ 357(2017):208–216

Ayalew AA, Aragaw TA (2020) Utilization of treated coffee husk as low-cost bio-sorbent for adsorption of methylene blue. Adsorpt Sci Technol 38(5–6):205–222

Ayu YS, Kasiamdari RS (2023) Biological treatment of naphthol yellow S and batik effluent using aspergillus tamarii and aspergillus sclerotiorum. Res J Chem Environ 27:1–7. https://doi.org/10.25303/2706rjce01007

Bahrodin MB, Zaidi NS, Kadier A, Hussein N, Syafiuddin A, Boopathy R (2022) A novel natural active coagulant agent extracted from the sugarcane bagasse for wastewater treatment. Appl Sci 12(16):7972

Balali-Mood M, Naseri K, Tahergorabi Z, Khazdair MR, Sadeghi M (2021) Toxic mechanisms of five heavy metals: mercury, lead, chromium, cadmium, and arsenic. Front Pharmacol. https://doi.org/10.3389/fphar.2021.643972

Bano A, Hussain J, Akbar A, Mehmood K, Anwar M, Hasni MS, Ullah S, Sajid S, Ali I (2018) Biosorption of heavy metals by obligate halophilic fungi. Chemosphere 199:218–222. https://doi.org/10.1016/j.chemosphere.2018.02.043

Benjawan L, Liamlaem W, Polprasert C (2022) Potential use of the constructed wetland system as an eco-technology for wastewater treatment in resort. Environ Asia 15(2):132–140. https://doi.org/10.14456/ea.2022.40

Bish DL (2013) Parallels and distinctions between clay minerals and zeolites. In: Developments in clay science, vol 5. Elsevier, pp 783–800

Borsatto F, Inglezakis VJ (2012) Natural zeolite markets and strategic considerations. In: Handbook of natural zeolites, 11–27

Bosha M, Abadi AL, Aini LQ (2022) Potential of indigenous yeast as a paraquat biodegradation agent in Sumberbrantas, Batu City, Indonesia. J Trop Plant Prot 3:1–10. https://doi.org/10.21776/ub.jtpp.2022.003.1.1

Brix H, Koottatep T, Fryd O, Laugesen CH (2011) The flower and the butterfly constructed wetland system at Koh Phi Phi-System design and lessons learned during implementation and operation. Ecol Eng 37(5):729–735. https://doi.org/10.1016/j.ecoleng.2010.06.035

Budianta W, Andriyani ND, Ardiana A, Warmada IW (2020) Adsorption of lead and cadmium from aqueous solution by Gunungkidul zeolitic tuff, Indonesia. Environ Earth Sci 79(8):172

Budihardjo MA, Wibowo YG, Ramadan BS, Serunting MA, Yohana E (2021) Mercury removal using modified activated carbon of peat soil and coal in simulated landfill leachate. Environ Technol Innov 24:102022

Chairin T, Nitheranont T, Watanabe A, Asada Y, Khanongnuch C, Lumyong S, Nitheranont T, Watanabe A, Asada Y, Khanongnuch C, Lumyong S (2013) Biodegradation of bisphenol A and decolorization of synthetic dyes by laccase from white-rot fungus, *Trametes polyzona*. Appl Biochem Biotechnol 169:539–545. https://doi.org/10.1007/s12010-012-9990-3

Chantawong V, Harvey NW, Bashkin VN (2003) Comparison of heavy metal adsorptions by Thai kaolin and ballclay. Water Air Soil Pollut 148:111–125

Chapman HD (1965) Cation-exchange capacity. Methods Soil Anal Part 2 Chem Microbiol Prop 9:891–901

Chen SH, Cheow YL, Ng SL, Ting ASY (2019) Biodegradation of triphenylmethane dyes by non-white rot fungus *Penicillium simplicissimum*: enzymatic and toxicity studies. Int J Environ Res 13:273–282. https://doi.org/10.1007/s41742-019-00171-2

Chen SH, Cheow YL, Ng SL, Ting ASY (2018) Influence of dyes on metal removal: a study using live and dead cells of *Penicillium simplicissimum* in single-metal and dye-metal mixtures. Water Air Soil Pollut, 229. https://doi.org/10.1007/s11270-018-3931-x

Cheng W, Sim H, Ahmad S, Syed M, Shukor M, Yusof M (2016) Characterization of an azo-dye-degrading white rot fungus isolated from Malaysia. Mycosphere 7:560–569. https://doi.org/10.5943/mycosphere/7/5/3

Chua LHC, Tan SBK, Sim CH, Goyal MK (2012) Treatment of baseflow from an urban catchment by a floating wetland system. Ecol Eng 49:170–180. https://doi.org/10.1016/j.ecoleng.2012.08.031

Crini G, Lichtfouse E, Wilson LD, Morin-Crini N (2019) Conventional and non-conventional adsorbents for wastewater treatment. Environ Chem Lett 17:195–213

Dan NP, Tam LTM, Le Quyen V, Chi DHL (2020) Application of constructed wetland for advanced treatment of industrial wastewater. Minist Sci Technol Vietnam 62(2):89–96. https://doi.org/10.31276/vjste.62(2).89-96

Dao ATN, Vonck J, Janssens TKS, Dang HTC, Brouwer A, de Boer TE (2019) Screening white-rot fungi for bioremediation potential of 2,3,7,8-tetrachlorodibenzo-p-dioxin. Ind Crops Prod 128:153–161. https://doi.org/10.1016/j.indcrop.2018.10.059

Debnath B, Singh W, Manna K (2019) Sources and toxicological effects of lead on human health. Indian J Med Spec 10:66. https://doi.org/10.4103/injms.injms_30_18

Dewi R, Mumpuni A, Yusiana R (2020) Chromium removal of batik wastewater using *Aspergillus* sp. and *Penicillium* sp. IOP Conf Ser Earth Environ Sci 593:012022. https://doi.org/10.1088/1755-1315/593/1/012022

Dewi RS, Hana (2021) Screening of microfungi from spent mushroom fordecolorizing and removing heavy metals from batikeffluent and its toxicity.pdf. In: IOP Conf Ser Earth Environ Sci, 012028

Dewi RS, Kasiamdari RS, Martani E, Purwestri YA (2018) Decolorization and detoxification of batik dye effluent containing Indigosol Blue-04B using fungi isolated from contaminated dye effluent. Indones J Biotechnol 23(54). https://doi.org/10.22146/ijbiotech.32332

Dewi RS, Ilyas M, Sari AA (2019a) Ligninolitic enzyme immobilization from Pleurotus ostreatus for dye and batik wastewater decolorization. J Pendidik IPA Indones 8:220–229. https://doi.org/10.15294/jpii.v8i2.19372

Dewi RS, Kasiamdari RS, Martani E, Purwestri YA (2019) Efficiency of *Aspergillus* sp. 3 to reduce chromium, sulfide, ammonia, phenol, and fat from batik wastewater. In: IOP conference series: earth and environmental science. Institute of Physics Publishing, p 012003. https://doi.org/10.1088/1755-1315/308/1/012003

Dickson UJ, Coffey M, Mortimer RJG, Di Bonito M, Ray N (2019) Mycoremediation of petroleum contaminated soils: progress, prospects and perspectives. Environ Sci Process Impacts 21:1446–1458. https://doi.org/10.1039/c9em00101h

Dologuin RB, Sheila M, Ramos K, Guihawan JQ, Amparado RF (2022) Subsurface vertical flow constructed wetland as potential landfill leachate treatment of the solid waste disposal facility of Iligan City, Philippines. J Bio Env Sci 4:140–153

Dugal S, Gangawane M (2012) Metal tolerance and potential of *penicillium* species for use in mycoremediation. J Chem Pharm Res 4:2362–2366

Eichlerová I, Baldrian P (2020) Ligninolytic enzyme production and decolorization capacity of synthetic dyes by saprotrophic white rot, brown rot, and litter decomposing basidiomycetes. J Fungi 6:1–22. https://doi.org/10.3390/jof6040301

El Bouraie M, El Din WS (2016) Biodegradation of Reactive Black 5 by *Aeromonas hydrophila* strain isolated from dye-contaminated textile wastewater. Sustain Environ Res 26:209–216. https://doi.org/10.1016/j.serj.2016.04.014

Esterhuizen-Londt M, Sani SB, Wang L, Pflugmacher S (2019) Improving mycoremediation of acetaminophen: effect of pH, nitrogen limitation, and co-cultivation. In: CEST 2019 conference proceedings, pp 1–2

Ezeonuegbu BA, Machido DA, Whong CM, Japhet WS, Alexiou A, Elazab ST, Qusty N, Yaro CA, Batiha GES (2021) Agricultural waste of sugarcane bagasse as efficient adsorbent for lead and nickel removal from untreated wastewater: biosorption, equilibrium isotherms, kinetics and desorption studies. Biotechnol Rep 30:e00614

Falah S, Sari NM, Hidayat A (2018) View of decolorization of Remazol Brilliant Blue R by laccase of newly isolated *Leiotrametes flavida* strain ZUL62 from Bangka Heath Forest, Indonesia. Biodiversitas 19:633–639. https://doi.org/10.13057/biodiv/d190235

Ghosh S, Rusyn I, Dmytruk OV, Dmytruk KV, Onyeaka H, Gryzenhout M, Gafforov Y (2023) Filamentous fungi for sustainable remediation of pharmaceutical compounds, heavy metal and oil hydrocarbons. Front Bioeng Biotechnol. https://doi.org/10.3389/fbioe.2023.1106973

Giap VD, Nghi DH, Cuong LH, Quynh DT (2022) Lignin peroxidase from the white-rot fungus *Lentinus squarrosulus* MPN12 and its application in the biodegradation of synthetic dyes and lignin. BioResources. https://doi.org/10.15376/biores.17.3.4480-4498

Giubilei MA, Leonardi V, Federici E, Covino S, Šašek V, Novotny C, Federici F, D'Annibale A, Petruccioli M (2009) Effect of mobilizing agents on mycoremediation and impact on the indigenous microbiota. J Chem Technol Biotechnol 84:836–844. https://doi.org/10.1002/jctb.2126

Gómez-Aguilar DL, Rodríguez-Miranda JP, Baracaldo-Guzmán D, Salcedo-Parra OJ, Esteban-Muñoz JA (2021) Biosorption of Pb (II) using coffee pulp as a sustainable alternative for wastewater treatment. Appl Sci 11(13):6066

Guerin TF (2000) The differential removal of aged polycyclic aromatic hydrocarbons from soil during bioremediation. Environ Sci Pollut Res 7:19–26. https://doi.org/10.1065/espr199910.004

Guino-o Ii RS, Aguilar AS, Oracion EG (2009) The efficiency and social acceptability of the constructed wetland of Bayawan City. Negros Oriental 50(1):1–26

Guo G, Tian F, Zhao Y, Tang M, Liu W, Liu C (2019) Aerobic decolorization and detoxi fi cation of Acid Scarlet GR by a newly isolated salt-tolerant yeast strain *Galactomyces geotrichum* GG. Int Biodeterior Biodegrad 145:104818. https://doi.org/10.1016/j.ibiod.2019.104818

Gupta N, Yadav KK, Kumar V, Kumar S, Chadd RP, Kumar A (2019) Trace elements in soil-vegetables interface: translocation, bioaccumulation, toxicity and amelioration—a review. Sci Total Environ 651:2927–2942. https://doi.org/10.1016/j.scitotenv.2018.10.047

Gupta VK, Carrott PJM, Ribeiro CMML, Suhas (2009) Low-cost adsorbents: growing approach to wastewater treatment—a review. Crit Rev Environ Sci Technol 39(10):783-842

Hadibarata T, Amalina Adnan L, Rahim Mohd Yusoff A, Yuniarto A, Mohd Fikri Ahmad Zubir M, Badr Khudhair A, Chuang Teh Z, Abu Naser M (2013) Microbial decolorization of an azo dye Reactive Black 5 using white-rot fungus *Pleurotus eryngii* F032. Water Air Soil Pollut 224:1595. https://doi.org/10.1007/s11270-013-1595-0

Hadibarata T, Kristanti RA, Bilal M, Al-Mohaimeed AM, Chen TW, Lam MK (2022) Microbial degradation and transformation of benzo[a]pyrene by using a white-rot fungus *Pleurotus eryngii* F032. Chemosphere 307:136014. https://doi.org/10.1016/j.chemosphere.2022.136014

Hanafiah MZ, Wan Mohtar WHM, Waaqi W-M, Bithi AS, Rohani R, Indarto A, Yaseen ZM, Sharil S, Binti Abdul Manan TS (2024a) Removal of pharmaceutical compounds and toxicology study in wastewater using Malaysian fungal *Ganoderma lucidum*. Chemosphere 358:142209. https://doi.org/10.1016/j.chemosphere.2024.142209

Hanafiah Z, Azmi AR, Wan-Mohtar WAAQI, Olivito F, Golemme G, Ilham , Jamaludin AA, Razali N, Halim-Lim SA, Wan Mohtar WHM (2024b) Water quality assessment and decolourisation of contaminated ex-mining lake water using bioreactor dye-eating fungus (BioDeF) system: a real case study. Toxics 12(60). https://doi.org/10.3390/toxics12010060

Hanapi SZ, Abdelgalil SA, Hatti-Kaul R, Aziz R, El Enshasy HA (2018) Isolation of a new efficient dye decolorizing white rot fungus *Cerrena* Sp. WICC F39. J Sci Ind Res

Hassan A, Khan A, Barros R (2023) Pharmaceuticals in water: risks to aquatic life and remediation strategies. Hydrobiol 2(2):395–409. https://doi.org/10.3390/hydrobiology2020026

Hefnawy MA, Gharieb MM, Shaaban MT, Soliman AM (2017) Optimization of culture condition for enhanced decolorization of direct blue dye by *Aspergillus flavus* and *Penicillium canescens*. J Appl Pharm Sci 7:83–092. https://doi.org/10.7324/JAPS.2017.70210

Hendrawan DI, Widanarko S, Moersidik SS, Triweko RW, Catholic P, Program ES, Program ES, Indonesia J (2013) The performance of subsurface constructed wetland for domestic wastewater treatment. Int J Eng Res Technol 2(6):3374–3382

Heng S, Boeut S, Khoeurn K, Eang KE (2022) Application of eggshell in neutralization of acid mine drainage produced by Artisanal mine tailing, Chong Phlah, Mondulkiri, Cambodia. In: International symposium on earth resources and geo-environment technology

Henny C, Jasalesmana T, Kurniawan R, Melati I, Suryono T, Susanti E, Yoga GP, Rosidah, Sudiono BT (2020) The effectiveness of integrated floating treatment wetlands (FTWs) and lake fountain aeration systems (LFAS) in improving the landscape ecology and water quality of a eutrophic lake in Indonesia. IOP Conf Ser Earth Environ Sci 535(1). https://doi.org/10.1088/1755-1315/535/1/012018

Henny C, Kurniawan R (2020) Evaluation of plant structure in urban lake ecosystem and their performance in treatment wetland systems (TWS) to maintain lake water quality in Megacity Jakarta, Indonesia. IOP Conf Ser Earth Environ Sci 477(1). https://doi.org/10.1088/1755-1315/477/1/012003

Henny C, Kurniawan R, Akhdiana I (2019a) Floating treatment wetlands and submerged vegetation for water quality improvement of an urban lake in megacity Jakarta, Indonesia. IOP Conf Ser Earth Environ Sci 308(1). https://doi.org/10.1088/1755-1315/308/1/012005

Henny C, Kurniawan R, Trisuryono (2019b) Application of floating treatment wetlands in a highly eutrophic lake, Indonesia: a new tool for lake restoration and provision of microhabitat. In: IOP conference series: earth and environmental science

Henny C, Rohaningsih D, Susanti E, Sumi F, Sudiono BT, Waluyo A (2022a) Application of constructed floating wetlands (CFWs) to treat textile effluent. IOP Conf Ser Earth Environ Sci 1108(1). https://doi.org/10.1088/1755-1315/1108/1/012036

Henny C, Rohaningsih D, Susanti E, Sumi F, Waluyo A (2022b) Evaluation of heliconia psittacorum in a horizontal flow constructed wetland (HFCW) system for the treatment of textile wastewater. IOP Conf Ser Earth Environ Sci 1062(1). https://doi.org/10.1088/1755-1315/1062/1/012026

Herath IS, Udayanga D, Jayasanka DJ, Hewawasam C (2024) Textile dye decolorization by white rot fungi—a review. Bioresour Technol Reports 25:101687. https://doi.org/10.1016/j.biteb.2023.101687

Hernahadini N, Suhandono S, Choesin DN, Chaerun SK, Kadarusman A (2014) Isolation, identification and characterization of indigenous fungi for bioremediation of hexavalent chromium, nickel and cobalt. AIP Conf Proc 1589:350–354. https://doi.org/10.1063/1.4868817

Hidayati N, Soeprobowati TR, Helmi M (2018) The evaluation of water hyacinth (*Eichhornia crassiper*) control program in Rawapening Lake, Central Java Indonesia. IOP Conf Ser Earth Environ Sci 142(1). https://doi.org/10.1088/1755-1315/142/1/012016

Hiền TTT, Hiền LT, Huynh NV, Hằng ĐTT, Hà ĐTC (2016) Study of suitable media for laccase synthesis of basidiomycete fungi from Bidoup—Nuiba park and its reactive dye decolorization ability. Vietnam J Biotechnol 14:313–325. https://doi.org/10.15625/1811-4989/14/2/9357

Hughes D, Thongkum W, Tudpor K, Turnbull N, Yukalang N, Sychareun V, Van Vo T, Win LL, Watkins A, Jordan S (2021) Pesticides use and health impacts on farmers in thailand, vietnam, and lao pdr: protocol for a survey of knowledge, behaviours and blood acetyl cholinesterase concentrations. PLoS One 16. https://doi.org/10.1371/journal.pone.0258134

Ibrahim NN, Talib SA, Ismail HN, Tay CC (2018) Decolorization of reactive red-120 by using macrofungus and microfungus. J Fundam Appl Sci 9:954. https://doi.org/10.4314/jfas.v9i6s.11

Indonesian Information Portal (2023) Indonesian textile exports grow

Irawati W, Parhusip AJN, Christian S, Yuwono T (2017) The potential capability of bacteria and yeast strains isolated from rungkut industrial sewage in Indonesia as a bioaccumulators and biosorbents of copper. Biodiversitas 18:971–977. https://doi.org/10.13057/biodiv/d180315

Jakovljevic V, Vrvic M (2016) The effect of ethoxylated oleyl-cetyl alcohol on metabolism of some fungi and their potential application in mycoremediation. Hem Ind 70:277–286. https://doi.org/10.2298/HEMIND141229034J

Jamali SAM, Badaluddin NA, Baharum SN, Sali JM, Ahmad A, Taib M (2020) *Trichoderma atroviride* isolated from mangroves of the East Coast of Peninsular Malaysia exhibited high tolerance against heavy metal cadmium. Malaysian Appl Biol 49:113–120. https://doi.org/10.55230/mabjournal.v49i4.1600

Kadlec RH (2009) Comparison of free water and horizontal subsurface treatment wetlands. Ecol Eng 35(2):159–174. https://doi.org/10.1016/j.ecoleng.2008.04.008

Katayon S, Fiona Z, Noor MJMM, Halim GA, Ahmad J (2008) Treatment of mild domestic wastewater using subsurface constructed wetlands in Malaysia. Int J Environ Stud 65(1):87–102. https://doi.org/10.1080/00207230601125192

Khatoon H, Rai J, Jillani A (2020) Role of fungi in bioremediation of contaminated soil. In: Sharma V, Shah MP, Kumar A (eds) Fungi bio-prospects in sustainable agriculture, environment and nano-technology volume 2: extremophilic fungi and myco-mediated enviromental management. Academic Press, p 498

Komunikasi M, Pengembangan D, Lingkungan T, Masriyanto PW, Yulistyorini A, Ariestadi D (2023) Jurnal presipitasi application of free water surface constructed wetland for reduction of Brantas River pollutants 20(1):93–103

Koottatep T, Polprasert C, Thi N, Oanh K, Surinkul N, Montangero A, Strauss M (2002) Constructed wetlands for septage treatment—towards effective faecal sludge management. In: IWA 8th international donference on wetlands systems for water pollution control

Kurniadie D (2011) Wastewater treatment using vertical subsurface flow constructed wetland in Indonesia. Am J Environ Sci 7(1):15–19. https://doi.org/10.3844/ajessp.2011.15.19

Kurniadie D, Wijaya D, Widayat D, Umiyati U, Iskandar (2018) Constructed wetland to treat tapioca starch wastewater in Indonesia. Asian J Water Environ Pollut 15(3):107–113. https://doi.org/10.3233/AJW-180049

Kurniati E, Arfarita N, Imai T, Higuchi T, Kanno A, Yamamoto K, Sekine M (2014) Potential bioremediation of mercury-contaminated substrate using filamentous fungi isolated from forest soil. J Environ Sci 26:1223–1231. https://doi.org/10.1016/S1001-0742(13)60592-6

Kurniawan SB, Abdullah SRS, Imron MF, Said NSM, Ismail NI, Hasan HA, Othman AR, Purwanti IF (2020) Challenges and opportunities of biocoagulant/bioflocculant application for drinking water and wastewater treatment and its potential for sludge recovery. Int J Environ Res Public Health 17(24):9312

Kurniawan T, Mandalagiri NL, Kustiningsih I (2022) Characterization of zeolites from Tasikmalaya, Indonesia, and its application for ammonium removal. ASEAN J Sci Eng Mater 1(1):37–48

Lee MY, Hong KJ, Kajiuchi T, Yang JW (2005) Synthesis of chitosan-based polymeric surfactants and their adsorption properties for heavy metals and fatty acids. Int J Biol Macromol 36(3):152–158

Li Y, Zhu G, Ng WJ, Tan SK (2014) A review on removing pharmaceutical contaminants from wastewater by constructed wetlands: design, performance and mechanism. Sci Total Environ 468:908–932

Lim LB, Priyantha N, Hei IC, Khairud DM, Tennakoon DTB, Zehra T, Suklueng M (2015) *Artocarpus odoratissimus* skin as a potential low-cost biosorbent for the removal of methylene blue and methyl violet 2B. Desalin Water Treat 53(4):964–975

Lueangjaroenkit P, Teerapatsakul, C, Sakka K, Sakka M, Kimura T, Kunitake E, Chitradon L (2019) Two manganese peroxidases and a laccase of *Trametes polyzona* KU-RNW027 with novel properties for dye and pharmaceutical product degradation in redox mediator-free system. https://doi.org/10.1080/12298093.2019.1589900

Lürling M, Kang L, Mucci M, van Oosterhout F, Noyma NP, Miranda M, Huszar VL, Waajen G, Marinho MM (2020) Coagulation and precipitation of cyanobacterial blooms. Ecol Eng 158:106032

Malyan SK, Yadav S, Sonkar V, Goyal VC, Singh O, Singh R (2021) Mechanistic understanding of the pollutant removal and transformation processes in the constructed wetland system. Water Environ Res 93(10):1882–1909

Marcharchand S, Ting ASY (2017) *Trichoderma asperellum* cultured in reduced concentrations of synthetic medium retained dye decolourization efficacy. J Environ Manage 203:542–549. https://doi.org/10.1016/j.jenvman.2017.06.068

Maxwell OI, Onyebuchukwu MG, Chinonso OE, Nnaemeka IC, Afamefuna EK (2023) Kinetics of bioremediation of oil contaminated water dispersed by environment-friendly bacteria (*Pseudomonas aeruginosa*) and fungi (*Aspergillus niger*). Adv Chem Eng Sci 13:19–35. https://doi.org/10.4236/aces.2023.131003

Melati I, Rahayu G, Henny C (2021a) The recent status of synthetic dyes mycoremediation: a review. In: IOP conference series: earth and environmental science, p 13. https://doi.org/10.1088/1755-1315/1062/1/012029

Melati I, Rahayu G, Surono, Effendi H, Henny C (2021b) Decolourization of congo red synthetic dyes by dark septate endophytes. In: IOP conference series: earth and environmental science. The 4th international conference on biosciences (ICoBio), p 10. https://doi.org/10.1088/1755-1315/948/1/012073

Melati I, Rahayu G, Surono, Effendi H, Henny C, Susanti E (2023a) Chromium (VI) bioremediation potential of dark septate endophytic (DSE) fungi. In: IOP conference series: earth and environmental science. IOP Publishing, p 012077. https://doi.org/10.1088/1755-1315/1201/1/012077

Melati I, Rahayu G, Surono EH, Henny C, Yanto DHY (2023b) Biodecolorization of anthraquinone and azo dyes by dark septate endophytic fungi. Bioresour Technol Reports 22:101427. https://doi.org/10.1016/j.biteb.2023.101427

Montiague E, Guzman J, Unciano N, Panerrio E, Bigol U, Castro I, Jose J, Mantaring S (2020) Biodecolourization of textile dye and wastewaters by crude laccase from Pleurotus florida ITDI 6003 cultivated in wheat grains. J Fungal Biol 10:167–175. https://doi.org/10.5943/cream/10/1/17

Mostafa AAF, Elshikh MS, Al-Askar AA, Hadibarata T, Yuniarto A, Syafiuddin A (2019) Decolorization and biotransformation pathway of textile dye by *Cylindrocephalum aurelium*. Bioprocess Biosyst Eng 42:1483–1494. https://doi.org/10.1007/s00449-019-02144-3

Mudhoo A, Ramasamy DL, Bhatnagar A, Usman M, Sillanpää M (2020) An analysis of the versatility and effectiveness of composts for sequestering heavy metal ions, dyes and xenobiotics from soils and aqueous milieus. Ecotoxicol Environ Saf. https://doi.org/10.1016/j.ecoenv.2020.110587

Muliyadi S, Elida TZMT, Hana HN, Mat NR (2022) Bioremediation of textile wastewater using *Pleurotus pulmonarius*. J Sustain Sci Manag 17:67–76. https://doi.org/10.46754/jssm.2022.02.007

Napaldet JT, Buot IE (2019) Treatment of Balili River in Benguet, Philippines with constructed wetland planted with dominant local macrophytes. Int J Phytoremediation 21(14):1463–1473. https://doi.org/10.1080/15226514.2019.1633268

Naseem M, Raghuwanshi R, Verma PC, Srivastava PK (2021) Mycoremediation-effective strategy to ameliorate arsenic toxicity. In: Fungi bio-prospects in sustainable agriculture, environment and nano-technology. Elsevier, pp 433–458. https://doi.org/10.1016/B978-0-12-821925-6.00019-8

Neeratanaphan L, Khamma S, Benchawattananon R, Ruchuwararak P, Appamaraka S, Intamat S (2017) Heavy metal accumulation in rice (*Oryza sativa*) near electronic waste dumps and related human health risk assessment. Hum Ecol Risk Assess 23:1086–1098. https://doi.org/10.1080/10807039.2017.1300856

Neoh CH, Lam CY, Lim CK, Yahya A, Bay HH, Ibrahim Z, Noor ZZ (2015) Biodecolorization of recalcitrant dye as the sole source of nutrition using *Curvularia clavata* NZ2 and decolorization ability of its crude enzymes. Environ Sci Pollut Res 22:11669–11678. https://doi.org/10.1007/s11356-015-4436-4

Nguyen TKX, Thayanukul P, Pinyakong O, Suttinun O (2017) Tiamulin removal by wood-rot fungi isolated from swine farms and role of ligninolytic enzymes. Int Biodeterior Biodegrad 116:147–154. https://doi.org/10.1016/j.ibiod.2016.10.010

Nguyen TLA, Dao ATN, Dang HTC, Koekkoek J, Brouwer A, de Boer TE, Van Spanning RJM (2022) Degradation of 2,4-dichlorophenoxyacetic acid (2,4-D) and 2,4,5-trichlorophenoxyacetic acid (2,4,5-T) by fungi originating from Vietnam. Biodegrad 33:301–316. https://doi.org/10.1007/s10532-022-09982-1

Nguyen V-A, Pham N-T, Nguyen TH, Morel A, Tonderski K (2007) Improved septic tank with constructed wetland, a promising decentralized wastewater treatment alternative in Vietnam. In: NOWRA 16th annual technial education conference and exposition, pp 1–17

Nguyen XC, Nguyen DD, Tran QB, Nguyen TTH, Tran TKA, Tran TCP, Nguyen THG, Tran TNT, La DD, Chang SW, Balasubramani R, Chung WJ, Yoon YS, Nguyen VK (2020) Two-step system consisting of novel vertical flow and free water surface constructed wetland for effective sewage treatment and reuse. Bioresour Technol 306. https://doi.org/10.1016/j.biortech.2020.123095

Nilratnisakorn S, Thiravetyan P, Nakbanpote W (2009) A constructed wetland model for synthetic reactive dye wastewater treatment by narrow-leaved cattails (*Typha angustifolia* Linn.). Water Sci Technol 60(6):1565–1574. https://doi.org/10.2166/wst.2009.500

Nkansah MA, Dua AB, Aryee GA and Adusei-Gyamfi J (2022) Evaluation of scales of Tilapia Sp. and Sciaenops ocellatus as low cost and green adsorbent for fluoride removal from water. Front Chem 10:813533

Noman E, Talip BA, Al-Gheethi A, Mohamed R, Nagao H (2020) Decolourisation of dyes in greywater by mycoremediation and mycosorption process of fungi from peatland; primary study. Mater Today Proc 31:23–30. https://doi.org/10.1016/j.matpr.2020.01.078

Noor M, Sidek M, Ghani A, Prof A, Environments B, South ML (2013) Tropical climate : an experience at humid tropic centre

Nur Asmaliza MN, Lariyah MS, Kok KH, Humaira HS, Hidayah B (2014) Floating treatment wetland as an alternative for water quality improvement: a preliminary study. Appl Mech Mater 567:68–73. https://doi.org/10.4028/www.scientific.net/AMM.567.68

Obanan S, Sadaba R, Alva-Gatchalian P, Cadiz A, Basilio E, Bela-ong D (2022) Decolorization of synthetic textile dyes by fungal endophytes isolated from the leaves of Philippine Mangrove (*Avicennia marina*). Biol Fac Publ

Othman M, Awang M, Samsudin N, Suif Z, Ahmad N, Nor MAM (2020) The performance of pilot-scale constructed wetland for treating stormwater. IOP Conf Ser Earth Environ Sci 498(1). https://doi.org/10.1088/1755-1315/498/1/012074

Pat-Espadas AM, Portales RL, Amabilis-Sosa LE, Gómez G, Vidal G (2018) Review of constructed wetlands for acid mine drainage treatment. Water (Switzerland) 10(11):1–25. https://doi.org/10.3390/w10111685

Patil R, Zahid M, Govindwar S, Khandare R, Vyavahare G, Gurav R, Desai N, Pandit S, Jadhav J (2021) Constructed wetland: a promising technology for the treatment of hazardous textile dyes and effluent. In: Development in wastewater treatment research and processes: removal of emerging contaminants from wastewater through bio-nanotechnology, pp 173–198. Elsevier. https://doi.org/10.1016/B978-0-323-85583-9.00016-8

Perdana MC, Hadisusanto S, Purnama ILS (2020) Implementation of a full-scale constructed wetland to treat greywater from tourism in Suluban Uluwatu Beach, Bali, Indonesia. Heliyon 6(10):e05038. https://doi.org/10.1016/j.heliyon.2020.e05038

Permana I, Awaluddin A, Saryono (2019) Methylene blue decolorization fungi from crude oil contaminated soils. Biodiversitas 20:2693–2697. https://doi.org/10.13057/biodiv/d200934

Phuengphai P, Singjanusong T, Kheangkhun N, Wattanakornsiri A (2021) Removal of copper (II) from aqueous solution using chemically modified fruit peels as efficient low-cost biosorbents. Water Sci Eng 14(4):286–294

Piyaviriyakul P, Boontanon N, Boontanon SK (2021) Bioremoval and tolerance study of sulfamethoxazole using whole cell *Trichoderma harzianum* isolated from rotten tree bark. J Environ Sci Heal Part A Toxic/Hazardous Subst Environ Eng 56:920–927. https://doi.org/10.1080/10934529.2021.1941558

Praipipat P, Ngamsurach P, Kosumphan S, Mokkarat J (2023) Powdered and beaded sawdust materials modified iron (III) oxide-hydroxide for adsorption of lead (II) ion and reactive blue 4 dye. Sci Rep 13(1):531

Prasetyo S, Anggoro S, Soeprobowati TR (2022) Water hyacinth Eichhornia crassipes (Mart) Solms management in Rawapening Lake, Central Java. AACL Bioflux 15(1):532–543

Purnomo AS, Prameswari AS, Rizqi HD, Alkas TR, Ediati R, Kusumawati Y (2022) Biotransformation of methylene blue by mixed fungal cultures of *Gloeophyllum trabeum* and *Aspergillus oryzae*. Int J Technol 13:1768. https://doi.org/10.14716/ijtech.v13i8.6111

Purnomo AS, Putra SR, Shimizu K, Kondo R (2014) Biodegradation of heptachlor and heptachlor epoxide-contaminated soils by white-rot fungal inocula. Environ Sci Pollut Res 21:11305–11312. https://doi.org/10.1007/s11356-014-3026-1

Qin P, Wu Y, Adil B, Wang J, Gu Y, Yu X, Zhao K, Zhang X, Ma M, Chen Q, Chen X, Zhang Z, Xiang Q (2019) Optimization of laccase from *Ganoderma lucidum* decolorizing remazol brilliant blue R and Glac1 as main laccase-contributing gene. Molecules 24. https://doi.org/10.3390/molecules24213914

Qu Y, Cao X, Ma Q, Shi S, Tan L, Li X, Zhou H, Zhang X, Zhou J (2012) Aerobic decolorization and degradation of Acid Red B by a newly isolated *Pichia* sp. TCL J Hazard Mater 223–224:31–38. https://doi.org/10.1016/j.jhazmat.2012.04.034

Quyen VT, Pham TH, Kim J, Thanh DM, Thang PQ, Van LQ, Jung SH, Kim T (2021) Biosorbent derived from coffee husk for efficient removal of toxic heavy metals from wastewater. Chemosphere 284:131312. https://doi.org/10.1016/j.chemosphere.2021.131312

Rahmadyanti E, Audina O (2020) The performance of hybrid constructed wetland system for treating batik wastewater. J Ecol Eng 21(3):94–103. https://doi.org/10.12911/22998993/118292

Rahmadyanti E, Wiyono A (2020) Constructed wetland with rice husk substrate as phytotechnology treatment for sustainable batik industry in Indonesia. J Phys Conf Ser 1569(4). https://doi.org/10.1088/1742-6596/1569/4/042018

Rahmadyanti E, Wiyono E (2017) The performance of vertical flow constructed wetland for grey water treatment as the efforts in preserving water resources. Ecol Environ Conserv 23(4):1909–1916

Ramadhan KP, Anita SH, Oktaviani M, Sari FP, Budi LRP, Nurhayat OD, Yuli YDH (2021) Biodecolorization of anthraquinone and azo dyes by newly isolated Indonesian white-rot fungi. Biosaintifika J Biol Biol Educ 13:16–25. https://doi.org/10.15294/biosaintifika.v13i1.26148

Rathankumar AK, Saikia K, Cabana H, Kumar VV (2022) Surfactant-aided mycoremediation of soil contaminated with polycyclic aromatic hydrocarbons. Environ Res 209:112926. https://doi.org/10.1016/j.envres.2022.112926

Ridtibud S, Suwannasai N, Sawasdee A, Champreda V, Phosri C, Sarp S, Pisutpaisal N, Boonyawanich S (2023) Screening of white-rot fungi isolates for decolorization of pulp and paper mill effluent and assessment of biodegradation and biosorption processes. Curr Microbiol, 80. https://doi.org/10.1007/s00284-023-03464-0

Riyanto R, Sari IN (2024) Treatment of dyes wastewater using bio-coagulant from rambutan seeds (Nephelium lappaceum L.). In: AIP conference proceedings, vol 3027, no 1. AIP Publishing

Rohaningsih D, Henny C, Susanti E, Sudiyono BT, Febrianti D, Waluyo A, Sugiarti, Lestari FS (2023) The efficacy of Vetiveria zizanioides in horizontal subsurface flow constructed wetland for treating textile wastewater. IOP Conf Ser Earth Environ Sci 1201(1). https://doi.org/10.1088/1755-1315/1201/1/012073

Rohmawati U, Sri Kasiamdari R (2022) Morphological, molecular characterization, and physicochemical analysis of *Trichoderma yunnanense* as Indigosol golden yellow dye-decolorizing fungus. Philipp J Sci 151:2459–2470. https://doi.org/10.56899/151.6B.11

Rosli MA, Daud Z, Awang H, Zainorabidin A, Halim AA (2017) The effectiveness of peat-AC composite adsorbent in removing SS, colour and Fe from landfill leachate. Int J Integr Eng 9(3)

Saeed T, Sun G (2012) A review on nitrogen and organics removal mechanisms in subsurface flow constructed wetlands: dependency on environmental parameters, operating conditions and supporting media. J Environ Manage 112:429–448. https://doi.org/10.1016/j.jenvman.2012.08.011

Saleh HEDM, Aglan RF (2018) Heavy metals. INTECH. https://doi.org/10.5772/intechopen.71185

Sarnthima R, Khammuang S, Svasti J (2009) Extracellular Ligninolytic Enzymes by *Lentinus polychrous* Lév. under solid-state fermentation of potential agro-industrial wastes and their effectiveness in decolorization of synthetic dyes. Biotechnol Bioprocess Eng 14:513–522. https://doi.org/10.1007/s12257-008-0262-6

Saroj S, Kumar K, Pareek N, Prasad R, Singh RP (2014) Biodegradation of azo dyes acid red 183, direct blue 15 and direct red 75 by the isolate *Penicillium oxalicum* SAR-3. Chemosphere 107:240–248. https://doi.org/10.1016/j.chemosphere.2013.12.049

Shah H, Yusof F, Alam MZ (2023) A new technique to estimate percentage decolorization of synthetic dyes on solid media by extracellular laccase from white-rot fungus. Bioremediat J. https://doi.org/10.1080/10889868.2021.1984197

Shaharuddin S, Chan NW, Zakaria NA, Ghani AA, Chang CK, Roy R (2014) Constructed wetlands as a natural resource for water quality improvement in Malaysia. Natural Resources 05(07):292–298. https://doi.org/10.4236/nr.2014.57027

Shaharuddin S, Ghani AAB, Zakaria NA, Mansor M (2011) Biodiversity in stormwater constructed wetlands : case study of USM Wetland. In: 3rd international conference on managing rivers in the 21st century: sustainable solutions for global crisis of flooding, pollution and water scarcity, pp 313–324

Shourie A, Vijayalakshmi U (2022) Fungal diversity and its role in mycoremediation. Geomicrobiol J 39:426–444. https://doi.org/10.1080/01490451.2022.2032883

Shukla A, Zhang YH, Dubey P, Margrave JL, Shukla SS (2002) The role of sawdust in the removal of unwanted materials from water. J Hazard Mater 95(1–2):137–152

Singh H (2006) Mycoremediation: fungal bioremediation, mycoremediation: fungal bioremediation. https://doi.org/10.1002/0470050594

Sitharam T (2020) Emerging issues in the water environment during anthropocene : a South East Asian perpective, 1st ed. Springer Singapore. https://doi.org/10.1007/978-981-32-9771-5_13

Sudiana IK, Sastrawidana IDK, Sukarta IN (2018) Decolorization study of remazol black B textile dye using local fungi of *Ganoderma* sp. and their ligninolytic enzymes. J Environ Sci Technol 11:16–22. https://doi.org/10.3923/jest.2018.16.22

Sulastri, Henny C, Nomosatryo S, Susanti E, Sulawesty F (2023) Monitoring planktonic cyanobacteria in Lake Maninjau, West Sumatra, Indonesia. IOP Conf Ser Earth Environ Sci 1260:012018. https://doi.org/10.1088/1755-1315/1260/1/012018

Sumandono T, Saragih H, Migirin WT, Amirta R (2015) Decolorization of remazol brilliant blue R by new isolated white rot fungus collected from tropical rain forest in east kalimantan and its ligninolytic enzymes activity. Procedia Environ Sci 28:45–51. https://doi.org/10.1016/j.proenv.2015.07.007

Susanti E, Wibowo H, Henny C, Rohaningsih D, Prihatinningtyas E, Fakhrurroja H, Febrianti D, Zulti F, Kurniawan R, Waluyo A, Lestari FS, Sudiyono BT (2023) Evaluation of ecotechnology performance in treating textile wastewater: constructed treatment wetlands and natural adsorbents. IOP Conf Ser Earth Environ Sci 1260(1). https://doi.org/10.1088/1755-1315/1260/1/012048

Suwandhi IA, Hendrasarie N, Rahyuni D (2022) Constructed wetland tipe horisontal subsurface flow Menggunakan Rumput Odot untuk Pengolahan Efluen IPAL Tahu. Jurnal Serambi Engineering 7(3):3252–3261. https://doi.org/10.32672/jse.v7i3.4350

Tan L, Li H, Ning S, Xu B (2014) Aerobic decolorization and degradation of azo dyes by suspended growing cells and immobilized cells of a newly isolated yeast *Magnusiomyces ingens* LH-F1. Bioresour Technol 158:321–328. https://doi.org/10.1016/j.biortech.2014.02.063

Tan L, Ning S, Zhang X, Shi S (2013) Aerobic decolorization and degradation of azo dyes by growing cells of a newly isolated yeast *Candida tropicalis* TL-F1. Bioresour Technol 138:307–313. https://doi.org/10.1016/j.biortech.2013.03.183

Tansengco M, Tejano J, Coronado F, Gacho C (2018) Heavy metal tolerance and removal capacity of *Trichoderma* species isolated from mine tailings in Itogon Benguet. Environ Nat Resour J 16: 39–57. https://doi.org/10.14456/ennrj.2018.5

Tegegn K, Yusuf Z, Sasikumar JM and Gorfu K (2022) Biosorbent efficacy of groundnut husk for the elimination of chromium from the effluent of mojo tannery industry, Ethiopia. Int J Biomater 2022(1):9997348

Thampraphaphon B, Phosri C, Pisutpaisal N, Thamvithayakorn P, Chotelersak K, Sarp S, Suwannasai N (2022) High potential decolourisation of textile dyes from wastewater by manganese peroxidase production of newly immobilised *Trametes hirsuta* PW17-41 and FTIR analysis. https://doi.org/10.3390/microorganisms10050992

Than MM, Aung NN, Zin AT (2020) Study on sorption properties of activated biosorbents (fishscale and seashell) for the removal of anionic surfactant. J Myanmar Acad Arts Sci 18(1A):77–88

Thoa LTK, Thao TTP, Hung NB, Khoo KS, Quang HT, Lan TT, Hoang VD, Park SM, Ooi CW, Show PL, Huy ND (2022) Biodegradation and detoxification of malachite green dye by extracellular laccase expressed from *Fusarium oxysporum*. Waste Biomass Valorization 13:2511–2518. https://doi.org/10.1007/s12649-022-01692-2

Thuy Anh T, Duc Ha T, Mai Huong VT (2020) Evaluating the efficiency of *Cyperus alternifolius* on constructed wetland for the domestic wastewater treatment in the peri-urban area of Northern Vietnam. IOP Conf Ser Mater Sci Eng 869(4). https://doi.org/10.1088/1757-899X/869/4/042038

Torres JMO, Cardenas CV, Moron LS, Guzman APA, dela Cruz TEE (2011) Dye decolorization activities of marine-derived fungi isolated from Manila Bay and Calatagan Bay, Philippines. Philipp J Sci 140:133–143

Tran HD, Vi HMT, Dang HTT, Narbaitz RM (2019) Pollutant removal by *Canna Generalis* in tropical constructed wetlands for domestic wastewater treatment. Glob J Environ Sci Manag 5(3):331–344. https://doi.org/10.22034/gjesm.2019.03.06

Tran T, Anh LD, Hoa V, Quang TL, Thi HNN, Huu LH, Khoi NN, Thi AHL, Hoang LT, Trung L, Duc T, Nam V (2020) Studies on some selected microorganisms for biodegradation of congo red. In: IOP conference series: materials science and engineering, p 022037. https://doi.org/10.1088/1757-899X/736/2/022037

Tran HQ, Tran VQ, Phan TKT, Tanaka S, Gogina E (2023) The ability of reusing municipal treated water for irrigation the landscaping in Vietnam. The research in constructed wetland in municipal wastewater treatment plant, Danang city. In: E3S web of conferences, p 403. https://doi.org/10.1051/e3sconf/202340306006

Tufail MA, Iltaf J, Zaheer T, Tariq L, Amir MB, Fatima R, Asbat A, Kabeer T, Fahad M, Naeem H, Shoukat U, Noor H, Awais M, Umar W, Ayyub M (2022) Recent advances in bioremediation of heavy metals and persistent organic pollutants: a review. Sci Total Environ. https://doi.org/10.1016/j.scitotenv.2022.157961

Tuxen N, Reitzel LA, Albrechtsen H, Bjerg PL (2006) Oxygen-enhanced biodegradation of phenoxy acids in ground water at contaminated sites. Groundwater 44:256–265. https://doi.org/10.1111/j.1745-6584.2005.00104.x

Ugya AY, Imam TS, Hua X and Ma J (2019) Efficacy of Eichhornia crassipes, Pistia stratiotes and Nymphaea lotus in the biosorption of Nickel from refinery wastewater. Appl Ecol Environ Res 17(6)

Van Thanh HT, Dang MP, Tu TL, Thien TH, Bach LG (2020) Assessing the ability to treat industrial wastewater by constructed wetland model using the *Brachiaria mutica*. Waste Biomass Valorization 11(10):5615–5626. https://doi.org/10.1007/s12649-020-01065-7

Van Thi Thanh Ho, Dang MP, Lien LT, Huynh TT, Van Hung T, Bach LG (2020) Study on domestic wastewater treatment of the horizontal subsurface flow wetlands (HSSF-CWs) using *Brachiaria mutica*. Waste Biomass Valorization 11(10):5627–5634. https://doi.org/10.1007/s12649-020-01084-4

Vásquez LAH, García FP, Méndez JP, Lass-man AA, Sánchez EMO (2022) Artificial wetlands and floating islands: use of macrophytes. South Fla J Dev 3(1):476–498. https://doi.org/10.46932/sfjdv3n1-036

Verma A, Agarwal M, Bhati D, Garg H (2019) Investigation on the removal of direct red dye using *Aspergillus Niger*. J Mater Sci Surf Eng 6:881–883. https://doi.org/10.jmsse/2348-8956/6-6.1

Vikrant K, Giri BS, Raza N, Roy K, Kim KH, Rai BN, Singh RS (2018) Recent advancements in bioremediation of dye: current status and challenges. Bioresour Technol 253:355–367. https://doi.org/10.1016/j.biortech.2018.01.029

Visoottiviseth, P., Panviroj, N., 2001. Selection of fungi capable of removing toxic arsenic compounds from liquid medium. ScienceAsia

Vymaza J (2022) The historical development of constructed wetlands for wastewater treatment. Land (Basel) 11(2). https://doi.org/10.3390/land11020174

Vymazal J (2009) The use of constructed wetlands with horizontal sub-surface flow for various types of wastewater. Ecol Eng 35(1):1–17. https://doi.org/10.1016/j.ecoleng.2008.08.016

Vymazal J (2011) Constructed wetlands for wastewater treatment: five decades of experience. Environ Sci Technol 45(1):61–69. https://doi.org/10.1021/es101403q

Vymazal J (2019) Constructed wetlands for wastewater treatment. Encyclopedia of ecology: volume 1–4, 2nd edn, pp 14–21. https://doi.org/10.1016/B978-0-12-409548-9.11238-2

Waajen G, van Oosterhout F, Douglas G, Lürling M (2016) Management of eutrophication in Lake De Kuil (The Netherlands) using combined flocculant–Lanthanum modified bentonite treatment. Water Res 97:83–95

Wang J, Chen C (2009) Biosorbents for heavy metals removal and their future. Biotechnol Adv 27:195–226. https://doi.org/10.1016/j.biotechadv.2008.11.002

Wararam W, Chunkao K, Phewnil O, Tangkananuruk N, Tangkananuruk K, Pattamapitoon T, Nimpee C, Soeihom C, Peumsin J (2015) Applicable VFCW technology in parallel with biochar-mixed soils for treating formaldehyde in ethylene glycol factory wastewater. Mod Appl Sci 9(12):154

Wijaya N, Priyani N, Munir E (2019) IOP conference series: earth and environmental science lead accumulation activity of fungi isolated from Batang Toru, South Tapanuli. North Sumatra 305:12064. https://doi.org/10.1088/1755-1315/305/1/012064

Williams CD (2018) Application of zeolites to environmental remediation. Urban Pollut Sci Manag 249–258

Wongputtisin P, Supo C, Suwannarach N, Honda Y, Nakazawa T, Kumla J, Lumyong S, Khanongnuch C (2021) Filamentous fungi with high paraquat-degrading activity isolated from contaminated agricultural soils in northern Thailand. Lett Appl Microbiol 72:467–475. https://doi.org/10.1111/lam.13439

Wu S, Lyu T, Zhao Y, Vymazal J, Arias CA, Brix H (2018) Rethinking intensification of constructed wetlands as a green eco-technology for wastewater treatment. Environ Sci Technol 52(4):1693–1694. https://doi.org/10.1021/acs.est.8b00010

Wulandari R (2020) Biodegradation of polycyclic aromatic hydrocarbons commonly found in coal slurry using white rot fungi isolated from East Kalimantan. Chulalongkom University

Xiao P, Kondo R (2020) Potency of Phlebia species of white rot fungi for the aerobic degradation, transformation and mineralization of lindane. J Microbiol 58:395–404. https://doi.org/10.1007/s12275-020-9492-x

Yanto DHY, Anita SH, Solihat NN (2022) Enzymatic degradation and metabolic pathway of acid blue 129 dye by crude laccase from newly isolated *Trametes hirsuta* EDN 082. Biocatal. Biotransformation. https://doi.org/10.1080/10242422.2022.2138360

Yanto DHY, Auliana N, Anita SH, Watanabe T (2019) Decolorization of synthetic textile dyes by laccase from newly isolated *Trametes hirsuta* EDN084 mediated by violuric acid. In: IOP conference series: earth and environmental science. Institute of Physics Publishing, p 012005. https://doi.org/10.1088/1755-1315/374/1/012005

Yanto DHY, Chempaka RM, Nurhayat OD, Argo BD, Watanabe T, Wibisono Y, Hung YT (2023) Optimization of dye-contaminated wastewater treatment by fungal Mycelial-light expanded clay aggregate composite. Environ Res, 231. https://doi.org/10.1016/j.envres.2023.116207

Yu J, Lai J, Neal BM, White BJ, Banik MT, Dai SY (2023) Genomic diversity and phenotypic variation in fungal decomposers involved in bioremediation of persistent organic pollutants. J Fungi 9:418. https://doi.org/10.3390/jof9040418

Zeng X, Huang JJ, Hua B (2021) Efficient phosphorus removal by a novel halotolerant fungus *Aureobasidium* sp. MSP8 and the application potential in saline industrial wastewater treatment. Bioresour Technol 334:125237. https://doi.org/10.1016/j.biortech.2021.125237

Zhang DQ, Jinadasa KBSN, Gersberg RM, Liu Y, Ng WJ, Tan SK (2014) Application of constructed wetlands for wastewater treatment in developing countries—a review of recent developments (2000–2013). J Environ Manage 141:116–131. https://doi.org/10.1016/j.jenvman.2014.03.015

Zhang DQ, Jinadasa KBSN, Gersberg RM, Liu Y, Tan SK, Ng WJ (2015) Application of constructed wetlands for wastewater treatment in tropical and subtropical regions (2000–2013). J Environ Sci (China) 30:30–46. https://doi.org/10.1016/j.jes.2014.10.013

Zhang Y, Zhao M, Cheng Q, Wang C, Li H, Han X, Fan Z, Su G, Pan D, Li Z (2021) Research progress of adsorption and removal of heavy metals by chitosan and its derivatives: a review. Chemosphere 279:130927. https://doi.org/10.1016/j.chemosphere.2021.130927

Zhao Y, Yi X (2010) Effects of soil oxygen conditions and soil pH on remediation of DDT-contaminated soil by laccase from white rot fungi. Int J Environ Res Public Health 7:1612–1621. https://doi.org/10.3390/ijerph7041612

Chapter 11
Trends and Feasibility

Tjandra Chrismadha, Livia Rossila Tanjung, Miratul Maghfiroh, and Djamhuriyah S. Said

The restoration of inland water ecosystems has become crucial as the deterioration of these ecosystems worsens with time due to human activities such as changes in land use and pollution (Boelee et al. 2017). These anthropogenic pressures have negative impacts on the quality of the water environment. The diminishing environmental conditions influence the quality of human existence, as humans depend on the environment for their well-being. The concepts of ecosystem management appear to connect sustainable environment, community benefit, and human wellness. Some common concepts utilized include Ecological Engineering, Green/Blue Infrastructure, Ecosystem Approach, Ecosystem-based Mitigation of Adaptation, and Ecosystem Services Approach/Framework.

In addition to those concepts, Nature-based Solutions (NbS) are considered inclusive since they incorporate ecosystem-based concepts to address societal challenges and promote human well-being and biodiversity benefits (Cohen-Shacham et al. 2016). NbS provide the foundation for sustainable actions that assist society respond to challenges from environmental, social, and economic perspectives (Nesshöver et al. 2017). The concept of "Nature" is a fundamental source of inspiration and support for sustainable actions. These actions encompass the protection of natural and semi-natural ecosystems, the integration of green and blue infrastructures, and the implementation of ecosystem-based principles (Seddon et al. 2020).

In the tropical region, where the human population is rapidly increasing and inland water ecosystems are being badly affected due to rapid urban activities, implementing NbS can be beneficial in supporting the restoration of inland water ecosystems, ultimately benefiting both biodiversity and human wellness. However, governmental restrictions and cultural variations might have hindered this implementation (Lechner et al. 2020).

T. Chrismadha (✉) · L. R. Tanjung · M. Maghfiroh · D. S. Said
Research Center for Limnology and Water ResourcesNational Research and Innovation Agency (BRIN), Jakarta, Indonesia
e-mail: tjan001@brin.go.id

© The Author(s), under exclusive license to Springer Nature Singapore Pte Ltd. 2025

A. C. Santos-Borja et al. (eds.), *Progress on Ecosystem Restoration of Tropical Inland Waters*, https://doi.org/10.1007/978-981-96-2284-9_11

Constructed wetland (CW) can be an example of NbS. Constructed wetlands remove excessive nutrients and other harmful pollutants, such as heavy metals, by being absorbed into the soil and taken up by microorganisms and plants (van Rees et al. 2023). Research conducted in the Putrajaya Wetlands in Malaysia found that the highest levels of nutrient removal were 84.32% for phosphate and 70.73% for nitrate, respectively (Sim et al. 2008). A more detailed explanation of CWs is described in the previous chapter.

Besides CWs, an implementation of peatland restoration might exemplify the practice of NbS. A peatland restoration in Indonesia had been conducted through the application of "Paludiculture," i.e., cultivation of selected plants in swamp, which used plant species that were relatively resistant to inundation and able to adapt in acidic soils such as sago (*Metroxylon sagu*), jelutung (*Dyera polyphylla*), ramin (*Gonystylus bancanus*), balangeran (*Shorea balangeran*), gelam (*Melaleuca cajuputi*), *Campnosperma coriaceum*, *Cratoxylum arborescens* (Budiman et al. 2020; Giesen 2021). This NbS practice aimed to restore biophysical as well as ecological properties and at the same time, provide economic benefits for the local communities (Giesen 2021). In Ogan Komering Ilir (OKI) regency, South Sumatra Province (Indonesia), Paludiculture was compromised with agro-sylvofishery and agri-silviculture which differed by the choice of the plants (dryland and native forest plants) and the involvement of fishery (Budiman et al. 2020). Nonetheless, challenges on Paludiculture include high cost of revegetation, dryland species dominance, lack of landscape approach, and information of potential Paludiculture species (Giesen 2021).

Another example of NbS is the adoption of Integrated Multi-trophic Aquaculture (IMTA). The following sections focus on analyzing this technique from the perspective of Nature-based Solutions (NbS).

11.1 Integrated Multi-Trophic Aquaculture (IMTA)

Integrated Multi-Trophic Aquaculture (IMTA) is the practice of cultivating different species of aquatic organisms collectively in a single system, considering their feeding characteristics. The principle of IMTA is to maximize the efficient use of space, water resources, and feed. The IMTA system follows an ecological strategy to promote the circular principle. This system employs scientific integration to utilize uneaten feed, waste, nutrients, and by-products generated by the species fed in fish farming activities to be converted into fertilizers, feed, and energy by other species at lower trophic levels. This system adds value to the products and comprehensively minimizes the adverse effects on the environment (Barrington et al. 2009; Brown et al. 2011). IMTA aims to enhance the rate and efficiency of biomass productivity per unit of water column within a specific time frame, such as kg/m^3 per year.

IMTA is a production strategy consisting of cultivating organisms with varied trophic statuses arranged in a production system that allows for internal circular mass and energy transfer. This arrangement creates an internal ecological balance, leading to greater production efficiency. Although IMTA has primarily been implemented in marine environments, its adaptation to freshwater systems represents a growing area of interest globally.

In freshwater environments, IMTA activity involves the simultaneous cultivation of multiple species that interact beneficially within the same aquaculture system. This approach typically integrates fish, such as Common Carp or Nile Tilapia, with species like freshwater shrimp, mussels, or aquatic plants (Hasan and Chakrabarti 2009). Scientific studies have demonstrated that IMTA systems can effectively utilize nutrients and organic matter generated by one species as inputs for others, thereby reducing waste and improving overall system efficiency (Chopin 2013).

As a promising strategy in freshwater aquaculture, IMTA utilizes ecological principles to optimize resource use, increase productivity, and reduce environmental impacts. Key publications highlight the ecological benefits of IMTA in freshwater aquaculture, emphasizing its potential to enhance water quality and biodiversity. Research conducted by Troell et al. (2009), Chopin (2013), Sukhdhane et al. (2018), and Checa et al. (2024) has demonstrated that the integration of filter-feeding species such as mussels or freshwater shrimp can effectively decrease nutrient loading and enhance water clarity by filtering out suspended particles and excess nutrients.

Furthermore, IMTA offers economic advantages by diversifying production and reducing reliance on external inputs. Studies have demonstrated that integrating multiple species can enhance farm profitability and resilience to market fluctuations (Whitmarsh et al. 2006; Hossain et al. 2022). This economic diversification is critical for the sustainability and long-term viability of freshwater aquaculture operations.

Despite its potential benefits, implementing IMTA in freshwater systems poses challenges related to system design, species compatibility, and regulatory frameworks. Scientific inquiry continues to address these challenges through interdisciplinary research collaborations, integrating fields such as aquaculture science, ecology, hydrology, and economics.

The scientific literature on IMTA in freshwater aquaculture underscores its role as a promising approach to addressing environmental and economic sustainability in aquaculture. Further exploration is into the evolving research and practical applications shaping IMTA's role in freshwater systems, offering solutions to meet growing global demands for freshwater food production while minimizing ecological footprint.

11.2 History of IMTA

The concept of Integrated Multi-trophic Aquaculture (IMTA) has its roots in ancient agricultural practices described in the 1639 book "Nong Zheng Quan Shu" (The Complete Book on Agriculture). This book provides comprehensive information on many techniques like irrigation, rotation of fish and plant production, integrated fish and livestock farming, and the use of manure in aquaculture. It also discussed the integrated cultivation of mulberry, paddy fields, and fish ponds (Chopin 2013).

Modern IMTA began its development in the 1970s with John Ryther's project on "integrated waste-recycling marine polyculture systems," which focused on using seaweed and bivalves to process waste from aquaculture (Troell et al. 2003). This concept evolved in the 2000s with various modifications like polyculture cultivation, integrated marine cultivation, and ecology-based aquaculture (Chopin 2013). The term "Integrated Multi-trophic Aquaculture" was coined in 2004 by Thierry Chopin and Jack Taylor. The basic concept of the IMTA system is the conversion of nutrients produced from aquaculture activities through species diversification. In the IMTA system, species from different trophic levels with different feeding habits are integrated into the same production system. These species can be placed in the same compartments or separate areas connected by waste streams. The primary goal of IMTA is to reuse waste produced by one species as inputs for others, thereby converting nutrients into valuable by-products like fertilizer, feed, and energy (Chopin et al. 2008; Barrington et al. 2009).

IMTA adheres to the "circular principle," promoting scientific integration in reusing uneaten feed, waste, nutrients, and by-products generated by the species being fed for conversion into fertilizer, feed, and energy by other species with lower trophic levels. The careful selection and proportioning of species help balance biological and chemical processes within the system. Species in IMTA acting as biofilters can be commercially harvested, contributing to both environmental sustainability and economic viability. The overarching aim of IMTA is to enhance ecosystem conditions in cultivation environments and achieve higher production outputs than monoculture systems (Neori et al. 2004). By utilizing diverse species, IMTA improves the ecological footprint of aquaculture and offers a more efficient and sustainable production model.

11.3 Current State of IMTA in Several Parts of the World

11.3.1 Canada

Integrated Multi-trophic Aquaculture (IMTA) systems in Canada have been developed for commercial salmon hatcheries. The flexible concept of IMTA permits the development of numerous variations, extending its applicability beyond open water systems (Chopin et al. 2016).

Atlantic salmon raised for commercial purposes spend the early life cycle in fresh water, often in land-based, enclosed hatcheries, before being moved to open seawaters. Consequently, IMTA principles can also be applied to land-based, closed-containment, and freshwater systems. The critical factor is carefully selecting appropriate co-cultivated organisms from different trophic levels, considering their complementary roles and economic value. Like in saltwater operations, fishes in freshwater operations also require attention due to the increased nutritional benefits of the water in which they live. Rather than disposing of and wasting the water, wastewater is repurposed as a valuable resource for irrigation, providing water, nutrients, and a growth medium for further crops.

A comprehensive field and laboratory feasibility study was started in 2013 to explore the possibility of creating land-based freshwater IMTA (or FIMTA) systems for Atlantic salmon hatcheries in New Brunswick, Canada (Scarratt 2015; Chopin et al. 2016). An evaluation was conducted on flow-through and recirculating facilities to determine the most suitable system design, considering water quality, nutrient content, flow rate, and discharge characteristics. Equally significant factors included the levels and accessibility of nutrients, temperature, light, space availability, plant-specific growth, absorption requirements, and economic and market viability. Analysis indicated that recirculating hatcheries were superior options for FIMTA systems compared to conventional flow-through hatcheries. Furthermore, a pilot-scale FIMTA system equipped with light and temperature controls was constructed at the University of New Brunswick in Saint John, New Brunswick. The purpose of this system is to evaluate the growth and nutrient removal capability of different plant species under certain seeding conditions. The effluent was obtained from a commercial salmon hatchery, and the levels of nutrients were monitored before and after they flowed through the crop-producing areas to assess the system's capacity to decrease nutrient levels.

Salmon hatcheries typically maintain water temperatures between 10 and 15 °C. Consequently, the plant species used for these hatcheries will differ from those utilized in modern aquaponic systems designed for tropical fish, like Nile Tilapia, in warmer water. The selected plant species must demonstrate resilience, thrive, and exhibit suitable nutrient absorption capacities at low temperatures, and should hold commercial value, such as being edible, ornamental, or medicinal. Based on these requirements, the plants tested so far with positive results (Chopin et al. 2016) include watercress (*Nasturtium officinale*), leaf lettuce (*Lactuca sativa*), Swiss chard (*Beta vulgaris cicla*), chives (*Allium schoenoprasum*), escarole (*Cichorium endivia*), tomato (*Solanum lycopersicum*), French marigold (*Tagetes patula*), nasturtium (*Tropaeolum majus*), yarrow (*Achillea millefolium*), German chamomile (*Matricaria chamomilla*), and mint (*Mentha* sp.).

The efficacy of the system in reducing nutrients from the collected salmon effluent has varied based on the plant species chosen and the biomass they generated. Additionally, it fluctuated depending on the nutrient concentrations in the initial effluent. Calculations were conducted to determine nutrient recovery and abatement scenarios based on the ratio of fish to extractive species and its scalability. The analysis will consider the inherent worth of the biomass, the ecosystem services provided, the significance of the FIMTA stage in eco-certification programs, and the public's views on the acceptability of these aquaculture activities. Finally, optimizing the pilot system on a broader scale will be examined to facilitate its commercial adoption and application in hatcheries.

11.3.2 Hungary

In Hungary, crop farming and animal husbandry were integrated with fish production technologies throughout the latter half of the twentieth century. Currently, the focus is on integrating different types of feed to reduce the amount of nutrients in the surrounding natural ecosystem and maximize the use of water resources. Several modern integrated freshwater aquaculture systems, such as intensive fish production with wetlands, recirculating aquaculture systems, and multifunctional aquaculture, have demonstrated their feasibility. Nevertheless, the potential prospects for these systems are often overlooked and not given due recognition when addressing the future of European aquaculture (Popp et al. 2018).

Hungary currently possesses multi-trophic integrated fish production systems built upon the latest research findings and the utilization of advanced techniques and technology. These systems effectively combine the benefits of conventional pond fish farming with intensive fish culture. Ongoing research is being conducted to build freshwater IMTA systems through collaboration between various Hungarian research centers and innovative fish farms.

The primary foundation for intensive fish production in Hungary is the usage of readily available geothermal water resources. By using this thermal water, it is feasible to maintain ideal temperatures for the target fish species regardless of the season, which in turn enables the cultivation of tropical species such as Tilapia (*Oreochromis* spp.) and African catfish (*Clarias gariepinus*).

The integration of intensive aquaculture systems with extensive ponds or wetlands became required due to the need for the safe and non-polluting disposal of the used thermal water. Nutrient-rich water from the intensive fish-rearing system, generally warmer than the recipient's, can be treated in extensive fish ponds or an adequately sized constructed wetland. Following this process, the wastewater from intensive farms can be discharged into the natural environment without a considerable increase in organic matter. The main product of the integrated system is the fish cultivated in the intensive system. The connected extensive system allows the production of additional aquatic animals or plants. The animals reared in extensive systems are mostly filter-feeding organisms that effectively utilize the waste nutrients discharged by the intensive systems, for example, Common Carp, Grass Carp, Tilapia, and freshwater mussels. In the final stage of treating effluent water, various plants grown in the constructed wetland, such as reed (*Phragmites communis*), cattail (*Typha latifolia*), and willow (*Salix* spp.) are used to eliminate the plant nutrients that are still present in the water. Wetlands of greater importance provide a habitat for diverse plant and animal species. As a result, this unique freshwater IMTA system offers significant ecological services. The primary function of the extensive fishpond and the constructed wetland is to reduce the organic matter content of the effluent water to levels that satisfy the criteria for its release into natural water bodies.

The latest innovation in modern pond-based IMTA systems in Hungary is Multifunctional Pond Farming (MFPF), which constitutes a complex wetland system. MFPF encompasses a broader range of activities than traditional pond farming. In addition to producing aquatic and terrestrial organisms, MFPF also involves cultivating non-food products and providing ecological and societal services. MFPF is a specific form of IMTA that utilizes a single water body and a complex wetland ecosystem. The fundamental component of the wetland is the fish pond, which is complemented by the natural and constructed environment of the fishpond. Practical results demonstrate that MFPF provides a higher and more secure income to farmers through diversification and complex utilization of resources while contributing to maintaining and increasing biodiversity, improving our knowledge of nature, and encouraging the social acceptance of fisheries (Popp et al. 2018).

11.3.3 Bangladesh

The potential of the IMTA system was demonstrated in a study in freshwater ponds in Bangladesh (Kibria and Haque 2018). This study highlights the ability of the IMTA system to improve fish survival, growth, and overall production while maintaining water quality through effective waste mitigation.

A six-month experimental study was conducted to evaluate the potential of IMTA in earthen freshwater ponds. The study utilized nine ponds, each 40 m^2 in size, randomly assigned to three treatments in triplicate: T1 (carps and stinging catfish), T2 (carps, stinging catfish, and snails), and T3 (carps, stinging catfish, snails, and water spinach) as IMTA.

The stocking density for Carps is 20,000 fingerlings per hectare with a species ratio of 3:1:2:2 for catla, silver carp, rohu, and mrigal, while the stocking density for stinging catfish is 24,700 fingerlings per hectare in a cage system in ponds for all treatments. The snail stocking density is 62 kg per hectare in T2 and T3.

Feeding regimes included supplementary feed (a mixture of rice bran and wheat bran in a 1:1 ratio) for carps at 3–5% of fish biomass and commercial feed for stinging catfish for the first 45 days, followed by snail-mixed pelleted feed for the remaining period, at 5–25% of body weight.

Results indicated the highest survival rates, particularly for silver carp, mrigal, and stinging catfish, in T3 ponds, which incorporated the entire IMTA system. Weight gains for silver carp, rohu, and stinging catfish were also highest in the IMTA ponds. Furthermore, the highest yields for carp and stinging catfish were observed in the IMTA ponds, demonstrating enhanced growth performance and survival compared to the other treatments.

The study emphasizes the role of snails as an extractive species and a feed ingredient for catfish. Additionally, it suggests that snails can generate extra income by being used to create feed for chickens and ducks. The inclusion of snails and water spinach in the IMTA ponds (T3) significantly contributed to the reduction of organic and inorganic waste, thereby maintaining water quality within optimal conditions for fish culture. This finding underscores the role of IMTA in promoting sustainable aquaculture practices by integrating species that enhance nutrient recycling and waste management (Kibria and Haque 2018).

11.3.4 India

Nath et al. (2021) investigated the effectiveness of IMTA systems in freshwater environments. The research focused on the famous Indian fish species *Labeo rohita* (Rohu), utilizing *Wolffia globosa* as an inorganic nutrient extractor and *Lamellidens marginalis* mussels as organic nutrient extractors.

The study hypothesized that the co-culture of fed fish with inorganic and organic extractive organisms would create a balanced ecosystem conducive to sustainable production. The primary objectives were to evaluate the impacts of incorporating *L. marginalis* and *W. globosa* on the growth performance and fish yield of *L. rohita*, the feed efficiency and overall welfare of the fish, and water quality parameters.

The results showed that the IMTA system demonstrated higher survival rates and productivity, reaching more than five tons of fish per hectare per year, significantly exceeding the national average fish productivity. In addition, the inclusion of *Wolffia* and *Lamellidens* contributed to better growth performance and higher yields of the fish cultured. The integrated system improved feed efficiency, leveraging the nutrient extraction capabilities of *Wolffia* and *Lamellidens* to optimize nutrient utilization.

The most important result is that IMTA can maintain optimum water quality, which reduces oxidative stress and improves the immunity and survival rates of the fish. The incorporation of inorganic and organic extractive components effectively remediated water quality, creating a more stable and sustainable aquaculture environment. This system balances nutrient loads, enhances fish growth, and maintains water quality effectively.

Finally, the IMTA approach generated economic returns through improved fish productivity and reduced the environmental impact of aquaculture practices by maintaining better water quality, leading to sustainable and economically viable aquaculture practices. Future research should explore the efficiency of other locally relevant edible aquatic plants and mollusks and quantify the nutrient sequestration in the biomass of co-cultured species to optimize further and assess the effectiveness of IMTA systems.

11.3.5 Thailand

In Thailand, IMTA has been increasingly explored as a sustainable strategy to address environmental challenges in aquaculture while promoting socio-economic development. Various freshwater IMTA models were examined to assess their viability. The main objective was to combine economically valuable species like Nile Tilapia with complementary organisms such as freshwater shrimp and aquatic plants. The research conducted by Sri-uam et al. (2016) emphasized the benefits of reducing environmental impacts through nutrient assimilation and waste management in integrated systems.

An innovation in the freshwater IMTA, which consists of a series of interconnected tanks, called an integrated multi-trophic recirculating aquaculture system (IMRAS), was designed by Sri-uam et al. (2016). The IMRAS consisted of a fish tank, phytoplankton tank, zooplankton tank, and aquatic plant tank. The water from the aquatic plant tank was transferred to the fish tank using a submersible pump. A surplus conduit was added to connect the fish tank with the phytoplankton and aquatic plant tanks. A valve was installed to control water overflow from the fish tank to the phytoplankton tank. The valve remained open until the phytoplankton tank reached its maximum capacity. Once this occurred, the valve was closed. Subsequently, the tank was operated in a batch mode to cultivate phytoplankton (*Chlorella* sp.). When the stationary growth phase was achieved, the phytoplankton biomass was moved to the zooplankton tank to be used as food for zooplankton (*Moina macrocopa*), and the overflow valve was reactivated. A portion of the water from the fish tank consistently spilled over into the aquatic plant tank before being returned to the fish tank to complete the cycle.

The aquatic plant tank played a crucial role in the treatment process as it effectively eliminated a significant amount of nitrogen and phosphorus compounds, which the aquatic plants then used for their growth. Additionally, aquatic plants are able to capture suspended particles from the fish tank, ensuring the clarity of the water in the system. The aquatic plants included in the study were *Hygrophila difformis*, *Vallisneria spiralis*, and *Vallisneria americana*, which were planted at the bottom of the tank, while *Egeria densa* and *Ceratophyllum demersum* were floated on the water surface (Sri-uam et al. 2016).

11.3.6 Indonesia

IMTA has been acknowledged in Indonesia as a potential approach to encourage sustainable aquaculture practices while supporting rural lives and ensuring food security. Similar to the IMTA implementation in other locations, the IMTA adopted in Indonesia likewise seeks to utilize nutrient-rich effluent from fish ponds to enhance plant growth and cultivate economically beneficial species, which contributes to increasing the overall efficiency and sustainability of the system (Chrismadha et al. 2014a). Several challenges have been documented regarding the implementation of IMTA for sustainable aquaculture in rural areas. These challenges include the introduction of new technologies, the establishment of efficient management systems, and the formation of institutional frameworks (Chrismadha and Said 2019).

- Awerange Bay Waters, South Sulawesi

Many IMTA experiments have been carried out, but most were small-scale. A specific example of this method of cultivation is the double-net cage system, which is employed to cultivate a variety of trophic species, including grouper, white shrimp, Rabbitfish, sea cucumbers, and seaweed in separate compartments within the cage (Putro et al. 2015).

The implementation of IMTA in the stratified double floating net cage (SDFNC) was initially conducted in the waters of Awerange Bay, South Sulawesi. Milkfish (*Chanos chanos*) and white shrimp (*Penaeus vannamei*) were used in fed aquaculture, while seaweed (*Kappaphycus alvarezii*) functioned as the inorganic extractive species. The trial effectively maintained the purity of the aquatic ecosystem in Awerange Bay and reduced the negative impacts linked to production in floating net cages. Moreover, it has shown the capacity to increase agricultural productivity in Awerange Bay (Putro et al. 2015).

- Sea Farming Karang Lebar in Kepulauan Seribu, Jakarta

Putro et al. (2015) advanced their research by including IMTA in the SDFNC system, which involved the addition of biomonitoring.

The research was conducted at Sea Farming Karang Lebar in Kepulauan Seribu, Jakarta. The primary objective was to sustain a functional aquaculture environment in offshore regions. The IMTA implementation and regular biomonitoring of SDNRC facilities resulted in a significant increase in aquaculture productivity. The biota cultivated for fed aquaculture during the trial consisted of Rabbitfish (*Siganus* sp.), Tiger grouper (*Epinephelus fuscoguttatus*), white-leg shrimp (*P. vannamei*), Pomfret (*Trachinotus blochii*), and Milkfish (*C. chanos*). Seaweed (*Eucheuma cottonii*) was employed as an inorganic extractive organism, while sea cucumbers (*Stichopus hermanii*) were utilized as non-fed components. Implementing SDRNC-IMTA floating cages alongside biomonitoring systems can effectively meet the requirements for sustainable aquaculture. The cultivation procedures adhere to existing regulations, primarily aiming to improve aquaculture productivity while considering the environmental carrying capacity. Integrating biomonitoring in aquaculture is crucial for effectively mitigating potential problems, minimizing environmental hazards, and facilitating proactive actions.

- Zhonegat Strait of the Banda Naira Islands, Moluccas

A study was conducted in Zhonegat Strait of the Banda Naira Islands, Moluccas, to investigate the implementation of the IMTA system in enhancing the productivity of marine aquaculture products and improve the overall condition of the aquatic environment in the Banda Naira Islands (Saimima et al. 2020).

The IMTA model utilized Rabbitfish (*Siganus* sp.) as fed aquaculture species and seaweed (*K. alvarezii*) as a species for inorganic extraction. Rabbitfish were reared in floating net cages, while seaweed was cultivated using floating rafts located 50 m away from the cages. The control floating raft was placed at a distance of 2–3 km from the cage. The cultivation of seaweed near the IMTA system exhibited a more favorable growth pattern, resulting in a 40.85% increase in productivity compared to the control seaweed farmed outside the system. This increase indicates the efficiency of utilizing nutrients from fish cage farming waste discarded into the seas. The IMTA system is highly appropriate for implementation in Banda Naira due to its relevance in land use, production outcomes, and water environmental conditions. Moreover, it has no detrimental impact on the coral reefs situated on the ocean floor, thus promoting the development of marine tourism in the Banda Islands.

- Gourami Cultivation Using IMTA Compared with Gourami Monoculture

The study conducted by Wilfart et al. (2020) evaluated the potential environmental benefits and resource efficiency of integrating red Azolla floating plants into traditional Gourami cultivation systems. This study compared two cultivation systems in a single earthen pond separated by nets. The first system involved the cultivation of Gourami species (*Osphronemus gourami*) in a monoculture setting where the fish were fed with artificial feed, referred to as GM. The second system involved farming Gourami combined with floating plants *Azolla* spp., referred to as IMTA-G.

The research evaluated the reduction in artificial feed usage by supplementing the Gourami diet with Azolla, assessed the decrease in nutrient emissions to water through nutrient recycling into Azolla production, and used life cycle assessment (LCA) to measure the environmental impacts of the IMTA system compared to the monoculture system.

The IMTA-G system demonstrated a significant reduction in the consumption of formulated feed per unit mass of fish. This reduction was achieved by incorporating Azolla into the fish diet, providing an additional source of nutrition. The integration of Azolla effectively recycled nutrients within the pond ecosystem, leading to lower nutrient emissions into the water. Azolla absorbed nutrients from fish waste and water, which were then reintroduced into the food chain when consumed by the fish. This nutrient recycling process resulted in a "virtuous" system that improved overall nutrient management and water quality.

The IMTA-G system decreased all selected environmental impact categories by 70% compared to the GM system. These impacts included greenhouse gas emissions, eutrophication potential, and resource use. The life cycle assessment results highlighted the lower environmental footprint of the IMTA system, showcasing its potential for sustainable aquaculture practices. The IMTA-G system demonstrated improved resource use efficiency by reducing the need for external feed inputs and maximizing the use of available nutrients within the pond. This system supported sustainable fish production and contributed to the economic value of the aquaculture operation through the additional biomass of Azolla.

- Lerang Village, Lanrisang District, Pinrang Regency, South Sulawesi

This study investigated the IMTA-Paddy system in the community ponds located in Lerang Village, Lanrisang District. The objective was to assess the viability of the organisms cultivated in the IMTA-Paddy system. Field experiments were carried out using a tarpaulin pond model to study three different biota combinations: Milkfish and paddy (K-2), Milkfish, Tiger Prawns, and paddy (K-3), and Milkfish, Tiger Prawns, mussels, and paddy (K-4). The biota population was reared for 80 days at a density of 30 individuals, each integrated with 30 clumps of paddy, using the floating (Hamsiah et al. 2021).

The results showed that the difference in the survival rate of Milkfish in K-2, K-3, and K-4 was in the range of 1.1–3.6%, which was not statistically significant ($P > 0.05$). The survival rate of Tiger Prawns in K-4 was 2.2% higher than in K-3. Additionally, the survival rate of mussels in K-4 was $85.6 \pm 1.90\%$. The water quality during the study remained within tolerable ranges for all biota. The cultivation of Milkfish, Tiger prawns, and mussels alongside paddy in the IMTA-Paddy system resulted in very high survival rates for each organism.

The combination in which mussels were included (K-4) resulted in the lowest NH_3 content and the highest overall survival rates. The presence of mussels is thought to play a crucial role in maintaining ammonia levels below 0.06 mg/L, thereby enhancing water quality and effectively supporting the survival of organisms in the IMTA system. The research suggested that more in-depth studies are needed to investigate the nutrient release and uptake dynamics among different biota and optimize the cultivation potential of the IMTA-Paddy system in brackish waters.

- Research at the Research Center for Limnology—Indonesian Institute of Sciences (LIPI)

The research conducted at the Research Center for Limnology—Indonesian Institute of Sciences by Chrismadha et al. (2014b) focused on the application of minute duckweed (*Lemna perpusilla*) in closed recirculating aquaculture systems. The experiment utilized minute duckweed for its dual functions as a phytoremediator to improve water quality and as an alternative feed to reduce the dependency on commercial feed. The research aimed to enhance water usage efficiency by implementing a water recirculation system in aquaculture. It sought to employ minute duckweed as a phytoremediator to effectively remove contaminants from the water and utilize its biomass to supplement the fish diet, thereby reducing overall production costs.

The trial involved four concrete ponds and eight tarpaulin ponds arranged in a serial water flow system using a submersible pump with a capacity of 6000 L/h (Fig. 11.1). Juvenile Common Carp (*Cyprinus carpio*) were reared in the concrete ponds (100 fish per pond) and fed a diet of commercial pellets and fresh duckweed biomass over 23 weeks. The duckweed was cultivated in the tarpaulin ponds at a controlled density of 700 g/pond. It was harvested regularly three times a week and was assessed for its ability to remove total inorganic nitrogen (TIN), total phosphorus (TP), and total suspended solids (TSS) from the water. The produced biomass was used to partially replace commercial feed, with 50% of the feed being substituted by fresh duckweed biomass. Figure 11.2 depicts the initially created IMTA system at the Research Center for Limnology LIPI.

The research found that minute duckweed significantly improved water quality through its phytoremediation capacity, with average uptake rates of 0.63 g/m^2/day for total inorganic nitrogen (TIN), 0.33 g/m^2/day for total phosphorus (TP), and 112.93 g/m^2/day for total suspended solids (TSS). These rates corresponded to removal efficiencies of 72.06% for TIN, 87.77% for TP, and 61.74% for TSS. The incorporation of duckweed biomass as a feed substitute notably enhanced fish growth and survival rates, reducing the feed conversion ratio (FCR) by up to 30% and leading to a 20% decrease in overall production costs. Despite the sub-optimal growth conditions during the trial, duckweed biomass productivity ranged from 6.14 to 11.61 tons dry weight (DW) per hectare per year. The duckweed also exhibited a high protein content, with a maximum observed value of 32% DW, underscoring its potential as a valuable feed ingredient.

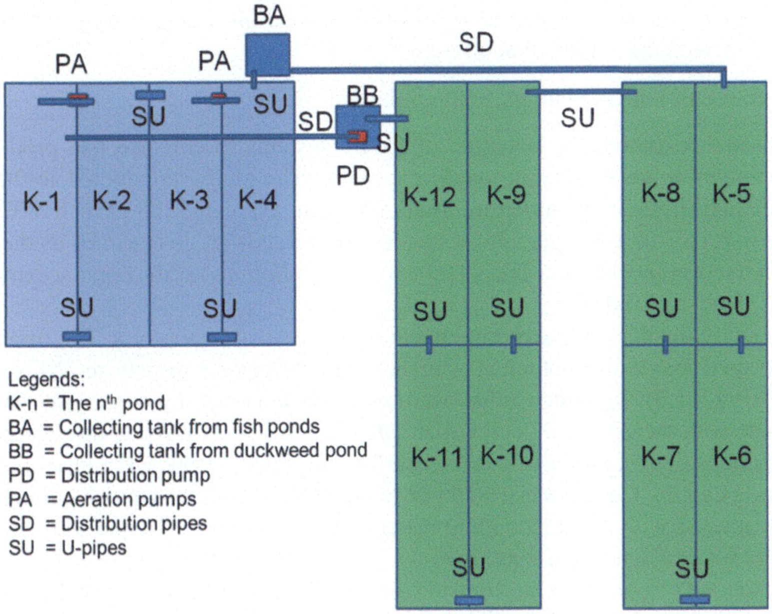

Fig. 11.1 Schematic of the experimental IMTA in closed-water recirculation (Chrismadha et al. 2014b)

Fig. 11.2 IMTA system that was first developed at the Research Center for Limnology LIPI. *Photo credit* BRIN/T. Chrismadha

This study by Chrismadha et al. (2014b) demonstrated that *L. perpusilla* could effectively perform the dual role in aquaculture. It improved water quality by removing significant amounts of waste from fish cultivation and as an alternative feed to reduce dependence on commercial feed, which can increase fish growth and survival while reducing production costs. Although the biomass productivity was relatively low, the high protein content of duckweed underscores its potential as a sustainable feed option in aquaculture.

- Enrekang Botanical Garden, South Sulawesi

The research explored the potential of IMTA systems to enhance fish production and maintain water quality in small reservoirs. The study, conducted in 2016 at the Enrekang Botanical Garden in South Sulawesi, evaluated the growth of Catfish (*Clarias* sp.) as the fed aquaculture species, duckweeds (*L. perpusilla*) as the inorganic extraction species, and Tilapia (*Oreochromis niloticus*) as the organic extraction species (Said et al. 2017).

The study involved two experiments in different small reservoirs. The first reservoir, located inside the botanical garden, was a fish pond measuring 20 × 25 × 1.5 m^3 (Fig. 11.3). Juvenile Catfish were cultured at a stocking density of 180 fish/m^2 inside nets measuring 2 × 5 × 1 m^3. *Lemna* was grown inside the two nets of equal dimensions on either side of the Catfish net. Plastic tarps were placed around the nets to capture Catfish waste and prevent the water from the Catfish culture from directly mixing with the surrounding water (Fig. 11.4), where Tilapia were cultured for 114 days at a density of 2000 fish.

The second reservoir, located outside the botanical garden, consisted of two ponds with dimensions of 30 × 30 × 1.5 m^3 and 20 × 40 × 1.5 m^3, respectively (Fig. 11.5). In the first pond, Catfish were cultured at a density of 188 fish/m^2 inside a net of 4 × 4 × 1 m^3. Tilapia were reared separately in the second pond, connected to the first, with an initial stock of 2000 fish. This experiment lasted for 63 days. Catfish in both reservoirs were fed with pellets equivalent to 5% of their body weight per day, while Tilapia were given 400 g of duckweed daily, grown within the system.

The results demonstrated significant growth rates, with the specific growth rate (SGR) of Catfish and Tilapia reaching 4.02% per day and 2.60% per day, respectively.

Fig. 11.3 First reservoir, located in the botanical garden. *Photo credit* BRIN/D. S. Said

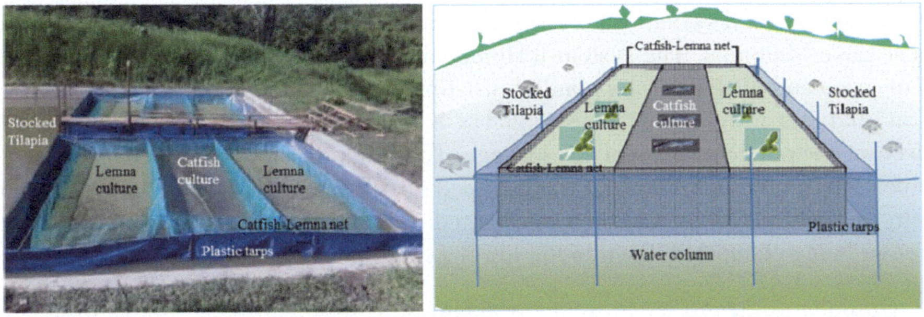

Fig. 11.4 First reservoir as an IMTA system. *Photo credit*: BRIN/D. S. Said; sketch: T. Chrismadha; Said et al. (2017)

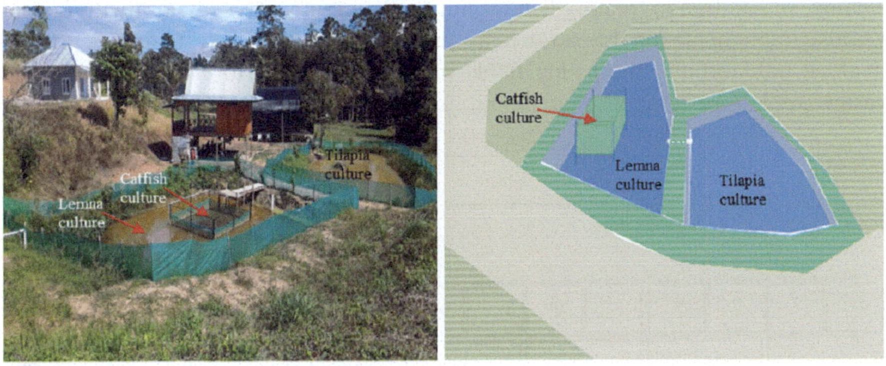

Fig. 11.5 Second reservoir as an IMTA system. *Photo credit* BRIN/D. S. Said; sketch: T. Chrismadha; Said et al. (2017)

L. perpusilla showed vigorous growth and proved a suitable food source for Tilapia. The IMTA systems effectively maintained water quality, achieving elimination rates of 15.63% for total nitrogen (TN) and 52.00% for total phosphorus (TP). These findings suggest that IMTA systems can establish a secure and effective setting for aquaculture, fostering sustainable fish production and water preservation.

- IMTA Trial in Lake Maninjau

Lake Maninjau has been experiencing eutrophication and requires immediate action to restore its condition. IMTA was introduced as a method that utilized lake water for inland aquaculture. This method employed an aquatic plant with a dual purpose: it acts as a water phytoremediator and provides natural feed for fish. This study investigated the capacity of *L. perpusilla* to facilitate the growth of Nile Tilapia (*O. niloticus*) in IMTA ponds for 40 days (September–October 2018). The IMTA system was implemented by utilizing lake water as a growth medium for *Lemna* (Said et al. 2020).

An IMTA pond system was constructed in the village of Sungai Batang, located near Lake Maninjau. The system featured five fish ponds with dimensions of 4 × 6 m^2 each and twelve *Lemna* ponds consisting of eight upper ponds measuring 4 × 4 m^2 and four lower ponds measuring 4 × 8 m^2. A plastic tarpaulin was employed to cover the pond and prevent any leakage. The water distribution system used PVC pipes. The water from the lake was pumped up, channeled, and equally distributed into the fish ponds. It was then discharged directly into the upper *Lemna* ponds before being channeled down to the lower *Lemna* ponds and finally returned to the lake (Fig. 11.6).

Each fish pond was stocked with 267 Tilapia fingerlings, or around 15 kg/pond. The fish was fed with fresh *Lemna* obtained from the twelve cultivation ponds (Fig. 11.7). The feeding quantity was adjusted to match the daily *Lemna* production rate from the available twelve ponds, delivered twice daily in the morning and evening. In addition to treating fish farming effluent, phytoremediation with *Lemna* was also implemented in the paddy fields uphill from the IMTA system. This activity involved transforming the paddy fields into *Lemna* ponds to cleanse agricultural wastewater before being discharged into Lake Maninjau (Fig. 11.8).

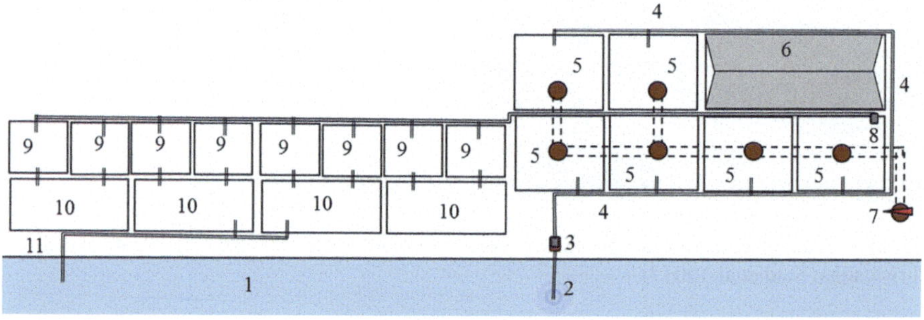

Fig. 11.6 Sketch of the IMTA system built on the shores of Lake Maninjau, drawn by T. Chrismadha (Said et al. 2020). *Notes* 1. Water inlet from the lake, 2. Intake pump, 3. Water distribution pipes, 4. Nile Tilapia ponds, 5. Water distribution pumps, 6. Monitoring house, 7. Pond spill way, 8. Tilapia pond water distribution system, 9. Upper *Lemna* ponds, 10. Lower *Lemna* ponds, 11. Outlet water

Fig. 11.7 IMTA ponds next to Lake Maninjau. *Photos* T. Chrismadha

Fig. 11.8 Converted paddy fields into *Lemna* ponds to purify agricultural water pollution. *Photos credit* BRIN/T. Chrismadha

- Research at the Research Center for Limnology and Water Resources—National Research and Innovation Agency (BRIN)

This IMTA research focused on using waste from Catfish aquaculture (*Clarias* sp.)—trophic status: omnivore and carnivore—to promote the growth of small floating aquatic plant *L. perpusilla*—trophic status: autotrophic or producer—which is subsequently consumed by Nile Tilapia (*O. niloticus*)—trophic status: omnivore and herbivore (Fig. 11.9). The advantage of this type of IMTA system is that *Lemna* converts waste material produced from catfish farming into biomass that can then be utilized as a valuable feed source in Tilapia farming. This application promotes the *Lemna* plant to serve a strategic dual function, specifically as a water phytoremediation agent that optimizes water conditions for fish growth requirements and as a supplementary feed source that decreases the expenses associated with fish feed (Chrismadha and Mulyana 2019; Chrismadha and Said 2019). Therefore, IMTA implementation will result in a simultaneous enhancement of production efficiency and the adoption of environmentally sustainable cultivation practices.

The experiment was undertaken to observe the growth and productivity of *Lemna* biomass utilizing catfish farming effluent as a substrate. It was conducted in closed-water recirculation ponds placed in open areas. These ponds consisted of catfish cultivation ponds, where the water medium was pumped into three sets of *Lemna* cultivation ponds before being recirculated back into the catfish ponds (Chrismadha and Said 2019). The ponds had dimensions of $2.4 \times 5\ m^2$ and a depth of 50 cm. Each pond contained 1500 catfish, meaning there were 125 fish per m^2 (Fig. 11.10). The catfish were fed with commercial feed containing 30–33% protein, with a quantity equivalent to 3% of their adult body weight. The amount of feed was systematically raised to get a predetermined growth rate of 6%.

The experiment, conducted during four consecutive cultivation periods from October 2017 to July 2018, demonstrated that catfish cultivation wastewater was

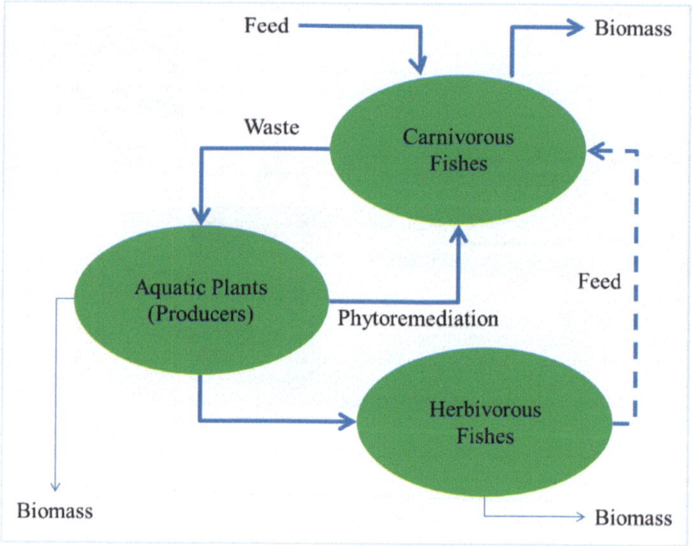

Fig. 11.9 Concept of IMTA implemented in the PRLSDA BRIN of Indonesia. *Sketch* T. Chrismadha

Fig. 11.10 IMTA trial ponds in the PRLSDA BRIN. *Photo* T. Chrismadha

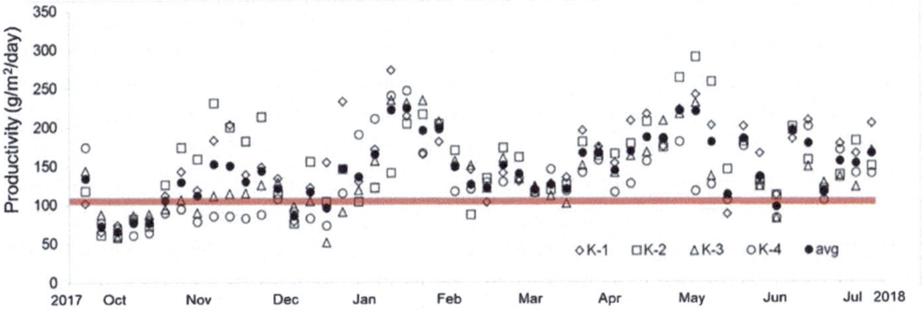

Fig. 11.11 Growth and biomass productivity of *L. perpusilla* in IMTA ponds using catfish aquaculture wastewater as the growth medium (Chrismadha and Said 2019)

well-suited as a growth medium for *Lemna*. This plant has the ability to regulate water quality, thereby reducing the need for water replacement throughout the cultivation period, except during the dry season when the evaporation rate was significantly high and there was no rainfall input. This capacity substantially reduced the usage of water by over 80%. Throughout the observation period, the average daily biomass productivity of *Lemna* consistently exceeded 100 g/m^2 or the equivalent of 328 tons/ha per year of *Lemna* fresh weight (Fig. 11.11). The optimal time frame for harvesting *Lemna* is within 3–4 days. The catfish culture yielded a productivity of 70–100 kg per pond per harvest, comparable to an annual production of 233–333 tons per hectare. The feed conversion ratio for this cultivation ranged from 0.8 to 1.4. This experiment also demonstrated that *Lemna* biomass may be dried, enabling it to be stored for extended durations and facilitating more flexible distribution. Nevertheless, this drying process may lead to a reduction in its nutritional value.

Another experiment involved using *Lemna* biomass as a substitute feed in aquaculture. In line with IMTA's principles, this directive aimed to utilize recently harvested organic material (fresh *Lemna* biomass) as animal feed. The biomass was promptly transported for fish feed, thus reducing the need for post-harvest processing and storage, which could involve extra energy and cost. Multiple efforts have been made to supply various fish species with freshly harvested *Lemna* biomass. The results suggested that almost all fish studied were ready to ingest this material. A group of fish had a strong preference for ingesting *Lemna*. This group consisted of Nile Tilapia, Common Carp, and Catfish. Other fish species, such as Gourami and Pomfret, consumed *Lemna* as their only food source in the trial.

Investigations were carried out to determine the growth performance of fish fed on *Lemna* biomass in two consecutive trials of Nile Tilapia cultivation. Firstly, the experiment was conducted in the nursery phase involving 2-week-old fish fries as the initial population (Sutrisno et al. 2021). These fries were raised for about eight weeks until they reached a size of 40–50 g per fish. Experiments at the nursery level were conducted in aquariums equipped with a closed-water recirculation system. Water quality management was primarily directed at regulating dissolved oxygen levels above 3 mg/L while maintaining a fish density of 100 fish/m^2. The *Lemna* used as feed was fresh biomass collected from nearby IMTA ponds specifically designated for cultivating catfish. This plant utilized wastewater from the catfish cultivation as a growth medium. The findings, as presented in Table 11.1, demonstrate the significant potential of *Lemna* biomass as a viable alternative to commercial pellet feed in nursery-level fish farming operations. Even though fish fed on *Lemna* biomass have a slower growth rate, simulations based on specific growth rate values showed that these fish required an additional 12–16 days to attain the same size as fish fed on commercial pellets. Based on the evidence, it can be concluded that *Lemna* has the potential to fully substitute commercial feed in the cultivation of tilapia fingerlings with the consequence of a more extended cultivation period. This finding presents a compelling benefit as the production of *Lemna* biomass is far more cost-effective than purchasing commercial feed.

In the second trial, a cultivation experiment was conducted on adult Tilapia in a 1 m^3 container equipped with a closed-water recirculation system. A submersible pump generates water circulation and maintains dissolved oxygen levels above 3 mg/L (Chrismadha and Mulyana 2019).

The experiment consisted of varying the cultivation density. At the lowest population density, with limited competition for space and resources, a single feeding of *Lemna* increased fish development by up to 3% per day (Table 11.2). The growth rate decreased as the density of culture increased.

This experiment demonstrates that the fish consumption rate on *Lemna* biomass is closely related to body weight development. The fish consumption rate ranges from 38.91 to 40.50% of body weight and follows a progressive pattern. Additionally, the feed conversion efficiency ratio decreases, forming a logarithmic model of body weight development. To determine the amount of *Lemna* biomass needed to grow fish to a specific size, one can plot the daily feed conversion ratio (FCR) alongside the growth of body weight and then multiply the daily FCR by the daily weight gain. To generate 250 g of fish, approximately 12.6–12.9 kg of fresh *Lemna* biomass is required, similar to an FCR value of 50.08–50.60. According to the FCR value, the annual production rate of 328 tons per hectare is adequate to cover the feed requirements to produce 7–8 tons of fish weighing 250 g each.

11.4 Integrated Multi-Trophic Aquaculture (IMTA) Practices Related to the Nature-Based Solutions (NbS)

The implementation of IMTA closely adheres to the ideas of NbS, which involve utilizing natural processes and biodiversity to address societal concerns sustainably. The IMTA methodology incorporates various fundamental principles of Nature-based Solutions, rendering it a pertinent and efficient method for promoting sustainable aquaculture and ecosystem management. IMTA utilizes these concepts to augment the sustainability of aquaculture, boost the health of ecosystems, and

Table 11.1 Growth performance of Nile Tilapia at nursery level (Sutrisno et al. 2021)

	Time				
	Initial	Week-2	Week-4	Week-6	Week-8
Pellet					
Length (cm)	35.00 ± 0.00	57.03 ± 1.59	87.03 ± 0.76	123.27 ± 3.62	138.60 ± 4.88
Weight (g/fish)	0.86 ± 0.00	3.42 ± 0.18	11.56 ± 0.28	30.96 ± 4.56	47.91 ± 5.62
CF (g/cm)	0.24 ± 0.00	0.59 ± 0.02	1.30 ± 0.03	2.49 ± 0.23	3.40 ± 0.35
SGR (%/day)		9.89 ± 0.38	7.62 ± 0.48	6.54 ± 0.68	3.36 ± 0.28
WG (g/fish/day)		0.18 ± 0.03	0.51 ± 0.03	1.13 ± 0.16	1.30 ± 0.19
TF (g)		120.00 ± 0.00	730.00 ± 0.00	670.00 ± 0.00	990.00 ± 0.00
FCR		0.47 ± 0.03	0.89 ± 0.05	0.81 ± 0.12	1.19 ± 0.18
Lemna					
Length (cm)	35.00 ± 0.00	51.07 ± 3.43	76.15 ± 2.68	106.40 ± 3.36	124.75 ± 3.65
Weight (g/fish)	0.86 ± 0.00	2.62 ± 0.24	8.42 ± 1.04	20.47 ± 2.31	32.18 ± 3.27
CF (g/cm)	0.24 ± 0.00	0.50 ± 0.03	1.07 ± 0.11	1.90 ± 0.16	2.57 ± 0.25
SGR (%day)		7.98 ± 0.64	7.27 ± 0.41	5.93 ± 0.25	3.48 ± 0.39
WG (g/fish/day)		0.13 ± 0.02	0.34 ± 0.03	0.69 ± 0.06	0.65 ± 0.06
TF (g)		191.67 ± 29.30	956.67 ± 40.41	983.33 ± 136.50	1100.00 ± 185.20
FCR		10.83 ± 0.30	17.56 ± 1.83	18.99 ± 0.95	27.78 ± 7.68

CF Condition factor; *SGR* specific growth rate; *WG* weight gain; *TF* total feed; *FCR* food conversion ratio

promote socio-economic development. To fully comprehend NbS for the International Union for Conservation of Nature (IUCN), it is crucial to consider a set of NbS principles, which should be studied alongside the NbS definition. The following are the eight proposed principles of NbS (Cohen-Shacham et al. 2016).

Nature-based Solutions:

1. embrace nature conservation norms (and principles);
2. can be implemented alone or in an integrated manner with other solutions to societal challenges (e.g., technological and engineering solutions);

Table 11.2 Growth performance of adult Nile Tilapia in the second trial (Chrismadha and Mulyana 2019)

Parameter	Cultivation density (fish/m^2)	Cultivation period (day)			
		0	40	75	115
Length (cm)	5	9.80 ± 0.00	14.22 ± 0.25	16.55 ± 0.27	18.30 ± 0.14
	10	9.92 ± 0.00	13.65 ± 0.07	15.79 ± 0.86	17.24 ± 0.61
	20	10.28 ± 0.25	13.16 ± 0.48	24.53 ± 1.91	16.28 ± 0.73
Weight (g/fish)	5	16.57 ± 0.63	55.30 ± 2.91	87.91 ± 4.99	126.36 ± 5.29
	10	16.83 ± 0.13	49.55 ± 3.71	76.29 ± 7.25	107.09 ± 5.85
	20	17.34 ± 0.20	43.26 ± 5.78	63.63 ± 13.66	88.03 ± 11.88
CF (g/cm)	5	1.69 ± 0.07	3.89 ± 0.14	5.31 ± 0.22	6.90 ± 0.24
	10	1.70 ± 0.01	3.63 ± 0.29	4.83 ± 0.20	6.21 ± 0.12
	20	1.69 ± 0.01	3.28 ± 0.32	4.35 ± 0.37	5.40 ± 0.49
SGR (%/day)	5		3.09 ± 0.04	1.32 ± 0.31	0.98 ± 0.04
	10		2.76 ± 0.21	1.23 ± 0.06	0.92 ± 0.11
	20		2.33 ± 0.39	1.08 ± 0.24	0.90 ± 0.22
WG (g/fish/day)	5		0.99 ± 0.06	0.84 ± 0.20	0.99 ± 0.01
	10		0.84 ± 0.10	0.86 ± 0.09	0.79 ± 0.04
	20		0.66 ± 0.16	0.52 ± 0.20	0.63 ± 0.05
TF (g)	5		3501 ± 634	4500 ± 1061	7450 ± 354
	10		6550 ± 2192	8800 ± 1556	12,100 ± 1131
	20		12,225 ± 884	13,800 ± 2687	18,400 ± 424
FCR	5		18.02 ± 2.22	27.62 ± 0.19	38.76 ± 2.14

(continued)

Table 11.2 (continued)

Parameter	Cultivation density (fish/m^2)	Cultivation period (day)			
		0	40	75	115
	10		20.56 ± 9.11	32.81 ± 1.47	39.41 ± 5.46
	20		24.04 ± 3.94	35.25 ± 7.05	37.78 ± 1.90

CF Condition factor; *SGR* specific growth rate; *WG* weight gain; *TF* total feed; *FCR* food conversion ratio

3. are determined by site-specific natural and cultural contexts that include traditional, local, and scientific knowledge;
4. produce societal benefits fairly and equitably in a manner that promotes transparency and broad participation;
5. maintain biological and cultural diversity and the ability of ecosystems to evolve;
6. are applied at a landscape scale;
7. recognize and address the trade-offs between the production of a few immediate economic benefits for development and future options for the production of the full range of ecosystem services and
8. are integral to the overall design of policies and measures or actions to address a specific challenge.

The principles of NbS (Cohen-Shacham et al. 2019) are directly relevant to the implementation of IMTA and are comprehensively explained as follows:

11.4.1 Principle 1

IMTA in freshwater environments offers a compelling example of how NbS can align with and support nature conservation norms and principles, as articulated in Principle 1 of NbS. This principle underscores that while NbS inherently embrace and benefit from nature conservation efforts, they are not a substitute for traditional conservation activities. Instead, NbS should be considered complementary to conservation, potentially enhancing and deriving benefits from conservation initiatives. This principle is particularly relevant in freshwater IMTA as it demonstrates how sustainable aquaculture practices can coexist and bolster conservation efforts.

Freshwater IMTA systems are designed to integrate multiple species from different trophic levels within a single aquaculture system. These systems typically include fish, aquatic plants, and invertebrates, creating a balanced ecosystem where the waste products of one species become resources for another. This approach enhances productivity and resource use efficiency and mitigates many environmental impacts of traditional monoculture aquaculture. By mimicking natural processes and promoting ecological balance, freshwater IMTA inherently aligns with nature conservation principles.

One of the primary ways freshwater IMTA supports nature conservation is through its potential to reduce nutrient pollution and eutrophication in natural water bodies. Traditional aquaculture systems often discharge significant amounts of nutrients, such as nitrogen and phosphorus, into surrounding waters, leading to harmful algal blooms and oxygen depletion. In contrast, IMTA systems utilize these nutrients by incorporating species like aquatic plants and filter-feeding invertebrates, which absorb and recycle these substances. This natural filtration process helps maintain water quality, protecting the health of aquatic ecosystems and supporting biodiversity. Thus, freshwater IMTA can complement conservation efforts by mitigating one of the major environmental impacts of traditional aquaculture.

Furthermore, freshwater IMTA systems can contribute to the conservation of wild fish populations and their habitats. IMTA reduces fishing pressure on natural fish stocks, often overexploited, by providing a sustainable alternative to wild fish capture. This reduction in fishing pressure allows wild populations to recover and ecosystems to maintain their ecological balance. Additionally, by promoting native species in aquaculture systems, IMTA helps preserve local biodiversity and prevents the introduction of potentially invasive species that could disrupt local ecosystems.

Freshwater IMTA also offers opportunities to enhance habitat conservation. For example, integrating IMTA systems with wetland restoration projects can create multifunctional landscapes that provide aquaculture products and critical ecosystem services. Restored wetlands can act as natural buffers, filtering runoff and reducing nutrient loads before they reach larger water bodies. These wetlands also provide habitat for various species, contributing to biodiversity conservation. By aligning aquaculture practices with habitat restoration, freshwater IMTA can support broader conservation goals and create synergies between food production and ecosystem health.

The principle that NbS should not replace traditional conservation but complement it is well-illustrated when freshwater IMTA is implemented near protected areas. For instance, a protected area established to conserve a particular species or habitat can provide ecological benefits that extend beyond its boundaries, supporting nearby IMTA systems. These systems, in turn, can enhance the conservation value of the protected area by reducing external pressures such as nutrient pollution and habitat degradation. This reciprocal relationship exemplifies how NbS interventions, like IMTA, can benefit from and contribute to conservation efforts across a landscape.

Moreover, IMTA can play a role in addressing broader biodiversity conservation priorities by promoting sustainable land and water use practices. By integrating multiple species and mimicking natural nutrient cycles, IMTA systems can enhance ecosystem resilience and productivity. This approach supports the conservation of aquatic biodiversity and contributes to the overall health of the watershed. By combining IMTA with other NbS interventions like reforestation and riverbank stabilization, a comprehensive approach can be developed to preserve and rehabilitate freshwater habitats effectively.

Implementing IMTA is a powerful demonstration of how NbS can embrace and support nature conservation norms and principles. By reducing nutrient pollution, alleviating fishing pressure, enhancing habitat conservation, and promoting sustainable land and water use practices, freshwater IMTA aligns closely with traditional conservation goals. It exemplifies the complementary relationship between NbS and nature conservation, highlighting how sustainable aquaculture practices can coexist with and bolster conservation efforts. As such, IMTA contributes to sustainable food production and plays a crucial role in preserving and enhancing the health and biodiversity of freshwater ecosystems, embodying the essence of Principle 1 of NbS.

11.4.2 Principle 2

Implementing freshwater IMTA is a compelling example of how NbS can be employed independently with technological and engineering solutions to address various societal challenges. Principle 2 of NbS emphasizes the versatility of NbS, highlighting its capacity to provide a comprehensive range of ecosystem services or to complement other approaches. This principle underscores the importance of policy coherence, linking it to NbS Principle 8, which focuses on inclusive governance and equitable implementation. Freshwater IMTA offers a unique perspective on how these principles can be actualized, promoting sustainability and resilience in aquaculture practices.

Freshwater IMTA involves the cultivation of multiple species from different trophic levels in a single, integrated system. For example, fish, aquatic plants, and invertebrates such as shrimp or mussels are grown together. Each species performs a specific ecological role, creating a balanced ecosystem where the waste products of one organism become resources for another. This system mimics natural processes, enhancing efficiency and sustainability by recycling nutrients and reducing environmental impacts commonly associated with conventional aquaculture.

One significant societal challenge freshwater IMTA addresses is nutrient pollution, a prevalent issue in traditional aquaculture systems. Excess nutrients from fish feed and waste can lead to eutrophication, causing algal blooms and oxygen depletion in water bodies. In a freshwater IMTA system, aquatic plants and invertebrates absorb these excess nutrients, effectively acting as biofilters. For instance, plants like water hyacinths or duckweeds can take up nitrogen and phosphorus, while mussels and other filter feeders can help clear suspended particles and organic matter from the water. This natural purification process complements existing technological solutions such as mechanical filters and aerators, providing an additional, sustainable layer of water quality management.

Moreover, freshwater IMTA systems enhance biodiversity and ecosystem health. By integrating multiple species, these systems create diverse habitats that support a variety of organisms. This biodiversity improves ecosystem resilience and provides multiple ecosystem services, such as water purification, carbon sequestration, and habitat provision. These services contribute to the overall health of the aquatic ecosystem, ensuring long-term sustainability and resilience against environmental changes and stresses.

The economic benefits of freshwater IMTA are also significant. By diversifying production, IMTA reduces the risks associated with market fluctuations and environmental changes that can affect monoculture systems. For example, if market prices for a particular fish species drop, farmers can still generate income from other products, such as aquatic plants or invertebrates. This economic resilience is crucial for the livelihoods of aquaculture farmers, particularly in regions where aquaculture is a primary source of income.

Freshwater IMTA also demonstrates the integration of NbS with engineered solutions. For example, constructed wetlands can be used with IMTA systems to enhance nutrient removal and water purification. These wetlands, engineered to mimic natural wetland processes, can provide additional filtration and habitat benefits, further improving water quality and ecosystem health. The combination of IMTA and constructed wetlands exemplifies how NbS can effectively integrate with technological solutions to address complex environmental challenges.

Policy coherence is essential for successfully implementing freshwater IMTA, aligning with Principle 2 and its connection to Principle 8 of NbS. Policymakers must ensure that regulatory frameworks support the integration of IMTA with other solutions and promote sustainable aquaculture practices, including developing policies that encourage the adoption of IMTA, providing financial incentives for sustainable practices, and facilitating research and innovation in the field. Coherent policies must also ensure that the benefits of IMTA are shared equitably among all stakeholders, including local communities, farmers, and consumers.

Inclusive governance is another critical aspect of successful IMTA implementation. Engaging local communities in the planning and managing IMTA systems ensures that their knowledge and needs are considered, leading to more effective and sustainable solutions. Participatory approaches that involve farmers, scientists, policymakers, and other stakeholders can foster collaboration and innovation, driving the successful adoption of IMTA practices.

In sum, freshwater IMTA implementation exemplifies Principle 2 of NbS by showcasing how NbS can be employed independently and in conjunction with other solutions to address societal challenges. Freshwater IMTA systems enhance ecosystem services, improve water quality, promote biodiversity, and provide economic benefits. By integrating these systems with engineered solutions and ensuring policy coherence and inclusive governance, freshwater IMTA can contribute significantly to sustainable and resilient aquaculture practices. This approach addresses immediate environmental and economic challenges and supports the long-term sustainability and health of aquatic ecosystems, aligning with the broader goals of Nature-based Solutions.

11.4.3 Principle 3

The implementation of freshwater IMTA vividly illustrates Principle 3 of NbS, which underscores the necessity of tailoring solutions to site-specific natural and cultural contexts, incorporating traditional, local, and scientific knowledge. This principle highlights that effective NbS are grounded in a comprehensive understanding of the unique characteristics of particular ecosystems and emphasizes the importance of participatory approaches in their development. IMTA exemplifies how these diverse knowledge systems can converge to create sustainable and context-sensitive aquaculture practices.

The design of IMTA systems is inherently site-specific, as the selection of species and the configuration of the system must align with the local environmental conditions and cultural practices. This site-specificity is a cornerstone of

Principle 3 of NbS, ensuring the solution is ecologically and culturally appropriate. In implementing freshwater IMTA, a thorough understanding of the local ecosystem is paramount. Scientific knowledge provides essential insights into the ecological interactions between species, nutrient cycles, and water quality parameters. For example, selecting suitable species for IMTA requires knowledge of their ecological roles, growth requirements, and compatibility with other species. Scientific research can inform the optimal stocking densities, feeding regimes, and system design to maximize productivity and minimize environmental impacts.

However, scientific knowledge alone is not sufficient. Incorporating traditional and local knowledge is equally critical to the success of IMTA systems. Local communities often possess a deep understanding of their natural environment, built over generations of interaction with the ecosystem. This knowledge includes practical insights into species behavior, seasonal variations, and effective management practices. For example, traditional aquaculture practices, such as integrating paddy and fish farming in Asian cultures, can offer valuable lessons for designing modern IMTA systems. By blending scientific research with traditional knowledge, freshwater IMTA can be tailored to fit the specific ecological and cultural contexts of the region.

The cultural context also plays a significant role in the acceptance and sustainability of IMTA. Aquaculture practices must align with local values, traditions, and livelihoods to be effectively adopted and maintained by communities. Engaging with local stakeholders and respecting their cultural practices ensures that IMTA systems are designed to support and enhance local livelihoods. For instance, in communities where fishing is a primary source of income, integrating IMTA can provide additional revenue streams through the cultivation of aquatic plants and invertebrates, thus enhancing economic resilience and food security.

Participatory approaches are essential for successful implementations of freshwater IMTA, as emphasized by Principle 3 of NbS. Involving local communities, farmers, scientists, policymakers, and other stakeholders in the planning and management of IMTA systems fosters collaboration and ensures that diverse perspectives and knowledge systems are integrated. This participatory process can include workshops, community meetings, and collaborative research projects, where stakeholders contribute to designing, implementing, and monitoring IMTA systems. Such inclusive governance ensures that the solutions are tailored to local needs and conditions, enhancing their effectiveness and sustainability.

An example of this participatory approach is the development of IMTA systems in rural communities, where local farmers are actively involved in designing and managing the aquaculture system. By engaging farmers in the process, they can share their knowledge of local environmental conditions, such as water quality and seasonal changes, which can inform the selection of species and system design. Additionally, involving farmers in monitoring and managing the system ensures that they have a vested interest in its success and are expected to adopt and maintain the practices.

As highlighted by Principle 3, evidence-based approaches are crucial for the continuous improvement and adaptation of freshwater IMTA systems. Monitoring and evaluating the performance of IMTA systems through scientific research and community observations can provide valuable data on their ecological and economic impacts. This evidence can inform adaptive management practices, allowing adjustments based on real-time feedback and changing conditions. For example, if water quality monitoring indicates an increase in nutrient levels, adjustments can be made to the stocking densities or feeding regimes to mitigate the impact.

Furthermore, the integration of traditional knowledge in monitoring practices can enhance the effectiveness of IMTA systems. Local communities often have established methods for observing environmental changes and assessing ecosystem health. By incorporating these traditional monitoring practices into scientific protocols, a more comprehensive understanding of the ecosystem can be achieved. This combined approach can enhance the resilience and adaptability of IMTA systems, ensuring their long-term sustainability.

The implementation of IMTA aligns closely with Principle 3 of NbS, which emphasizes the importance of site-specific natural and cultural contexts, incorporating traditional, local, and scientific knowledge. Freshwater IMTA systems are inherently site-specific, requiring a thorough understanding of local ecosystems and cultural practices. By integrating scientific research with traditional and local knowledge and fostering participatory approaches, IMTA systems can be tailored to fit the unique characteristics of the region. This holistic approach enhances the sustainability and effectiveness of IMTA, supporting the broader goals of Nature-based Solutions in promoting ecological and cultural resilience. As such, IMTA contributes to sustainable aquaculture practices and embodies the principles of NbS by embracing and integrating diverse knowledge systems and participatory governance.

11.4.4 Principle 4

The implementation of freshwater IMTA offers a robust demonstration of Principle 4 of NbS, which underscores the necessity of producing societal benefits fairly and equitably, promoting transparency, and ensuring broad participation. This principle is critical as NbS interventions often provide essential services such as food and water security or disaster risk reduction, benefiting distant communities and governments. However, it is essential to manage these interventions to avoid the loss of opportunities for those living in or near the source of these services. IMTA systems exemplify how equitable and inclusive approaches can be embedded in NbS, ensuring that the benefits are shared broadly and fairly among all stakeholders.

Integrating multiple species from different trophic levels in a single aquaculture system within freshwater IMTA enhances resource efficiency, reduces environmental impacts, and increases productivity. However, the successful IMTA implementation goes beyond ecological and economic benefits; it requires a fair and equitable approach that involves local communities and stakeholders at all levels.

One of the primary societal benefits of IMTA is its contribution to food security. Diversifying aquaculture products and improving system resilience can provide a stable and sustainable food source for local communities, which is particularly important in regions where traditional aquaculture practices are vulnerable to environmental changes and market fluctuations. By ensuring that local communities have access to and benefit from the products of IMTA, food security is enhanced fairly and equitably. Furthermore, the diversification inherent in IMTA can create additional income streams for local farmers, thereby contributing to economic stability and resilience.

Transparency and broad participation are fundamental to achieving these benefits. Implementing freshwater IMTA requires the involvement of local communities, farmers, scientists, policymakers, and other stakeholders in the planning and decision-making processes. This participatory approach ensures that the diverse needs and perspectives of different stakeholders are considered, leading to more inclusive and practical solutions. For example, organizing community workshops and meetings to discuss the design and management of IMTA systems can foster collaboration and build trust among stakeholders. This participatory process can also help identify and address potential conflicts or concerns, ensuring that the benefits of IMTA are distributed fairly.

Ensuring fair compensation and avoiding loss of local opportunities is another critical aspect of implementing freshwater IMTA in line with Principle 4. In some cases, introducing IMTA systems may require changes in land or water use that could impact local communities. It is essential to establish fair and transparent processes for negotiating these changes, including mechanisms for compensating affected individuals or communities. Payment for Ecosystem Services (PES) schemes can be an effective tool, providing financial incentives to local communities for maintaining and managing IMTA systems that deliver broader ecosystem services. For instance, communities that manage wetlands or riparian zones as part of IMTA could be compensated for maintaining water quality and biodiversity, benefiting downstream users, and contributing to regional water security.

The equitable distribution of benefits also addresses potential disparities in access to resources and opportunities. Ensuring that marginalized or vulnerable groups, such as small-scale farmers or indigenous communities, are included in IMTA initiatives is crucial. This goal can be achieved by providing capacity-building programs, technical support, and access to financial resources, enabling these groups to participate fully in and benefit from IMTA. IMTA can help bridge social and economic gaps by promoting inclusivity and equity contributing to broader societal well-being.

Moreover, transparency in the implementation and management of IMTA systems is essential for building trust and accountability. Clear communication about the goals, processes, and expected outcomes of IMTA initiatives helps ensure that all stakeholders understand their roles and responsibilities. Regular monitoring and reporting on the performance and impacts of IMTA systems can provide valuable feedback, allowing for adaptive management and continuous improvement. This transparency enhances the credibility of IMTA initiatives and fosters a sense of ownership and commitment among stakeholders.

Freshwater IMTA also has the potential to contribute to broader societal benefits beyond local communities. For example, by improving water quality and enhancing ecosystem health, IMTA systems can provide critical services such as flood regulation, carbon sequestration, and biodiversity conservation. These services benefit local communities and regional and national stakeholders, including infrastructure managers, private sector actors, and government agencies. Ensuring that the benefits of these ecosystem services are recognized and valued at all levels is essential for sustaining IMTA initiatives and promoting broader support for NbS.

In summary, the implementation of freshwater IMTA aligns closely with Principle 4 of NbS, which emphasizes producing societal benefits fairly and equitably, promoting transparency, and ensuring broad participation. By involving local communities and stakeholders in the planning and managing of IMTA systems, ensuring fair compensation, avoiding loss of local opportunities, and promoting transparency and accountability, freshwater IMTA can deliver significant societal benefits while upholding the principles of equity and inclusivity. This approach enhances the sustainability and effectiveness of IMTA systems and contributes to the broader goals of NbS, supporting resilient and equitable societies and healthy ecosystems.

11.4.5 Principle 5

IMTA exemplifies the core tenets of Principle 5 of NbS, which emphasizes the necessity of maintaining biological and cultural diversity and the ability of ecosystems to evolve. This principle is foundational to ensuring that NbS interventions are effective in the short term and sustainable and resilient in the face of future environmental changes. Freshwater IMTA systems, by their very design, inherently support these goals, integrating ecological, economic, and cultural dimensions harmoniously and adaptively.

IMTA promotes biodiversity within the aquaculture environment. The presence of multiple species creates a more complex and resilient system capable of withstanding environmental fluctuations and disturbances. For example, if a disease only affects one species, the impact on the overall system is mitigated because other species continue to thrive, thereby maintaining the ecological and economic functions of the system.

This biodiversity is crucial for the long-term sustainability of aquaculture operations. Diverse systems are more resilient to pests, diseases, and environmental changes, reducing the need for chemical interventions and thereby minimizing ecological footprints. Additionally, integrating species like filter-feeding invertebrates and aquatic plants helps maintain water quality by recycling nutrients and reducing waste, which supports the health of the entire aquaculture system. This nutrient recycling is a crucial aspect of IMTA that enhances the ability of ecosystems to evolve and adapt over time, as it fosters a self-sustaining and balanced environment.

Cultural diversity is another critical dimension supported by freshwater IMTA. Aquaculture practices are often deeply embedded in local traditions and livelihoods, and successful implementation of IMTA requires an understanding and respect for these cultural contexts. Integrating traditional knowledge with scientific approaches can enhance the effectiveness and acceptance of IMTA systems. For instance, local communities may have long-standing practices and observations about species interactions, seasonal changes, and resource management that can inform and improve IMTA practices. By valuing and incorporating this traditional knowledge, IMTA not only supports cultural diversity but also fosters community engagement and ownership, which are essential for the long-term success and sustainability of the system.

Moreover, IMTA systems can contribute to preserving and revitalizing cultural practices related to aquaculture and resource management. In many regions, traditional aquaculture practices have been eroded by the introduction of industrial monoculture systems. By promoting diverse and integrated approaches, IMTA can help revive these traditional practices, ensuring that cultural heritage is maintained and passed down through generations. This cultural continuity is vital for the resilience of communities, as it strengthens social cohesion and provides a sense of identity and belonging.

The adaptability of IMTA systems is a critical factor in their ability to support the evolving dynamics of ecosystems. As environmental conditions change due to climate change, pollution, and habitat degradation, IMTA systems can be adjusted to maintain their functionality and benefits. For example, changes in water temperature or quality may require modifications in the species composition or management practices within the IMTA system. By continuously monitoring and responding to environmental changes, IMTA systems can remain resilient and productive, ensuring their ecosystem services are sustainable over time.

Furthermore, the flexibility of IMTA allows for innovation and experimentation, which are essential for adapting to future challenges. Researchers and practitioners can explore new combinations of species, management techniques, and technologies to enhance the efficiency and resilience of IMTA systems. This continuous learning and adaptation process is critical for maintaining the relevance and effectiveness of IMTA in the face of evolving environmental and socio-economic conditions.

Implementing IMTA also contributes to the broader goals of biodiversity conservation. By reducing the environmental impacts of aquaculture, such as nutrient pollution and habitat destruction, IMTA helps protect surrounding natural ecosystems. Healthy and diverse aquaculture systems can act as buffers, absorbing and mitigating the effects of human activities on the environment. This protective role is significant in regions where natural habitats are under pressure from urbanization, agriculture, and other forms of development. By supporting the health of these ecosystems, IMTA contributes to biodiversity conservation at a landscape scale.

Thus, the implementation of freshwater IMTA aligns closely with Principle 5 of NbS, which emphasizes the maintenance of biological and cultural diversity and the ability of ecosystems to evolve. By promoting ecological balance, supporting traditional knowledge and practices, and fostering adaptability and resilience, IMTA systems provide a sustainable and robust approach to aquaculture that can withstand future environmental changes. This alignment ensures that the services provided by IMTA are effective in the present, sustainable and resilient in the long term, contributing to the broader goals of NbS, which are to create harmonious and enduring relationships between human societies and the natural world.

11.4.6 Principle 6

The implementation of IMTA is a compelling illustration of Principle 6 of NbS, which underscores the necessity of applying NbS at a landscape scale. This principle is pivotal in ensuring that the benefits of NbS extend beyond localized interventions to influence broader ecological, social, and economic systems. Freshwater IMTA, with its holistic and integrative approach, aligns closely with this principle by fostering ecological connectivity, enhancing resource efficiency, and promoting resilience across multiple ecosystems within a landscape.

The multispecies integration of IMTA maximizes resource use efficiency and enhances ecological functions within the aquaculture system. However, the true potential of freshwater IMTA is realized when it is scaled up and integrated into more extensive landscape-level planning and management strategies.

One of the primary benefits of implementing freshwater IMTA at a landscape scale is the promotion of ecological connectivity. Landscapes comprise interconnected ecosystems, such as rivers, lakes, wetlands, forests, and agricultural lands. Freshwater IMTA can link between these ecosystems, facilitating nutrient cycling, habitat connectivity, and biodiversity conservation. For example, IMTA system integration along river corridors can enhance water quality and provide habitat for diverse aquatic and terrestrial species, contributing to the overall health and resilience of the watershed. By enhancing ecological connectivity, freshwater IMTA helps maintain the integrity and functionality of landscapes, supporting the long-term sustainability of ecosystem services.

Applying freshwater IMTA at a landscape scale also enhances resource efficiency and resilience. In a landscape context, IMTA systems can be strategically located to optimize available resources, such as water, nutrients, and land. For instance, integrating IMTA with agricultural landscapes can create synergies between crop production and aquaculture, where nutrient-rich water from IMTA systems can be used to irrigate and fertilize crops, reducing the need for chemical inputs. This integration not only improves resource efficiency but also enhances the resilience of agricultural systems to environmental stresses, such as droughts and floods. Additionally, by diversifying production systems at a landscape scale, communities can build economic resilience, reducing their vulnerability to market fluctuations and environmental changes.

The landscape-scale application of IMTA also promotes the broader goals of sustainable development and climate change adaptation. As landscapes are often the units of planning and management for natural resource use, implementing IMTA at this scale can contribute to integrated water resource management, sustainable land use planning, and biodiversity conservation. For example, IMTA systems can be integrated into watershed management plans to enhance water quality, reduce nutrient loading, and support aquatic biodiversity. In coastal regions, IMTA can be linked with coastal zone management strategies to mitigate the impacts of land-based activities on marine ecosystems. By addressing multiple objectives simultaneously, IMTA supports the holistic and multifunctional approach required for sustainable landscape management.

Moreover, the transboundary nature of many landscapes highlights the importance of international cooperation and coordination in the implementation of freshwater IMTA. River basins, for instance, often span multiple countries, requiring collaborative efforts to manage shared water resources sustainably. Freshwater IMTA can serve as a model for transboundary cooperation, where countries work together to implement integrated aquaculture practices that benefit all riparian communities. By fostering cross-border collaboration, IMTA can contribute to peacebuilding and regional stability while enhancing the resilience of shared ecosystems to climate change and other environmental pressures.

The landscape-scale application of freshwater IMTA is also linked to the principles of transparency, broad participation, and policy coherence. Implementing IMTA at this scale requires the involvement of diverse stakeholders, including local communities, farmers, policymakers, scientists, and private sector actors. Transparent and participatory processes are essential for building trust, ensuring equitable benefit-sharing, and addressing potential conflicts. For example, engaging local communities in the planning and managing of IMTA systems can help incorporate traditional knowledge, address local concerns, and foster a sense of ownership and stewardship. By promoting inclusive governance, freshwater IMTA supports the principles of equity and social justice, which are fundamental to the success of NbS at any scale.

Furthermore, policy coherence is critical for scaling up freshwater IMTA across landscapes. Integrating IMTA into national and regional policies, such as agricultural, water, and biodiversity strategies, can create an enabling environment for widespread adoption and implementation. For instance, policies supporting sustainable aquaculture practices, providing financial incentives, and facilitating access to technical assistance can encourage farmers to adopt IMTA systems. By aligning policies across sectors and scales, governments can create synergies and leverage resources to maximize the benefits of freshwater IMTA.

Therefore, the implementation of freshwater IMTA aligns closely with Principle 6 of NbS, which emphasizes the application of NbS at a landscape scale. By promoting ecological connectivity, enhancing resource efficiency, and supporting sustainable development, freshwater IMTA contributes to the resilience and sustainability of landscapes. The transboundary nature of many landscapes highlights the importance of international cooperation and coordination, while the principles of transparency, broad participation, and policy coherence ensure that the benefits of IMTA are shared equitably and sustainably. As such, freshwater IMTA supports the goals of NbS and contributes to the broader objectives of sustainable development, climate change adaptation, and biodiversity conservation at a landscape scale.

11.4.7 Principle 7

Freshwater IMTA embodies the essence of Principle 7 of NbS, which emphasizes recognizing and addressing trade-offs between immediate economic benefits and the future sustainability of ecosystem services. This principle is crucial in ensuring that NbS interventions do not compromise long-term ecological health and resilience for short-term gains. IMTA balances these trade-offs, promoting a holistic approach that integrates economic viability with ecological sustainability and social equity.

Diversification in the IMTA system enhances resource use efficiency, reduces environmental impacts, and supports a broader range of ecosystem services. However, implementing IMTA at a landscape scale requires a nuanced understanding of the trade-offs involved and a commitment to managing these trade-offs transparently and equitably.

One of the primary trade-offs in aquaculture is between maximizing short-term economic returns and maintaining the long-term health of the ecosystem. Traditional monoculture aquaculture systems often prioritize the production of a single species, which can lead to environmental degradation, such as nutrient pollution, habitat destruction, and biodiversity loss. These systems may provide immediate economic benefits but can undermine the ecosystem's capacity to provide other essential services in the future. On the other hand, freshwater IMTA systems distribute and recycle nutrients throughout different trophic levels, reducing the environmental footprint and promoting ecosystem health. This comprehensive strategy promotes the long-term sustainability of the ecosystem by ensuring its ability to deliver a diverse variety of services, including water purification, habitat provision, and climate regulation.

However, transitioning from monoculture to IMTA involves trade-offs that must be carefully managed. For example, IMTA systems may require more complex management practices and initial investments than traditional monoculture systems. Farmers and stakeholders need to weigh these immediate economic costs against the long-term benefits of a more sustainable and resilient system. Transparent processes and fair negotiations are essential in addressing these trade-offs. Engaging stakeholders, including local communities, farmers, policymakers, and scientists, in decision-making ensures that diverse interests and perspectives are considered and the benefits and costs are equitably distributed.

Additionally, the trade-offs in IMTA are not solely economic and environmental but also social. Implementing IMTA can impact local livelihoods and cultural practices. For instance, traditional monoculture practices may be deeply embedded in the local culture and economy, and transitioning to IMTA might be perceived as a threat to these established ways of life. It is crucial to incorporate traditional knowledge and practices into the design and management of IMTA systems to address these social trade-offs. This inclusion not only respects and preserves cultural heritage but also enhances the acceptance and success of IMTA by aligning it with local values and practices.

Negotiating trade-offs in IMTA also involves addressing potential conflicts between different stakeholder groups. In landscapes where resources are shared by multiple users, such as farmers, fishers, and conservationists, there may be competing demands and priorities. For example, while farmers may prioritize maximizing aquaculture production, conservationists may focus on preserving biodiversity and ecosystem health. These complex and sometimes conflictual relationships must be identified and managed through inclusive and transparent negotiation processes. Establishing multi-stakeholder platforms or councils can facilitate dialogue, foster mutual understanding, and build consensus on balancing different interests and objectives.

Moreover, IMTA systems can provide a framework for adaptive management, where ongoing monitoring and evaluation inform continuous improvement and adaptation to changing conditions. This adaptive approach allows stakeholders to reassess and renegotiate trade-offs as new information and circumstances arise, ensuring the system remains sustainable and resilient. For instance, if monitoring indicates that a particular species in the IMTA system is underperforming or causing adverse environmental impacts, stakeholders can collaboratively decide on adjustments or replacements to optimize the system's performance and sustainability.

Implementing freshwater IMTA also involves considering broader landscape-scale trade-offs. While IMTA can enhance the sustainability of local aquaculture practices, its impacts on the broader landscape and connected ecosystems must be considered. For example, the nutrient recycling benefits of IMTA can improve water quality downstream, benefiting other aquatic and terrestrial ecosystems. However, if not properly managed, there could be unintended consequences, such as introducing non-native invasive species or changes in local hydrology. Therefore, landscape-scale planning and coordination are essential to ensure that the benefits of IMTA are maximized while minimizing potential negative impacts on other parts of the ecosystem.

Therefore, the implementation of freshwater IMTA closely aligns with Principle 7 of NbS, which highlights the importance of recognizing and addressing trade-offs between immediate economic benefits and the long-term sustainability of ecosystem services. By promoting ecological balance, integrating traditional knowledge, and fostering inclusive and transparent decision-making processes, IMTA provides a model for balancing these trade-offs. The comprehensive and adaptive nature of IMTA ensures that it not only enhances the sustainability and resilience of aquaculture systems but also supports broader landscape-scale conservation and development goals. Through careful management of trade-offs and equitable distribution of benefits, IMTA contributes to the sustainable development of both human and natural systems, exemplifying the principles of Nature-based Solutions.

11.4.8 Principle 8

Freshwater IMTA epitomizes Principle 8 of NbS, which asserts that NbS should be an integral part of the overall design of policies and measures to address specific challenges. This principle emphasizes embedding NbS within policy frameworks to ensure scalability, sustainability, and effectiveness. By aligning IMTA with policy design and implementation, we can maximize its ecological, economic, and social benefits, addressing broader challenges such as food security, environmental sustainability, and climate resilience.

Freshwater IMTA, involving the co-cultivation of multiple species from different trophic levels within a single aquaculture system, optimizes resource use, minimizes environmental impacts, and enhances the resilience and productivity of aquaculture operations. However, the successful implementation and scaling of IMTA require supportive policy frameworks that recognize and promote its multifaceted benefits.

To begin with, incorporating IMTA into national and regional policies can significantly enhance food security and economic development. Aquaculture is a critical component of global food systems, providing a substantial source of protein and livelihoods for millions of people. Policies that support IMTA can promote sustainable aquaculture practices that increase productivity and profitability while reducing environmental impacts. For instance, governments can provide financial incentives, technical assistance, and capacity-building programs to encourage farmers to adopt IMTA systems. By integrating IMTA into agricultural and fisheries policies, policymakers can create a conducive environment for widespread adoption, contributing to food security and rural development.

Moreover, embedding IMTA within environmental policies can help address key challenges such as water quality management, biodiversity conservation, and climate change mitigation. IMTA systems inherently enhance water quality by recycling nutrients and reducing waste outputs, thereby mitigating the impacts of nutrient pollution on freshwater ecosystems. Policies that mandate or incentivize the use of IMTA can help achieve water quality targets and support the health of aquatic ecosystems. Additionally, by promoting biodiversity within aquaculture systems, IMTA can contribute to the conservation of genetic resources and the resilience of aquatic habitats. Policymakers can integrate IMTA into biodiversity strategies and action plans to support these conservation goals.

Furthermore, IMTA can significantly influence climate change adaptation and mitigation efforts. Aquaculture systems are vulnerable to climate-related impacts such as temperature fluctuations, extreme weather events, and changes in water availability. IMTA enhances the resilience of aquaculture operations by diversifying species and optimizing resource use, thereby reducing vulnerability to climate variability. Policies that support IMTA can be integrated into climate adaptation strategies to enhance the resilience of food production systems. Additionally, IMTA systems can contribute to climate mitigation by sequestering carbon by cultivating aquatic plants and shellfish. Policymakers can incorporate IMTA into carbon offset programs and climate mitigation plans to leverage its potential for reducing greenhouse gas emissions.

Incorporating IMTA into policy frameworks also supports the principle of adaptive management, which is crucial for the sustainability and scalability of NbS. Adaptive management involves continuously monitoring and evaluating the outcomes of interventions and adjusting management practices based on new information and changing conditions. By embedding IMTA within policy frameworks, governments can establish mechanisms for monitoring and reporting on the performance of IMTA systems. This data can inform policy adjustments and improvements, ensuring that IMTA practices remain effective and responsive to environmental, social, and economic changes. For instance, policies can include provisions for regular assessments of IMTA systems, stakeholder consultations, and adaptive management plans that allow for iterative improvements.

Additionally, integrating IMTA into policy frameworks facilitates broader stakeholder engagement and collaboration. Effective NbS implementation requires the involvement of diverse stakeholders, including farmers, local communities, researchers, policymakers, and private sector actors. Policies that support IMTA can establish platforms for multi-stakeholder dialogue and collaboration, fostering partnerships and knowledge exchange. For example, governments can create advisory committees or working groups that include representatives from different sectors and communities to guide the development and implementation of IMTA policies. This inclusive approach ensures that policies are informed by diverse perspectives and expertise, enhancing their relevance and effectiveness.

Embedding IMTA within policy frameworks can help address socio-economic inequalities and promote social equity. Aquaculture provides livelihoods for many marginalized and vulnerable communities, and policies that support IMTA can enhance their economic opportunities and resilience. For instance, targeted support for small-scale and community-based IMTA projects can empower local communities, improve food security, and increase adaptive capacity. Policymakers can design inclusive policies that prioritize support for disadvantaged groups, promote gender equality, and ensure equitable access to resources and benefits.

Thus, the implementation of freshwater IMTA closely aligns with Principle 8 of NbS, which emphasizes the integration of NbS within policy frameworks to address specific challenges. By embedding IMTA into national and regional policies, governments can enhance food security, promote environmental sustainability, and support climate adaptation and mitigation efforts. The integration of IMTA within policy frameworks also facilitates adaptive management, stakeholder engagement, and social equity, ensuring the sustainability and scalability of NbS interventions. As such, freshwater IMTA provides a model for how NbS can be effectively incorporated into policy design and implementation, contributing to the broader goals of sustainable development and environmental resilience.

11.5 Practical Challenges in IMTA

IMTA offers numerous benefits but faces several practical challenges and gaps for improvement. These challenges span technical, environmental, regulatory, and economic aspects, which need to be addressed to enhance the scalability, efficiency, and sustainability of IMTA systems. Some of the main challenges and gaps for improvement in IMTA will be explained in detail below.

IMTA implementation presents an innovative approach to sustainable aquaculture, aiming to mimic natural ecosystems by integrating species from different trophic levels in the same aquaculture system. Despite its ecological and economic potential, practical challenges may hinder the widespread and efficient implementation of IMTA. These difficulties are explored, and gaps for improvement and future development are identified.

11.5.1 Technical and Operational Challenges

One of the primary challenges in IMTA implementation is the technical complexity involved in managing multi-species systems. Each species within an IMTA system has specific biological and environmental requirements, such as water quality, temperature, and nutrient levels. Balancing these needs to optimize growth and health for all species simultaneously can be intricate and demands sophisticated management practices. Furthermore, the design and operation of IMTA systems often require specialized infrastructure, such as integrated water recirculation systems, which can be costly and technically demanding to establish and maintain.

Moreover, the lack of standardized protocols and guidelines for IMTA systems poses a significant barrier. Although numerous pilot projects and case studies exist, a universally accepted framework for IMTA design, operation, and management is still lacking. This absence of standardization complicates the replication and scaling up of successful IMTA models, as each system must be tailored to specific environmental conditions and species combinations.

11.5.2 Economic and Market Barriers

Economic viability is a critical concern for IMTA implementation. Initial setup costs for IMTA systems can be substantially higher than for traditional monoculture systems due to the need for specialized infrastructure and technology. Additionally, the economic benefits of IMTA, such as increased productivity and resource efficiency, may not be immediately apparent, creating financial risk for investors and farmers. The market for products from IMTA systems is also underdeveloped. Consumer awareness and demand for sustainably produced aquaculture products are growing but remain insufficient to drive widespread market acceptance and premium pricing for IMTA products.

Furthermore, there is a need for a better economic valuation of the ecosystem services provided by IMTA, such as nutrient cycling, water purification, and carbon sequestration. These services often go unrecognized in traditional economic assessments, leading to an undervaluation of IMTA's tangible benefits. Integrating these ecosystem services into economic models and policy frameworks is essential for demonstrating the total value of IMTA and attracting investment.

11.5.3 Regulatory and Policy Constraints

The regulatory landscape for aquaculture, including IMTA, is often fragmented and inconsistent. Regulations governing water use, species introduction, nutrient discharge, and environmental impacts vary widely between regions and countries. This regulatory complexity can hinder the establishment and operation of IMTA systems, as operators must navigate a maze of permits and compliance requirements. In some cases, existing regulations designed for traditional monoculture systems may not be applicable or appropriate for IMTA, creating additional hurdles.

Policy support for IMTA is also generally limited. While there is growing recognition of the benefits of sustainable aquaculture, specific policies, and incentives promoting IMTA are rare. Policy frameworks must evolve to provide targeted support for IMTA through financial incentives, technical assistance, and streamlined regulatory processes. Encouraging public–private partnerships and fostering collaboration between government agencies, research institutions, and industry stakeholders can also accelerate the development and adoption of IMTA.

11.5.4 Knowledge and Capacity Gaps

A critical barrier to IMTA implementation is the knowledge and capacity gap among farmers, researchers, and policymakers. Effective IMTA management requires a deep understanding of species interactions, ecosystem dynamics, and integrated system design. However, this expertise is still relatively scarce, particularly in regions where traditional aquaculture practices dominate. Building capacity through education, training, and extension services is essential to equip stakeholders with the skills and knowledge to implement and manage IMTA systems effectively.

Research gaps also exist, particularly in understanding the ecological interactions and optimal configurations of IMTA systems. More research is needed to identify suitable species combinations, develop best practices for system design and management, and assess the long-term sustainability and resilience of IMTA systems. Collaborative research initiatives that involve stakeholders from academia, industry, and government can help bridge these knowledge gaps and drive innovation in IMTA practices.

11.5.5 Social and Cultural Challenges

Social and cultural factors can also influence the adoption of IMTA. In many communities, traditional aquaculture practices are deeply entrenched, and shifting to IMTA may face resistance due to perceived risks, lack of familiarity, or cultural preferences. Engaging local communities and stakeholders in the planning and implementation of IMTA is crucial to overcoming these barriers. Participatory approaches involving stakeholders in decision-making can build trust, foster acceptance, and ensure that IMTA systems are tailored to local needs and conditions.

Additionally, addressing the social and economic impacts of IMTA on local communities is important. While IMTA can enhance sustainability and resilience, it may also lead to changes in labor requirements, income distribution, and access to resources. Ensuring that the benefits of IMTA are equitably distributed and potential adverse impacts are mitigated is essential for the long-term success and social acceptance of IMTA systems.

11.6 Future Directions and Gaps for Improvement

Several gaps for improvement and development are identified to overcome the challenges and promote the widespread adoption of IMTA.

1. *Standardization and Best Practices*: Developing standardized protocols and guidelines for IMTA system design, operation, and management can facilitate replication and scaling up. Best practices should be based on rigorous scientific research and field-tested experiences.
2. *Economic Valuation and Incentives*: Integrating the economic value of ecosystem services provided by IMTA into economic models and policy frameworks can enhance its attractiveness to investors and policymakers. Financial incentives, such as subsidies, grants, and tax breaks, can also support the establishment and operation of IMTA systems.
3. *Regulatory Reforms*: Streamlining regulatory processes and harmonizing regulations across regions can reduce barriers to IMTA implementation. Policies should be adapted to support the unique requirements of IMTA systems and encourage sustainable aquaculture practices.
4. *Capacity Building and Training*: Investing in education, training, and extension services can build the knowledge and skills needed to implement and manage IMTA effectively. Collaborative research initiatives and knowledge-sharing platforms can also enhance capacity and innovation.

5. *Community Engagement and Social Equity*: Engaging local communities and stakeholders in planning and implementing IMTA can build trust and acceptance. Ensuring that the benefits of IMTA are equitably distributed and addressing potential social impacts are crucial for its long-term success.

11.7 Conclusion

Freshwater Integrated Multi-Trophic Aquaculture (FIMTA) technology has made significant strides, particularly in Asian countries. By integrating fish, plants, and invertebrates, these systems enhance productivity and promote environmental sustainability and economic resilience in aquaculture practices. This approach is aligned with Nature-based Solutions (NbS) principles, imitating natural ecosystems to optimize resource use, increase biodiversity, and provide ecosystem services, so as to enhance resilience to environmental changes and offer social and economic benefits. In order to achieve its full potential, continual research, innovation, and strategic collaborations are essential, in particular to find out resolutions for some practical obstacles such as technical challenges, economic constraints, regulatory issues, knowledge gaps, and social barriers is crucial for its widespread adoption and long-term success. By fostering collaborative and inclusive approaches, stakeholders can raise the full potential of IMTA to contribute to food security, environmental sustainability, and climate resilience.

The future of IMTA lies in integrating it into broader policy frameworks, deepening our understanding of its benefits and challenges, and developing the necessary resources and support for effective implementation. Despite the significant challenges, IMTA offers considerable prospects for promoting sustainable aquaculture and ecosystem management.

IMTA's applicability and feasibility in inland waters are particularly promising. By enhancing water quality and providing a balanced ecosystem for aquatic organisms, IMTA systems in freshwater environments can lead to improved fish growth and survival rates, reduced feed conversion ratios, and lower production costs. This makes IMTA a viable and attractive option for inland aquaculture. In addition to its success in Asian countries, IMTA is also being implemented in other regions, either as full-scale systems or in pilot projects aimed at scaling up. These implementations demonstrate IMTA's versatility and effectiveness across diverse aquatic ecosystems. Continued efforts to overcome barriers and promote IMTA through policy support, research, and innovation will be crucial in harnessing its full potential to meet the growing demand for fish production in a responsible and sustainable manner.

References

Barrington K, Chopin T, Robinson S (2009) Integrated multi-trophic aquaculture (IMTA) in marine temperate waters. In: Soto D (ed) Integrated mariculture: a global review. FAO Fisheries and Aquaculture, Rome, pp 7–46

Boelee E, Janse J, Le Gal A, Kok M, Alkemade R, Ligtvoet W (2017) Overcoming water challenges through nature-based solutions. Water Policy 19(5):820–836. https://doi.org/10.2166/wp.2017.105

Brown N, Eddy S, Plaud S (2011) Utilization of waste from a marine recirculating fish culture system as a feed source for the polychaete worm, Nereis virens. Aquaculture 322–323:177–183. https://doi.org/10.1016/j.aquaculture.2011.09.017

Budiman I, Sari EN, Hadi EE, Siahaan H, Januar R, Hapsari RD (2020) Progress of paludiculture projects in supporting peatland ecosystem restoration in Indonesia. Glob Ecol Conserv 23:e01084

Checa D, Macey BM, Bolton JJ, Brink-Hull M, O'Donohoe P, Cardozo A, Poersch LH, Sánchez I (2024) Circularity assessment in aquaculture: the case of integrated multi-trophic aquaculture (IMTA) systems. Fishes 9(5). https://doi.org/10.3390/fishes9050165

Chopin T (2013) Aquaculture, integrated multi-trophic (IMTA). In: Christou P, Savin R, Costa-Pierce BA, Misztal I, Whitelaw CBA (eds) Sustainable food production. Springer Science+Business Media, New York, pp 184–205

Chopin T, Robinson SMC, Troell M, Neori A, Buschmann AH, Fang J (2008) Multitrophic integration for sustainable marine aquaculture. Ecol Eng 2463–2475

Chopin T, Murray S, Bakhsh HK (2016) Freshwater IMTA. Hatch Int RECIRC ACTION 31:2

Chrismadha T, Mulyana E (2019) Laju Konsumsi Tumbuhan Air Mata Lele (*Lemna perpusilla*) oleh Ikan Nila (*Oreochromis* sp.) dengan Padat Tebar Berbeda. LIMNOTEK Perair darat Trop di Indones 26(1):39–46

Chrismadha T, Said D (2019) Pengembangan alternatif budidaya perikanan jaring apung: Meningkatkan efisiensi produksi kolam perikanan darat dengan tumbuhan mata lele (*Lemna perpusill*a Torr.)

Chrismadha T, Sulawesty F, Awalina A, Mardiati Y, Mulyana E, Widoretno MR (2014a) Growth performance of minute duckweed (*Lemna perpusilla*) in an integrated common carp (*Cyprinus carpio*) closed recirculation aquaculture. In: International conference of aquaculture Indonesia (ICAI), pp 16–26

Chrismadha T, Sulawesty F, Awalina, Mardiati Y, Mulyana E, Widoretno MR (2014b) Phytotechnology application for enhancing water conservation: use of minute duckweed (*Lemna perpusilla*) for phytoremediator and alternative feed in a water closed recirculation aquaculture. In: Sutapa IDA, Zhou D, Kobayashi K, Hidayat, Subehi L, Apip, Rustini HA (eds) International conference on ecohydrology (ICE): ecohydrology approaches facing the global water environment challenges. Asia Pacific Centre for Ecohydrology—UNESCO Research Center for Limnology—LIPI, Yogyakarta, pp 153–166

Cohen-Shacham E, Walters G, Janzen C, Maginnis S (2016) Nature-based solutions to address global societal challenges. IUCN, Gland, Switzerland

Cohen-Shacham E, Andrade A, Dalton J, Dudley N, Jones M, Kumar C, Maginnis S, Maynard S, Nelson CR, Renaud FG, Welling R (2019) Core principles for successfully implementing and upscaling nature-based solutions. Environ Sci Policy 98:20–29

Giesen W (2021) Tropical peatland restoration in Indonesia by replanting with useful indigenous peat swamp species: paludiculture. In: Tropical peatland eco-management, pp 411–441

Hamsiah CI, Heriansah KW, Kabangnga A (2021) Kelulushidupan Biota pada Sistem integrated multi trophic aquaculture (IMTA)-Padi. J Sumberd Akuatik Indopasifik 5(2):127–136

Hasan MR, Chakrabarti R (2009) Use of algae and aquatic macrophytes as feed in small-scale aquaculture

Hossain A, Senff P, Glaser M (2022) Lessons for coastal applications of IMTA as a way towards sustainable development: a review. Appl Sci 12(23):11920. https://doi.org/10.3390/app122311920

Kibria ASM, Haque MM (2018) Potentials of integrated multi-trophic aquaculture (IMTA) in freshwater ponds in Bangladesh. Aquac Reports 11(Aug):8–16. https://doi.org/10.1016/j.aqrep.2018.05.004

Lechner AM, Gomes RL, Rodrigues L, Ashfold MJ, Selvam SB, Wong EP, Raymond CM, Zieritz A, Sing KW, Moug P, Billa L, Sagala S, Cheshmehzangi A, Lourdes K, Azhar B, Sanusi R, Ives CD, Tang YT, Tan DT, Chan FKS, Nath TK, Sabarudin NAB, Metcalfe SE, Gulsrud NM, Schuerch M, Campos-Arceiz A, Macklin MG, Gibbins C (2020) Challenges and considerations of applying nature-based solutions in low- and middle-income countries in Southeast and East Asia. Blue-Green Syst 2(1):331–351. https://doi.org/10.2166/bgs.2020.014

Nath K, Munilkumar S, Patel AB, Kamilya D, Pandey PK, Banerjee Sawant P (2021) *Lamellidens* and *Wolffia* canopy improves growth, feed utilization and welfare of *Labeo rohita* (Hamilton,1822) in integrated multi-trophic freshwater aquaculture system. Aquaculture 534(Nov 2020):736207. https://doi.org/10.1016/j.aquaculture.2020.736207

Neori A, Chopin T, Troell M, Buschmann AH, Kraemer GP, Halling C, Shpigel M, Yarish C (2004) Integrated aquaculture: rationale, evolution and state of the art emphasizing seaweed biofiltration in modern mariculture. Aquaculture 231(1–4):361–391. https://doi.org/10.1016/j.aquaculture.2003.11.015

Nesshöver C, Assmuth T, Irvine KN, Rusch GM, Waylen KA, Delbaere B, Haase D, Jones-Walters L, Keune H, Kovacs E, Krauze K, Külvik M, Rey F, van Dijk J, Vistad OI, Wilkinson ME, Wittmer H (2017) The science, policy and practice of nature-based solutions: an interdisciplinary perspective. Sci Total Environ 579:1215–1227. https://doi.org/10.1016/j.scitotenv.2016.11.106

Popp J, Váradi L, Békefi E, Péteri A, Gyalog G, Lakner Z, Oláh J (2018) Evolution of integrated open aquaculture systems in Hungary: results from a case study. Sustain 10(1):177. https://doi.org/10.3390/su10010177

Putro SP, Widowati S, Muhammad F (2015) The application of integrated multi trophic aquaculture (IMTA) using stratified double net rounded cage (SDFNC) for aquaculture sustainability. Int J Sci Eng 9(2):85–89

Said DS, Chrismadha T, Triyanto (2017) Integrated multi trophic aquaculture as safe environment for fish production in small reservoirs. In: Maghfiroh M, Dianto A, Jasalesmana T, Melati I, Samir O, Kurniawan R (eds) The 16th World lake conference. Lake ecosystem health and its resilience: diversity and risks of extinction. Research Center for Limnology, Indonesian Institute of Sciences, Kuta, Bali, pp 388–394

Said DSS, Chrismadha T, Mayasari N, Badjoeri M (2020) Integrated multitrophic aquaculture in Maninjau Lake: converting eutrophic water into fish meal. IOP Conf Ser Earth Environ Sci 535(1):0–11. https://doi.org/10.1088/1755-1315/535/1/012006

Saimima A, Basir AP, Marjud T, Wasman A (2020) Penerapan Sistem integrated multi-trophic aquaculture (IMTA) untuk Peningkatan Performa Komoditas Budidaya Laut dan Kualitas Lingkungan Perairan di Kepulauan Banda Naira, Maluku. MUNGGAI J Ilmu Perikan Masy Pesisir 6(Feb):19–28

Scarratt D (2015) Freshwater IMTA. Hatc Int. https://www.hatcheryinternational.com/freshwater-imta-1448/. Accessed 26 June 2024

Seddon N, Chausson A, Berry P, Girardin CAJ, Smith A, Turner B (2020) Understanding the value and limits of nature-based solutions to climate change and other global challenges. Philos Trans R Soc B Biol Sci 375(1794):20190120. https://doi.org/10.1098/rstb.2019.0120

Sim CH, Yusoff MK, Shutes B, Ho SC, Mansor M (2008) Nutrient removal in a pilot and full scale constructed wetland, Putrajaya city, Malaysia. J Environ Manag 88(2):307–317. https://doi.org/10.1016/j.jenvman.2007.03.011

Sri-uam P, Donnuea S, Powtongsook S, Pavasant P (2016) Integrated multi-trophic recirculating aquaculture system for Nile Tilapia (*Oreochromis niloticus*). Sustain 8(7). https://doi.org/10.3390/su8070592

Sukhdhane KS, Kripa V, Divu D, Vase VK, Mojjada SK (2018) Integrated multi-trophic aquaculture systems: a solution for sustainability. Aquac Asia 22(4):26–29

Sutrisno S, Chrismadha T, Mayasari N (2021) Potensi *Wolffia globosa* dan *Lemna perpusilla* (Lemnaceae) sebagai Pakan Ikan Nila (*Oreochromis niloticus*). LIMNOTEK Perair darat Trop di Indones 28(1):1–11. https://doi.org/10.14203/limnotek.v28i1.346

Troell M, Halling C, Neori A, Chopin T, Buschmann AH, Kautsky N, Yarish C (2003) Integrated mariculture: asking the right questions. Aquaculture 226(1–4):69–90. https://doi.org/10.1016/S0044-8486(03)00469-1

Troell M, Joyce A, Chopin T, Neori A, Buschmann AH, Fang JG (2009) Ecological engineering in aquaculture—potential for integrated multi-trophic aquaculture (IMTA) in marine offshore systems. Aquaculture 297(1–4):1–9. https://doi.org/10.1016/j.aquaculture.2009.09.010

van Rees CB, Jumani S, Abera L, Rack L, McKay SK, Wenger SJ (2023) The potential for nature-based solutions to combat the freshwater biodiversity crisis. PLOS Water 2(6):e0000126. https://doi.org/10.1371/journal.pwat.0000126

Whitmarsh DJ, Cook EJ, Black KD (2006) Searching for sustainability in aquaculture: an investigation into the economic prospects for an integrated salmon-mussel production system. Mar Policy 30(3):293–298. https://doi.org/10.1016/j.marpol.2005.01.004

Wilfart A, Favalier N, Metaxa I, Platon C, Pouil S, Caruso D, Slembrouck J, Aubin J (2020) Integrated multi-trophic aquaculture in ponds: what environmental gain? An LCA point of view. In: 12th International conference on life cycle assessment of food 2020 (LCA Food 2020) "Towards sustainable agri-food systems", pp 206–208

Part IV
Monitoring and Measurement

Chapter 12
Monitoring and Measurement Tools

Bastari, Chang Wan Kim, Ick Hwan Ko, Abdul Ghoni Majdi, Bagus Adi Rahmadi, Khalil Mujtahid, Dewinta Ramdaniah, Solehudin, Muzzalifah Abd Hamid, Siti Norasikin Ismail, and Wan Maznah Wan Omar

Monitoring can be seen as an arranged process which incorporates numerous activities such as recording, sample collection, and measurement of variables. Monitoring benefits in controlling qualities and determining whether they are consistent or comply to the standard. In the case of water quality monitoring, the parameters to observe can be physical, chemical, or biological properties within spatial or temporal dimension. The following discussion describes principle aspects of monitoring along with selected implementation in some regions.

Monitoring Objectives

Water quality monitoring is important in safeguarding public health, protecting ecosystems, and ensuring the sustainable use of water resources. Monitoring data is used to identify potential sources of pollution, track changes in water quality over time, and inform decision-making on water management and policy. Monitoring can

Bastari · A. G. Majdi · B. A. Rahmadi (✉) · K. Mujtahid · D. Ramdaniah · Solehudin
Balai Besar Wilayah Sungai Citarum Ditjen SDA Kementerian PUPR, Jakarta, Indonesia
e-mail: bbwscitarum@yahoo.com

C. W. Kim · I. H. Ko
Yooshin Engineering, Seoul, South Korea

H2R Inc., Seoul, South Korea

S. N. Ismail
Institute of Tropical Biodiversity and Sustainable Development, Universiti Malaysia Terengganu, Kuala Nerus, Terengganu, Malaysia

M. A. Hamid
Institute of Oceanography and Environment (INOS), Universiti Malaysia Terengganu, Kuala Nerus, Terengganu, Malaysia

W. M. W. Omar
School of Biological Sciences, Universiti Sains Malaysia, George Town, Pulau Pinang, Malaysia

© The Author(s), under exclusive license to Springer Nature Singapore Pte Ltd. 2025
A. C. Santos-Borja et al. (eds.), *Progress on Ecosystem Restoration of Tropical Inland Waters*, https://doi.org/10.1007/978-981-96-2284-9_12

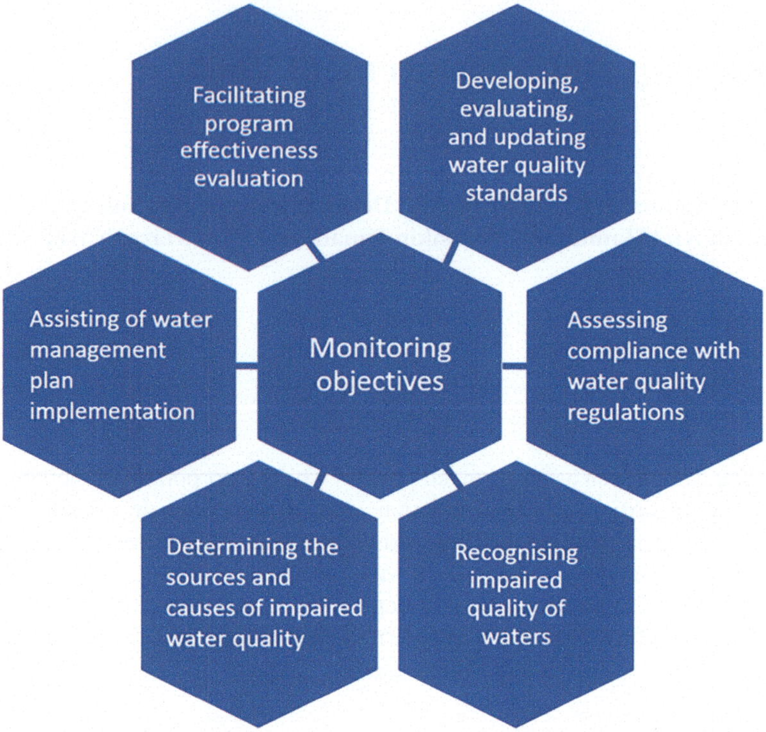

Fig. 12.1 Objectives of monitoring efforts

be conducted by government agencies, environmental organizations, research institutions, and private companies using various tools and techniques such as field-testing kits, water quality sensors, and laboratory analysis. Regular monitoring is essential for identifying potential risks to water quality and taking appropriate actions to address them.

It is essential to identify monitoring objectives so that efficient and successful monitoring programs can be established for all water resources decision-making. Identifying impaired water quality standards, assessing trends and status, identifying causes and sources of problems with water quality, putting water quality management programs into action, and assessing the success of those initiatives are a few examples of monitoring objectives (Fig. 12.1).

Water Quality Monitoring

Water quality monitoring should be able to answer the following questions (REMA 2010):

1. What is water quality as a whole?

   Determines the extent to which waters meet applicable water quality criteria while also protecting and propagating healthy fish and wildlife populations.

2. How much does water quality change over time?

    - Assess and report on the extent to which management programs have improved or will enhance water quality in order to protect and propagate a balanced population of fish and wildlife, as well as recreational activities in and on the water.
    - Evaluate and report on non-point source loading reductions and resulting improvements in water quality.
    - Evaluate and report on the state and trends of water quality in lakes and wetlands.

3. Identify troubled regions and locations that need protection.

    - To evaluate damaged waters.
    - Identify present high-quality waters that should be preserved from degradation.
    - Monitoring and evaluation programs must define the causes and sources of impaired waterways in order to protect and restore them.

4. Which protection level is required?

    To determine the degree of water quality security that is being monitored. Data from monitoring programs can be used to conduct triennial reviews of water quality standards, use attainability analyses, construct and implement revised designated uses and water quality standards, establish water quality-based effluent limits, establish total maximum daily loads, and determine the best management practices for non-point sources.

5. What is the feasibility of clean water initiatives and programs?

    To assess the efficacy of specific projects and programs, such as non-point source control, total maximum daily loads, water quality standards, compliance programs, and overall management measures.

Biomonitoring

Definition and Role

Biomonitoring, also known as biological monitoring, is a novel approach to assessing environmental pollution. Unlike traditional methods that rely on estimates and analyses of soil, air, water, and sediments, biomonitoring uses living organisms to evaluate contamination levels (Madheshiya et al. 2022). This innovative method involves systematically using various taxa, such as plants and animals, or their responses to assess environmental conditions and changes (Li et al. 2010). By observing the impact of external factors on ecosystems through temporal and spatial variations, biomonitoring provides a more dynamic and accurate picture of environmental health (Markert et al. 2003; Li et al. 2010).

Biomonitoring in Freshwater Ecosystems

Freshwater ecosystems, such as rivers and lakes, are particularly vulnerable to human activities. Since the Industrial Revolution, these habitats have faced significant challenges, necessitating robust environmental management strategies. Biomonitoring in freshwater ecosystems involves collecting individual organisms for taxonomic identification and using inventories to estimate environmental conditions at specific sites (Keck et al. 2017). Traditional monitoring techniques like physical, chemical, and bacteriological measurements provide valuable data. However, biomonitoring offers an additional approach that is especially useful in dynamic environments like running waters, where rapid changes require a more thorough evaluation (Li et al. 2010). As a result, biomonitoring provides a holistic view of ecosystem health and becomes a vital component of sustainable environmental management efforts, ensuring the long-term health of freshwater ecosystems.

Aquatic Pollution: Bioaccumulation and Biomagnification as Biomonitoring Strategies

The increasing levels of chemical pollution in aquatic environments have become a significant concern affecting organisms and the entire food chain. Heavy metal pollution is a notable environmental, ecological, and economic issue resulting from both geological and human activities, including industrial, urban, and agricultural practices (Ali and Khan 2018). These metals contaminate air, water, and soil, severely impacting plants, animals, and human health. Aquatic ecosystems, in particular, act as receivers of pollutants, leading to the accumulation of harmful substances throughout the food chain. These pollutants ultimately reach marine environments, impacting seafood species and, consequently, human health (Yarsan and Yipel 2013).

Biomarkers and Bioindicators

Biomarkers are essential tools for detecting exposure to and the effects of pollutants in aquatic organisms. They provide comprehensive and biologically relevant information on the health status of organisms, revealing the potential impact of pollutants. Due to their feeding and living environments, aquatic organisms are heavily exposed to pollution, making them effective bioindicators for evaluating water quality. In biomonitoring, bioindicator organisms such as fish, crustaceans, algae, protozoa, macrophytes, bacteria, and plankton are used to determine water pollution levels, helping to assess chemical contamination in aquatic environments (Yarsan and Yipel 2013).

Bioaccumulation and Biomagnification

Bioaccumulation and biomagnification are key biomonitoring strategies that provide crucial information for understanding how pollutants affect marine ecosystems and human health (Yarsan and Yipel 2013). Heavy metals can move from the environment into living organisms, accumulating in biota and contaminating food chains (Ali and Khan 2018). Bioaccumulation refers to the build-up of contaminants in an organism's tissues from its surrounding environment (Yarsan and Yipel 2013). This is particularly

Table 12.1 Bioaccumulation versus biomagnification

Bioaccumulation	Aspect	Biomagnification
A process where toxic chemicals accumulate in the tissues of an individual organism	Definition	A process where toxic chemicals become more concentrated in organisms higher up in the food chain
Occurs within an individual organism at a specific trophic level	Description	Occurs across multiple organisms between different trophic levels in a food chain
Increased substance concentration within an organism	Pollutant concentration	Increased concentration from one trophic level to the next as organisms move up the food chain
Mercury build-up in fish and aquatic animals	Example	Transfer of pollutants and toxins from microscopic aquatic organisms to small fish, which are then consumed by larger aquatic animals

concerning when it comes to toxic heavy metals in fish, a common food source for humans (Ali and Khan 2018). Higher concentrations of these metals in fish can pose significant health risks, especially when wild fish from industrially contaminated water bodies are consumed regularly (Ali and Khan 2018). When bioaccumulation involves the organism's diet, leading to higher concentrations of pollutants up the food chain, it is termed biomagnification. This process results in heavy metals becoming more concentrated at higher trophic levels (Yarsan and Yipel 2013).

The transfer of heavy metals through trophic levels, along with processes like bioaccumulation and biomagnification, has profound implications for wildlife and human health (Ali and Khan 2018). The following table illustrates the distinctions between bioaccumulation and biomagnification (Table 12.1).

Advancement in Biomonitoring

Biomonitoring traditionally involves identifying indicator species based on physical features, requiring high taxonomical expertise. Conventional biomonitoring methods are time-consuming, error-prone, and face challenges in standardization (Lefrançois et al. 2018). This process includes sampling, identifying, enumerating, counting, and documenting biological organisms across multiple sites and extended periods, resulting in large amounts of data (Keck et al. 2017).

Transition to DNA Barcoding

Recent advancements in biomonitoring are shifting from reliance on morphological criteria, to using DNA barcoding for species identification (Hebert et al. 2003; Keck et al. 2017). DNA barcoding employs short and standardized regions of genetic material (DNA fragments about 300 base pairs) to distinguish individual species. These DNA fragments, known as genetic markers, are linked to species taxonomy through databases that focus on specific genes such as CO1, 18S, and rbcL.

Development of Molecular Operational Taxonomic Units (MOTUs)

Molecular operational taxonomic units (MOTUs) are clusters formed based on similar DNA sequences identified through DNA barcoding. These MOTUs serve as or proxies (indicators) for the presence or absence of species in an environment. The effectiveness of MOTUs in biomonitoring depends on accurate ecological profiles, which can be derived from traditional indices or directly from sequencing data. Additionally, databases that link MOTUs to environmental responses are essential for further developing MOTU-based indices in biomonitoring practices (Keck et al. 2017).

Metabarcoding: A Breakthrough

While DNA barcoding is effective for identifying individual specimens, it becomes challenging with community-level samples that require sorting and isolating individuals. Metabarcoding addresses this challenge by using high-throughput DNA sequencing (HTS) to detect the DNA barcodes of all species present in a sample simultaneously. It uses the same principles as DNA barcoding but handles more complex samples, such as water containing various organisms. Bioinformatics tools then compare sequences to a reference database to identify species (Keck et al. 2017). Despite challenges like DNA spatial transfer, PCR amplification biases, and sequencing errors, metabarcoding is rapidly advancing and expected to become crucial in biomonitoring (Coissac et al. 2012; Shokralla et al. 2012; Keck et al. 2017).

Environmental DNA (eDNA)

Recent advancements in DNA sequencing, coupled with extended reference databases and efficient bioinformatics tools, enable reliable and cost-effective community inventories from environmental DNA (eDNA). Environmental DNA (eDNA) is DNA collected from environmental samples such as water, soil, or air, rather than directly from an organism (Taberlet et al. 2012). It includes DNA from microorganisms, free-floating DNA, and fragments from various organisms. eDNA can detect invasive species and monitor rare or indicator species by identifying their genetic material in the environment using metabarcoding techniques (Ficetola et al. 2008; Mächler et al. 2014). Longer DNA barcodes from microorganisms like bacteria, fungi, and diatoms can identify species that are difficult or impossible to grow in a lab (uncultured taxa) or observe through traditional methods (Taberlet et al. 2012). For instance, diatom DNA from a river can be analyzed to provide insights into the health of the river's ecosystems (Kermarrec et al. 2014; Visco et al. 2015).

Phylogenetic Approaches

DNA sequences, including those used in DNA barcoding, are employed to study evolutionary relationships between species, creating 'family trees' that illustrate their historical connections. By integrating phylogenetic methods, biomonitoring is enhanced to capture diversity uncovered through DNA metabarcoding. These methods estimate ecological profiles by utilizing evolutionary relationships, offering insights into how species respond to environmental stressors. Successful implementation requires robust phylogenies and predictive models to encompass rare and lesser-studied taxa in ecological assessments.

Machine Learning for Ecological Assessment

The rise of big data in biomonitoring has led to the adoption of machine learning techniques. These techniques use computer algorithms to analyze extensive DNA datasets, identify patterns, and make predictions, thereby enhancing species identification and analysis. Methods such as random forest and support vector machines enable the analysis of large datasets without the need for explicit models, improving the understanding and management of complex ecological systems. Effective collaboration between biologists and data scientists is essential to fully harness the potential of machine learning in biomonitoring applications.

Strategic Approaches to Monitoring System and Network: Designing and Implementing

In monitoring efforts, critical aspects include a strategic approach that emphasizes real-time monitoring, stakeholder engagement, and continuous improvement. Implementing such a network can significantly enhance the ability to detect and respond to environmental changes, ultimately contributing to the health and sustainability of aquatic ecosystems.

Real-Time Monitoring

Real-time monitoring involves continuous or near-continuous acquisition and analysis of data, minimizing the delay between data collection and the availability of results. This approach provides a constant stream of information, enabling immediate awareness and response to environmental changes. It is particularly vital for assessing water quality, where prompt actions may be necessary to prevent or mitigate environmental impacts and ensure the sustainability of water resources. The key components to real-time monitoring:

- Continuous Measurement: Systems continuously measure parameters such as water temperature, dissolved oxygen, and pH at predetermined intervals.
- Immediate Data Transmission: Data is transmitted in real time using wireless communication systems (e.g., GSM) to ensure prompt availability to users.
- Immediate Access: Data is disseminated through web-based portals and mobile devices in graphical and tabular formats, facilitating quick interpretation by end-users.

Case Study: Wireless Sensor Network (WSN) for Water Quality Monitoring (Demetillo et al. 2019).

Researchers have developed various systems for remote water quality monitoring. One notable example is the wireless sensor network (WSN) for real-time monitoring of physical and chemical water characteristics. This system is designed to be cost-effective and efficient, utilizing off-the-shelf electrochemical sensors, a microcontroller, wireless communication, and a customized buoy. The system features are:

- Reliability and Validation: The system's reliability was validated through stability tests and efficiency assessments, including buoy stability, energy consumption, data transmission efficiency, and web-based display effectiveness.
- Remote Monitoring: Data on dissolved oxygen, pH, and temperature is transmitted via GSM for remote accessibility, accessible via a web portal and mobile phones.
- Scalability: The system's flexibility and reproducibility make it suitable for widespread deployment. Efforts to expand coverage using autonomous surface vehicles and incorporating heavy metal ion monitoring are ongoing.

Stakeholder Engagement and Training

To foster effective monitoring, collaboration among various stakeholders, including scientists, environmental managers, and biotechnological companies is crucial. Raising awareness and providing training on new monitoring tools are essential steps. Stakeholder engagement strategies may include:

- Workshops: Organize workshops for students, experts, and the public to discuss findings and applications of monitoring tools.
- Economic and Sustainability Goals: Support the local economy, ensure project sustainability beyond government funding, and develop commercial plans for new methods.
- Environmental Management: Enhance management practices for lakes and rivers.
- Expertise Building: Foster expertise in environmental monitoring.

Incorporating DNAqua-Net Framework (Leese et al. 2016)

The DNAqua-Net initiative provides a comprehensive framework for addressing challenges in DNA-based biomonitoring through five working groups (WGs):

- WG 1: DNA Barcode Reference Databases: Collaborates with national and international projects to fill gaps in DNA barcode databases for aquatic species and provides guidance for DNA barcoding campaigns.
- WG 2: Biotic Indices and Metrics: Assesses the ecological value of new bioindicators identified by DNA barcoding and improves existing biotic indices to better suit high-throughput sequencing (HTS) data.
- WG 3: Field and Lab Protocols: Establishes gold-standard protocols for assessing biological quality elements (BQEs) using eDNA/metabarcoding.
- WG 4: Data Analysis and Storage: Identifies best practices for analyzing HTS data and develops guidelines for data analysis and long-term storage.
- WG 5: Implementation Strategies and Legal Issues: Develops a framework for large-scale assessments and addresses legal aspects.

Online Platform for Biomonitoring Networks

An online platform is crucial for modern biomonitoring, acting as a central hub for collaboration, data sharing, and standardization. This significantly enhances the efficiency and effectiveness of biomonitoring efforts. Essential elements of the online platform involve:

- Centralized Data Repository: Aggregates biomonitoring data in one place, ensuring comprehensive analyses and easy access for stakeholders.
- Standardization of Methods: Hosts standardized protocols to ensure consistency and reliability in sampling, DNA barcoding, sequencing, and data analysis.
- Training and Capacity Building: Offers online educational resources, preparing researchers with the latest genomic tools and techniques.
- Collaborative Research and Networking: Includes features for seamless communication, encouraging the sharing of ideas and best practices among researchers.
- Real-Time Data Sharing and Visualization: Provides tools for immediate data upload and advanced visualizations, aiding informed decision-making.
- Integration with Existing Systems: Designed to combine data from various sources, offering a comprehensive view of ecosystem health.
- Privacy and Security: Implements strong security measures to protect user data and ensure access control for authorized personnel only.

Future Directions

Advancements in technology and global collaborations present exciting future directions for monitoring:

- Advancements in Technology: Innovations in noninvasive sampling and high-throughput screening methods will enhance data collection and analysis capabilities.
- Integration with Other Data Sources: Combining biomonitoring data with environmental monitoring and health data will provide more comprehensive assessments.
- Global Initiatives: International collaborations and standardization efforts, such as the WHO's Human Biomonitoring Program, will promote consistency and comparability across regions.

Case Study: Flood Forecasting Warning System (FFWS) at Citarum Watershed and River Area

Water is an important and inseparable part of the life of all living creatures. As a tropical country with high rainfall, Indonesia has a large water resource potential. Based on the FAO report in 2020, Indonesia holds potential surface water resources reaching 2.02 trillion cubic meters per year and is the largest in Southeast Asia. Ministry of Public Works and Public Housing through the Directorate General of Water Resources carries out the duties of managing surface water resources at the national level. Considering that water is a natural object that crosses administrative boundaries, the management boundaries for Indonesia's water resources are determined based on river area (WS). River areas-based water resources management could increase the efficiency and effectiveness of its implementation (https://www.kompasiana.com/henkthomas/62f46e823555e45b16751404/mengenal-wilayah-sungai-ws-dan-daerah-aliran-sungai). Based on the Regulation of the Minister of Public Works and Public Housing of the Republic of Indonesia (Peraturan Menteri Pekerjaan Umum dan Perumahan Rakyat No. 04/PRT/M/2015) concerning Criteria and Determination of River Area Status, a river area is defined as unitary water resources management area in one or more river basins and/or small islands in which area is less than or equal to 2000 km^2. All land areas in Indonesia, including waters that are still influenced by land activities such as estuaries, lakes, rivers, and other waters, are completely divided into river areas. Currently, Indonesia is divided into 128 river areas, consisting of 5 cross-country river areas, 31 cross-provincial river areas, 28 national strategic river areas, 52 cross-regency/city river areas, and 12 river areas within districts/cities. Among the 128 river areas, there are 64 river areas managed by the Directorate General of Natural Resources including Citarum River area. The Citarum River area is a national strategic river area and is managed by the Citarum River Area Center (BBWS Citarum). BBWS Citarum is an agency under the Directorate General of Natural Resources, Ministry of Public Works and Public Housing. This agency has the main task to carry out water resource management in the Citarum River Region which includes planning, construction implementation, operation, and maintenance in the context of conservation and utilization of water resources and controlling the destructive power of water in rivers, beaches, dams, lakes, ponds, reservoirs and other water storage, irrigation, ponds, groundwater and raw water as well as managing urban main drainage (Undang-undang Sumber Daya Air Nomor 17 Tahun 2019). The organizational structure starts from the Head of the Center as the leader of BBWS which oversees 5 fields/sections and 1 group of functional positions. The Citarum River area is the largest in West Java Province with an area of 11,323.3 km^2 that consists of 19 river basins (DAS) and covers 12 districts/cities. The total area of the Citarum watershed is 6617 km^2 or around 60% of the total area of the river area. This river stretches for 297 km, originating from Mount Wayang to the south of Bandung City, flows north through the central part of West Java Province, and empties into Muara Gembong, Bekasi Regency (Fig. 12.2).

The population in the Citarum watershed consists of 12.25 million people (West Java Statistics Agency, 2022). Bandung Regency, West Bandung Regency, and Bandung City are the districts/cities with the largest populations. Due to the high population and need of built-up land, this watershed consists of high use of developed areas. Based on MoEF land cover data in 2022, the built-up area is more than 10%, while forest area is only 5.6% of the watershed. Judging from the topographic aspect, the elevation of Citarum watershed ranges between 0–3000 m above sea level. The lowland areas are located on the North Coast and the highland in the south has a role as upstream of several tributaries of the Citarum River. Viewed from a geological aspect, the Citarum watershed has complex and varied geological conditions, ranging from alluvial deposits on the north coast to volcanic material in the highlands of Bandung and Cianjur. There are seven types of soil starting from alluvial on the

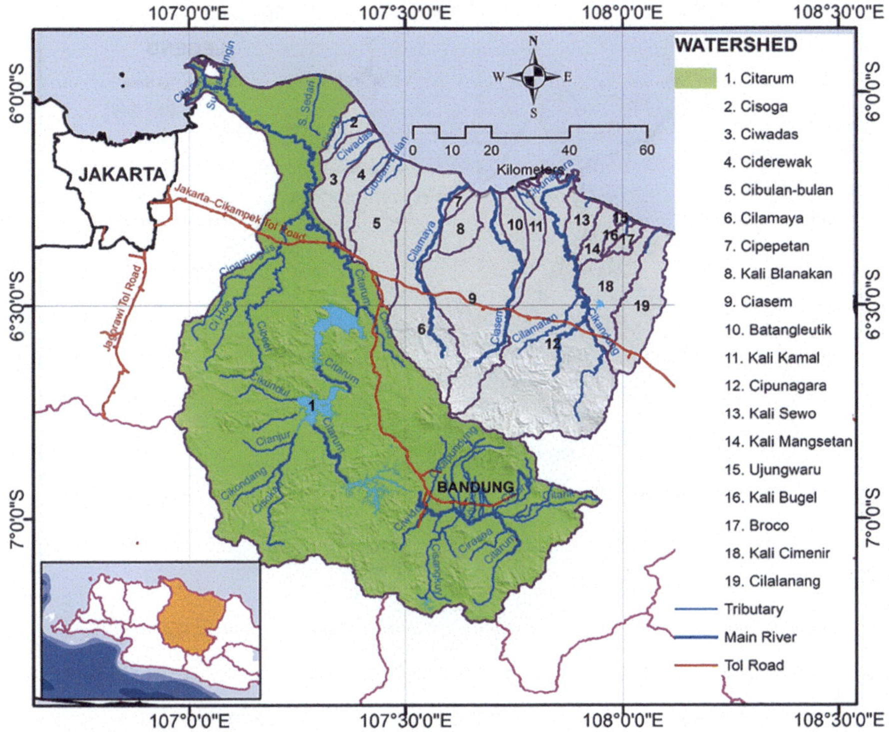

Fig. 12.2 Position of Citarum watershed to Citarum River Area. *Source* BBWS Citarum, 2024

north coast and Bandung basin, andosol in the mountainous areas in the south and north of Bandung, and a small part in the west of Cianjur. The Latosol soil type is the most dominant, and widespread in the Central Citarum, Lower Citarum, and Upper Citarum. Several other types of soil in this watershed are Glei, Grumosol, Red and Yellow Podzol, and Regosol. In terms of management, the Citarum watershed is divided into 3 zones: Upper, Central, and Lower Citarum. Upstream Citarum starts from the upstream tributaries of the Citarum River to the inlet of the Saguling Reservoir in Nanjung. Central Citarum covers the river channel and watershed between the Saguling Reservoir and the Jatiluhur Reservoir. Meanwhile, Lower Citarum starts from the Jatiluhur Reservoir outlet to Muara Gembong in Bekasi (Fig. 12.3).

Problems, Challenges, and Potential

Citarum is the largest watershed in West Java Province, as well as the economic lifeblood for West Java Province and Indonesia. This river system possesses 3 cascade dams (Saguling, Cirata, and Jatiluhur) which are very important for Indonesia's food, energy, and water security. The three reservoirs are provided to irrigate more than 250 thousand hectares of agricultural area, hydroelectric power plants (PLTA) for Java and Bali supplies, fisheries, recreation, and the provision of raw water for

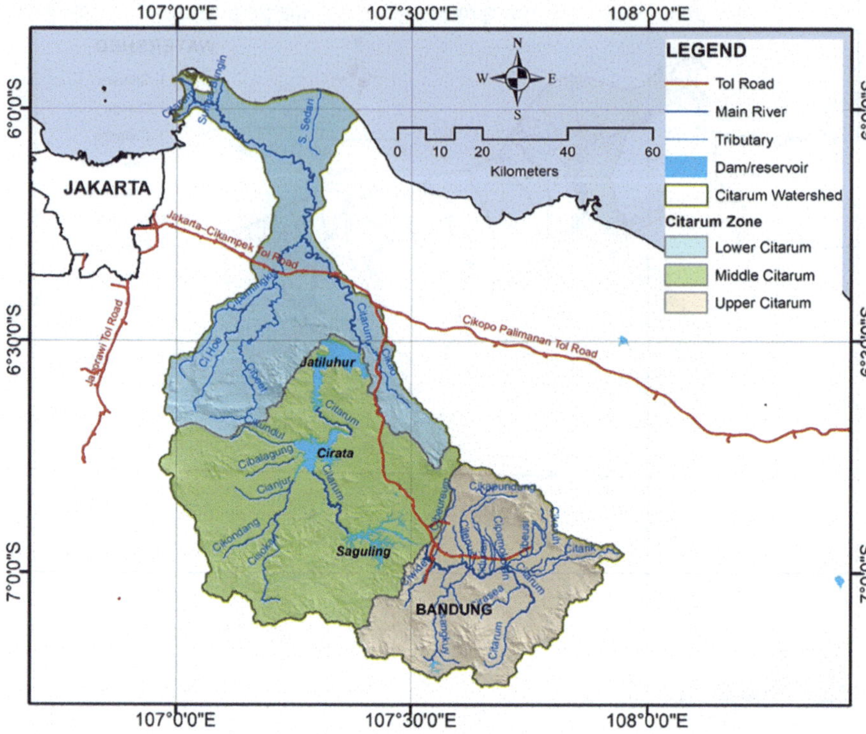

Fig. 12.3 Citarum Watershed zonation. *Source* BBWS Citarum, 2024

Jakarta Megapolitan via the West Tarum Canal with an average capacity of around 12 m^3/second. Apart from that, there are hundreds of industrial areas in Bekasi and Karawang which rely on their raw water source from the Citarum River. However, despite these great benefits, the Citarum River has a high destructive power. High urbanization and intense agricultural activities have caused the Citarum watershed ecosystem to be ecologically degraded. As a result, droughts in the dry season are getting worse, flooding in the rainy season is inevitable, and global climate change worsen this situation. Flooding in the Citarum River is generally concentrated in Upper Citarum and Central Citarum, while in Central Citarum it is relatively safe. Every year, nearly 10 million residents along the Citarum River are affected by floods, which are caused by overflows or embankment breaches. Based on the West Java disaster management agency report, in the last 10 years (2013–2023), around 1007 flood events recorded in the Citarum watershed, where the highest cases occurred in 2013 (267 incidents), 2021 (142 incidents), and 2016 (110 incidents) (Figs. 12.4 and 12.5).

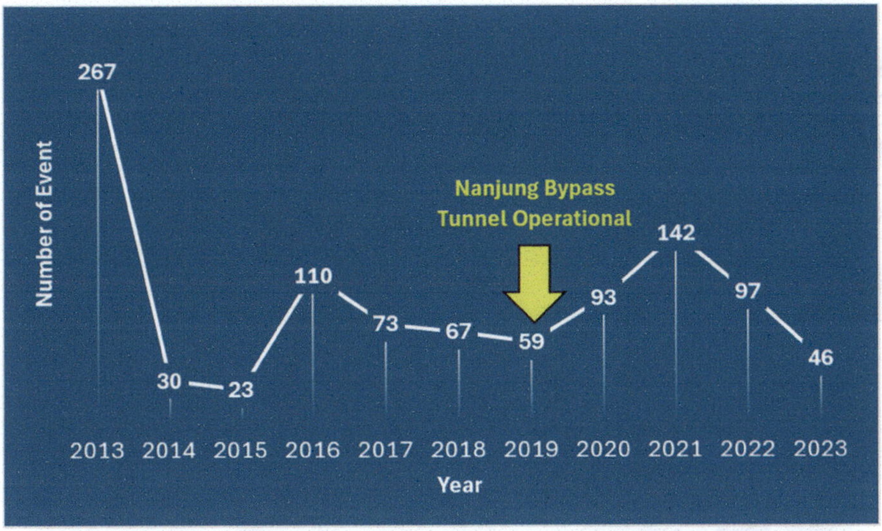

Fig. 12.4 Number of flood events in Citarum watershed in 2013–2023. *Source* FFWS Phase II KOICA, 2024

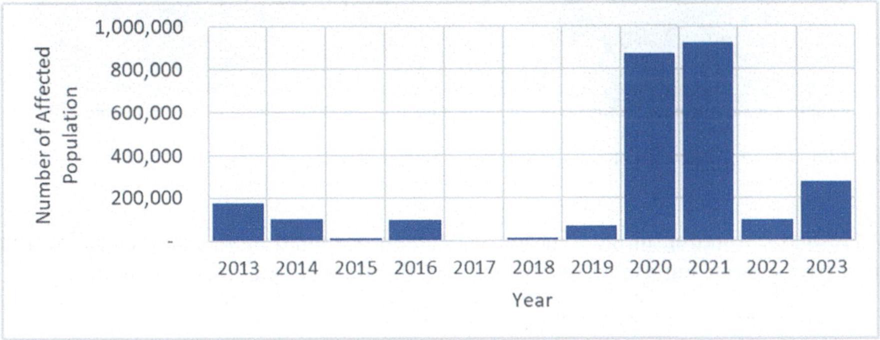

Fig. 12.5 Flood-affected population. *Source* FFWS Phase II KOICA, 2024

Dayeuhkolot area in Upper Citarum is one of the flood-prone area where flood threatens this area every single year. The unfavorable geographical conditions indicated by the estuary of several tributaries of the Citarum River, coupled with land subsidence, make flood control in this location quite difficult (Fig. 12.6).

Apart from Dayeuhkolot, flood-prone areas also appear in Citarum Hilir every time heavy rain occurs. The location is Karangligar Village, West Telukjambe District, Karawang Regency. Flooding at this location is caused by backwater from the Cibeet River flowing through the Cidawolong and Cikaranggelam Drain Channels. As a result, inundation of residential areas and hundreds of hectares of productive rice fields is unavoidable (Fig. 12.7).

Fig. 12.6 Flood in Dayeuhkolot. *Source* Various Social Media

However, flooding in the Citarum watershed is spread in several places, both in main rivers and tributaries. This disaster is caused by several factors, including:

1. Topographical characteristics
   - Approximately 40% of coastal areas are located in low-lying areas below sea level
   - As a source of flow to the Java Sea, the flow will stagnate when the sea level rises
   - When flow stagnates, flooding occurs in nearby areas due to increased flow.
2. Insufficient river capacity
   - Garbage disposal issue

   Based on West Java in Figures for 2016, the total daily waste volume in the Greater Bandung area (Bandung City, Cimahi City, Bandung Regency, and West Bandung Regency) reached 24,988.95 m^3/day. Some of waste generation has not been managed properly due to the lack of ecological behavior from community causing rubbish to enter water bodies, including rivers. As a result,

Fig. 12.7 Flood in Karangligar Village, West Telukjambe District, Karawang Regency in 2024. *Source* BBWS Citarum, 2024

the river volume has decreased due to the accumulation of rubbish, including several places in Bandung City and Cimahi City (Fig. 12.8).

- Sedimentation

  Based on data quoted from www.citarum.org (2014), the annual sedimentation volumes in the 3 main dams of the Citarum River are: (1) 8.2 million m^3/year in Saguling, (2) 6.4 million m^3/year in Cirata and 1.6 million m^3/year in Jatiluhur.

3. Built-up area development

   - Mangroves are being cut down due to downstream development, causing increased coastal erosion, saltwater intrusion, and flooding
   - More runoff flows into the river during rainfall due to increased developed areas.

Fig. 12.8 Rubish-covered under the bridge in Bandung City. *Source* BBWS Citarum, 2022

4. Flood control infrastructure

   The failure due to poor maintenance and flood mainly occurs at river structures in curved channel junction.

The large ecosystem services, high population density, and high potential for destructive power are some of big challenges for Citarum watershed managers, in this case BBWS Citarum. This is in line with Law no. 17 of 2019 concerning Water Resources, where one of the main tasks that must be carried out by the Directorate General of Natural Resources, represented by BBWS Citarum, is controlling the destructive power of water by preventing, overcoming, and restoring damage to environmental quality.

Hydrological Station System

List of Managed Stations

As one of the implementing units that manage water resources, the BBWS Citarum pointed Hydrology and Water Quality (HWQ) unit is to manage hydrological data in the Citarum WS. This unit was formed in 2010 to carry out five-year annual program planning, operations, maintenance, data analysis, publication, and dissemination of hydrological data. In 2024, HWQ unit has 8 members and chaired by a Unit Head. The number of hydrological stations managed by the HWQ Unit is 67 stations,

Table 12.2 Rainfall station and water inspection post managed by the Hidrology and Water Quality BBWS Citarum Unit

No	Station type	Recording data		Total
		Manual	Manual + Telemetry	
1	Rainfall station	7	31	38
2	Water level station	16	13	29
Total		23	44	67

Source BBWS Citarum

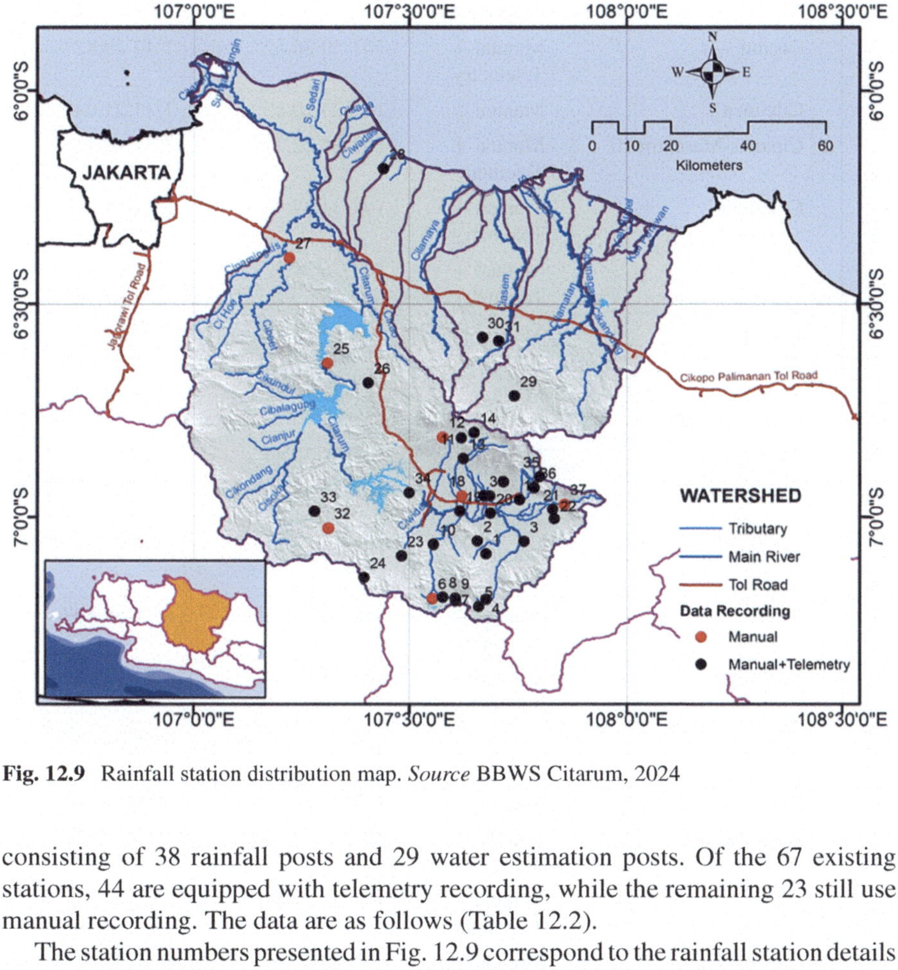

Fig. 12.9 Rainfall station distribution map. *Source* BBWS Citarum, 2024

consisting of 38 rainfall posts and 29 water estimation posts. Of the 67 existing stations, 44 are equipped with telemetry recording, while the remaining 23 still use manual recording. The data are as follows (Table 12.2).

The station numbers presented in Fig. 12.9 correspond to the rainfall station details presented in Table 12.3. The map of the distribution of stations for recording water level can be seen in Fig. 12.10 and the details of them are presented in Table 12.4.

Table 12.3 Rainfall station

No	Rainfall station	Data recording	Coordinate	
			X (East)	Y (South)
1	Ancolmekar	Manual + Telemetry	107°40′37.96″	7° 5′7.07″
2	Arjasari	Manual + Telemetry	107°39′25.884″	7°3′17.652″
3	Paseh	Manual + Telemetry	107°45′49.85″	7° 3′23.82″
4	Kertasari	Manual + Telemetry	107°40′36.78″	7°11′30.84″
5	Cisanti	Manual + Telemetry	107°39′36.62″	7°12′29.82″
6	Cileunca	Manual	107°33′10.89″	7°11′21.04″
7	Cipanas-Margamukti	Manual + Telemetry	107°36′22.27″	7°11′42.11″
8	Cinchona	Manual + Telemetry	107°34′37.53″	7°11′11.11″
9	Kertamanah	Manual + Telemetry	107°36′21.27″	7°11′19.92″
10	Hantap	Manual + Telemetry	107°33′21.21″	7° 3′47.58″
11	Meteo	Manual + Telemetry	107°37′12.64″	6°48′48.78″
12	Cipeusing	Manual	107°34′37.44″	6°48′39.02″
13	Dagopakar	Manual + Telemetry	107°37′28.25″	6°51′42.55″
14	Margahayu	Manual + Telemetry	107°38′58.61″	6°48′4.35″
15	Cibiru	Manual + Telemetry	107°43′0.84″	6°54′58.59″
16	Rancaekek	Manual + Telemetry	107°45′14.04″	6°57′31.11″
17	Cinambo	Manual + Telemetry	107°41′8.88″	6°56′58.08″
18	Pasirluyu	Manual	107°37′17.94″	6°56′52.40″
19	Dayeuhkolot	Manual + Telemetry	107°36′59.09″	6°59′5.15″
20	Sapan	Manual + Telemetry	107°41′14.11″	6°59′23.57″
21	Cicalengka	Manual + Telemetry	107°49′47.96″	6°58′52.57″
22	Cikancung	Manual + Telemetry	107°50′0.15″	7° 0′11.63″

(continued)

Table 12.3 (continued)

No	Rainfall station	Data recording	Coordinate	
			X (East)	Y (South)
23	Cisondari	Manual + Telemetry	107°28′54.21″	7° 5′25.25″
24	Rancaupas	Manual + Telemetry	107°23′40.95″	7°8′27.55″
25	Cikao-Galumpit	Manual	107°18′32.44″	6°38′16.27″
26	Cisomang	Manual + Telemetry	107°24′15.72″	6°41′5.03″
27	Cibeet-Bendung	Manual	107°13′17.22″	6°23′28.80″
28	Tempuran	Manual + Telemetry	107°26′23.60″	6°10′55.53″
29	Cisalak	Manual + Telemetry	107°44′31.24″	6°42′55.03″
30	Ciherang	Manual + Telemetry	107°40′8.118″	6°34′38.84″
31	Cisampih	Manual + Telemetry	107°42′21.53″	6°35′8.247″
32	Gunung Halu	Manual	107°18′41.17″	7° 1′27.47″
33	Ngamprah	Manual + Telemetry	107°16′46.736″	6°59′7.699″
34	Rongga	Manual + Telemetry	107°29′58.38″	6°56′29.63″
35	Tanjungsari	Manual + Telemetry	107°47′58.97″	6°54′15.24″
36	Jatiroke	Manual + Telemetry	107°47′13.60″	6°55′47.37″
37	Dampit	Manual	107°51′25.36″	6°58′7.59″
38	Cidurian-Balai	Manual + Telemetry	107°40′18.13″	6°56′56.20″

Source BBWS Citarum

Operating System

The equipment systems managed by the Hydrology and Water Quality Unit are divided into manual and telemetry. Manual equipment is managed by hydrological data recording station officers in the field spread across the Citarum. Reporting of rainfall recording data is carried out once every 24 h at 07.00 AM. Meanwhile, water level reporting is provided 3 times a day at 07.00 AM, 12.00 PM, and 05.00 PM. Previously, reporting was conducted via WhatsApp group and manual data recording form (rain and water level). However, reporting using this method experiences several obstacles, including late reporting and report fabrication by postal officers due to the loose reporting rules. Based on these conditions, starting in 2024, data reporting will

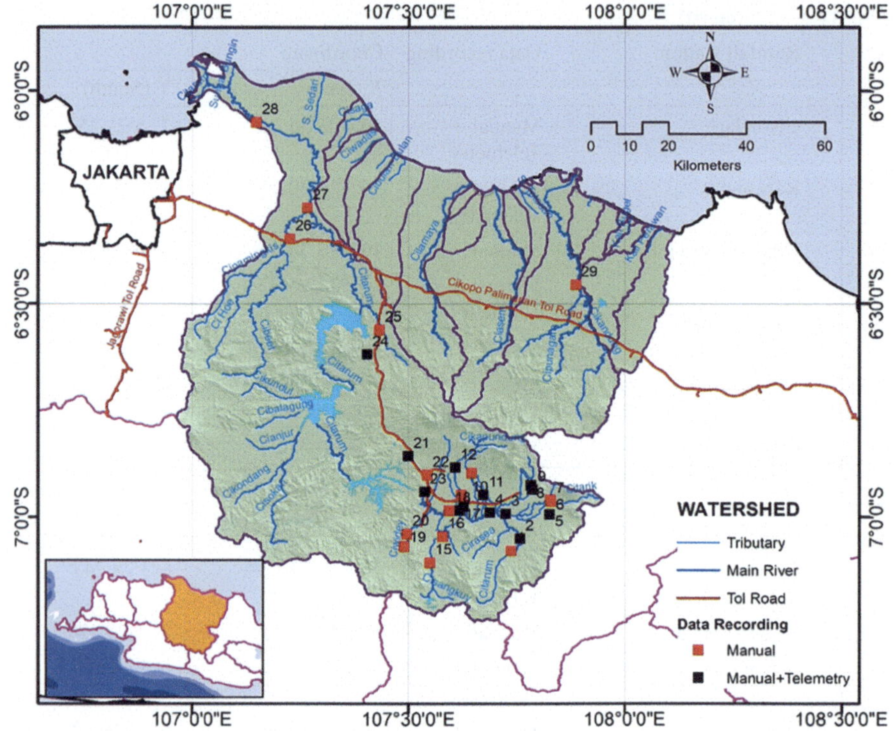

Fig. 12.10 Water level station distribution map. *Source* BBWS Citarum,2024

be carried out through the APOLO application (Hydrology Reporting Application). The APOLO application was developed by the HWQ in cooperation with local consultants. This app is available on Playstore and can be downloaded directly by postal officers. In their reports, officers must include a picture of the water level for water estimation post officers and the rain station for rainfall post officers. In addition, the report could be posted via APOLO within a radius of one kilometer from the station. The interface of the APOLO application can be seen in the following images (Fig. 12.11).

Meanwhile, telemetry equipment for reading and recording is carried out by machine tools or equipment. Water level recording at telemetry posts in the field uses sensors installed on water level observation buildings. There are two main vendors of post-hydrology telemetry equipment used by the HWQ Unit, namely Higertech Telemetry and KOICA Telemetry. Water level telemetry posts are recorded every 10 min, while rainfall telemetry posts are recorded hourly. The telemetry equipment was obtained through procurement (purchase) and grants from overseas company including KOICA. To support water quality observations, samples are taken at all water level station to test water quality and floating sediment. Sampling is routinely carried out, between 3–5 times a year. The results of water quality laboratory tests are

Table 12.4 Water level station

No	Water level station	Data recording	River	Coordinate	
				X (East)	Y (South)
1	Wangisagara Weir	Manual	Citarum	107°44′9.08″	07°04′46.4″
2	Majalaya	Manual + Telemetry	Citarum	107°45′23.92″	7° 3′2.23″
3	Solokan Jeruk	Manual + Telemetry	Citarik	107°43′25.00″	6°59′36.50″
4	Sapan	Manual + Telemetry	Citarum	107°41′14.11″	6°59′23.57″
5	Peundeuy	Manual + Telemetry	Cijalupang	107°49′29.57″	6°59′36.79″
6	Cimanggung	Manual	Cimande	107°49′35.51″	6°57′33.54″
7	Sindangpakuon	Manual	Ciburaleng	107°49′39.47″	6°57′48.05″
8	Cikuda	Manual + Telemetry	Cikeruh	107°46′53.63″	6°55′33.22″
9	Jatiroke	Manual + Telemetry	Cikeruh	107°47′6.00″E	6°56′10.10″
10	Sukapada	Manual	Cidurian	107°38′39.98″	6°53′49.40″
11	Balai AWLR	Manual + Telemetry	Cidurian	107°40′17.84″	6°56′56.14″
12	Teras Cikapundung	Manual + Telemetry	Cikapundung	107°36′22″	6°53′3″
13	Pasirluyu	Manual	Cikapundung	107°37′17.94″	6°56′52.40″
14	Komplek Radio	Manual + Telemetry	Cikapundung	107°37′33.14″	6°58′35.84″
15	Pataruman	Manual	Cisangkuy	107°32′54.95″	7° 6′27.66″
16	Kamasan	Manual	Cisangkuy	107°34′40.46″	7° 2′45.88″
17	Dayeuhkolot	Manual + Telemetry	Citarum	107°36′59.09″	6°59′5.15″
18	Rancamanyar	Manual	Citarum	107°35′33.2″	6°59′07.0″
19	Cukanggenteng	Manual	Ciwidey	107°29′23.14″	7° 4′8.70″
20	Sadu	Manual	Ciwidey	107°29′47.40″	7° 2′23.65″
21	Kertajaya	Manual + Telemetry	Ciburandul	107°29′54.83″	6°51′28.41″
22	Cimahi-Ciujung	Manual	Cimahi	107°32′32.11″	6°54′7.59″
23	Nanjung	Manual + Telemetry	Citarum	107°32′9.94″	6°56′30.02″
24	Cilalawi	Manual + Telemetry	Cilalawi	107°24′14.94″	6°37′7.97″
25	Cikao Sasak Beusi	Manual	Cikao	107°25′59.78″	6°33′41.37″
26	Siphon	Manual	Cibeet	107°13′35.70″	6°20′48.80″

(continued)

Table 12.4 (continued)

No	Water level station	Data recording	River	Coordinate	
				X (East)	Y (South)
27	Tanjungpura	Manual	Citarum	107°16′1.33″	6°16′26.62″
28	Cabang Bungin	Manual	Citarum	107° 8′58.52″	6° 4′23.34″
29	Kiarapayung	Manual	Cipunagara	107°53′9.74″	6°27′17.71″

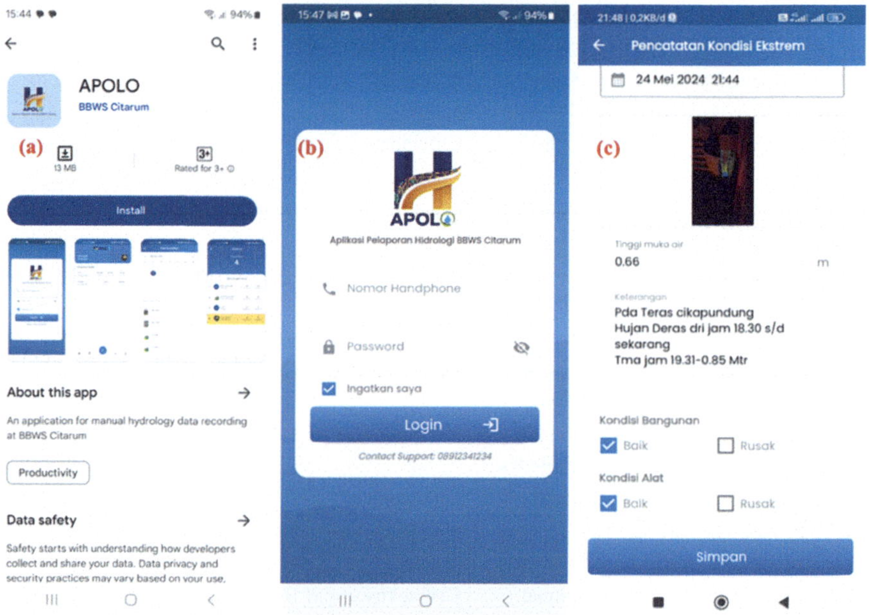

Fig. 12.11 **a** APOLO app di Playstore, **b** login interface, and **c** report from station officers. *Source* BBWS Citarum, 2024

then analyzed for quality standards, referring to the regulations in force in Indonesia. The results of the drift sediment testing are then analyzed into a sediment rating curve, presenting the relationship between the amount of discharge and the sediment. Field documentation of water level recording stations and manual rainfall posts can be seen in Fig. 12.12.

Documentation of KOICA telemetry equipment for recording rainfall is seen in Fig. 12.13 and the water level can be seen in Fig. 12.14. Meanwhile, the rainfall telemetry post installed by Higertech can be seen in Fig. 12.15.

Challenges, Problems, and Obstacles Faced in Managing Hydrological Stations

In managing hydrological and water quality data, there are certainly challenges, problems, and obstacles faced, including the following:

Fig. 12.12 Manual water level recording station (left) and manual rainfall recording tool (right). *Source* BBWS Citarum, 2024

Fig. 12.13 Rainfall telemetry equipment by KOICA. *Source* BBWS Citarum, 2017

Fig. 12.14 Water level telemetry equipment by KOICA. *Source* BBWS Citarum, 2017

Fig. 12.15 Rainfall telemetry equipment by Higertech. *Source* BBWS Citarum, 2024

1. Hydrology post managers at Citarum WS are also tasked in several other agencies (not only BBWS Citarum), required the high need of synchronization and coordination with other agencies.
2. Many hydrological monitoring equipments are in damage condition required special funding to fix them.
3. Telemetry equipment spare parts are difficult to obtain
4. Manual data reading on some posts needs validation for accuracy.
5. There is no integration of hydrological data and information with related agencies or stakeholders.

Development of the Flood Forecasting Warning System (FFWS) for the Citarum Watershed

1. BBWS Citarum—KOICA collaboration

   FFWS is a grant activity from South Korea through Korea International Cooperation Agency (KOICA) and the Ministry of Public Works and Public Housing through the Citarum River Region Center, Directorate General of Water Resources. This activity is an early warning and prediction system for flood events and consists of 2 stages, FFWS Phase I and II. The hope to be achieved in this FFWS activity is to save property and people's lives affected by flooding in the Citarum River.
2. Flood Forecasting Warning System (FFWS)

   a. FFWS Phase I was implemented from 2015 to April 2017, over a period of 28 months. The objectives of FFWS Phase I are:

      - Flood risk management capacity building for Citarum Basin Water Managers & Engineers
      - Reducing vulnerability of people and properties in the flood-prone area in the Upper Citarum (Bandung Basin)

During this phase, equipment was installed in the upstream Citarum watershed including 13 rainfall stations; 8 water supply stations; 5 warning posts; 2 CCTVs, and 1 LED Display installed on Cikapundung Terrace, Bandung City. All equipment installed during FFWS Phase I is entirely located in Citarum Hulu, where the distribution is as follows.

The data recorded by the equipment in the field is then processed by the system and transmitted by the telecommunications modem to the receiving antenna at BBWS Citarum. The data is then transferred to the communication server, processed in the server database, and presented on the monitor screen in the command center. Apart from being connected to the flyover in the command center, telemetry data is also available on several computers in the HWQ Unit which are connected to the server.

Automatically, rain and water level telemetry data is also connected to the FFWS software that has been developed. This rain and water level data is the basis for calibrating the river discharge forecasting model. If the model calibration results are good enough, then the model can be used for discharge forecasting activities,

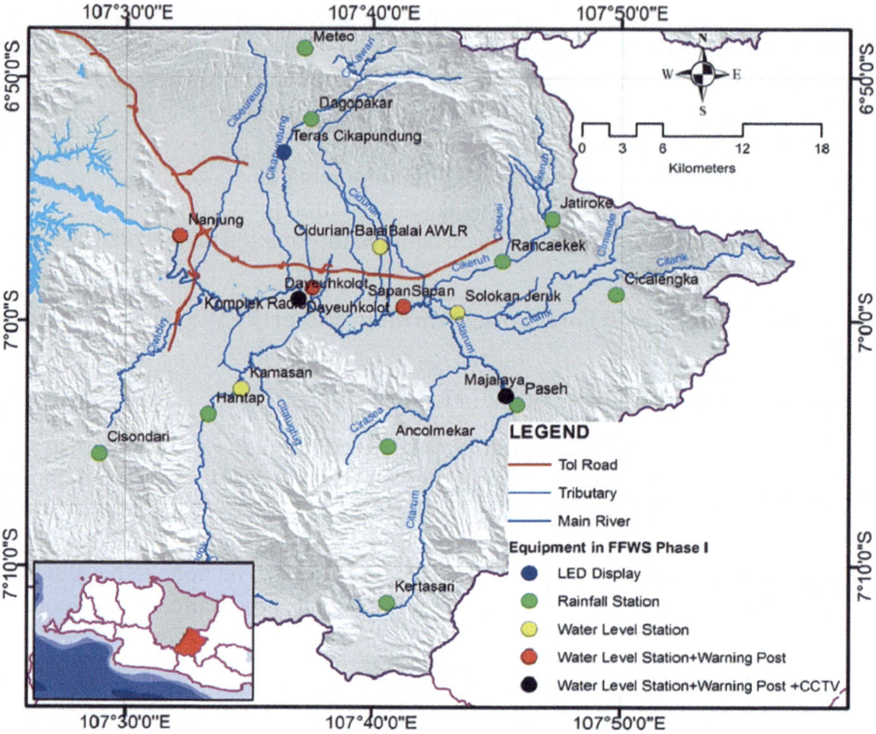

Fig. 12.16 Installed equipment during FFWS Phase I. *Source* BBWS Citarum, 2024

where the rain data used for forecasting comes from another agency, in this case, Indonesia Meteorology Agency (BMKG). The data acquisition process can be seen in Fig. 12.16 while the forecasting model process can be seen in Figs. 12.17 and 12.18.

Other Monitoring Cases

Advanced geospatial technologies and remote sensing for environmental management have been applied. Some projects using remote sensing technology have been developed in Philippines for environmental monitoring, particularly in the context of rivers and lakes. One of those projects is the Multi-Sensor Cross-Platform Water Quality Monitoring (MCWQM) Project, which integrates various sensors to monitor water quality in real time (Blanco 2015). This project aims to enhance the accuracy and efficiency of water quality assessments, which is crucial for managing aquatic ecosystems.

In addition, the BlueCARES Project, which focuses on the conservation of blue carbon ecosystems, further integrating technology with environmental conservation efforts. Moreover, the Phil-LiDAR 2 Program, which assessed the country's natural resources, and the Multi-Sensor Cross-Platform Water Quality Monitoring

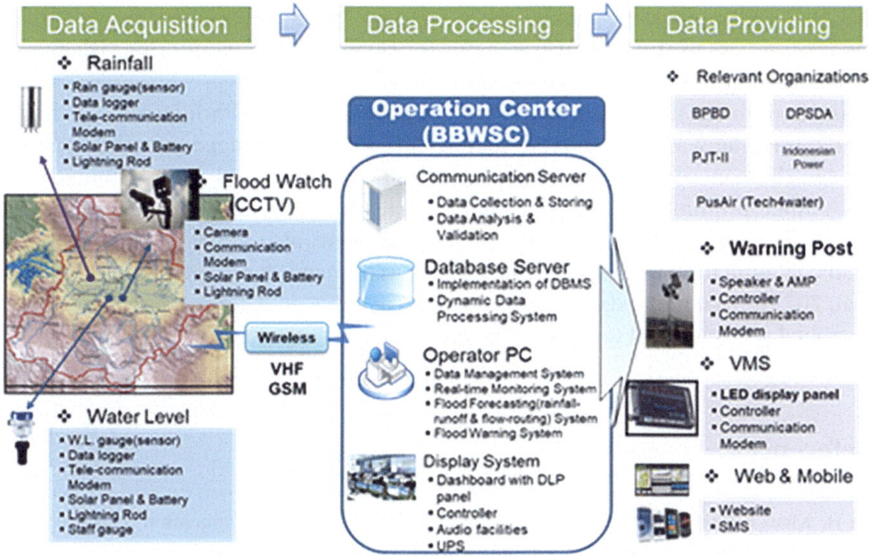

Fig. 12.17 Developed system configuration in FFWS Phase I. *Source* Presentation FFWS File from Dr. Ko, Ick Hwan. Reproduced with permission

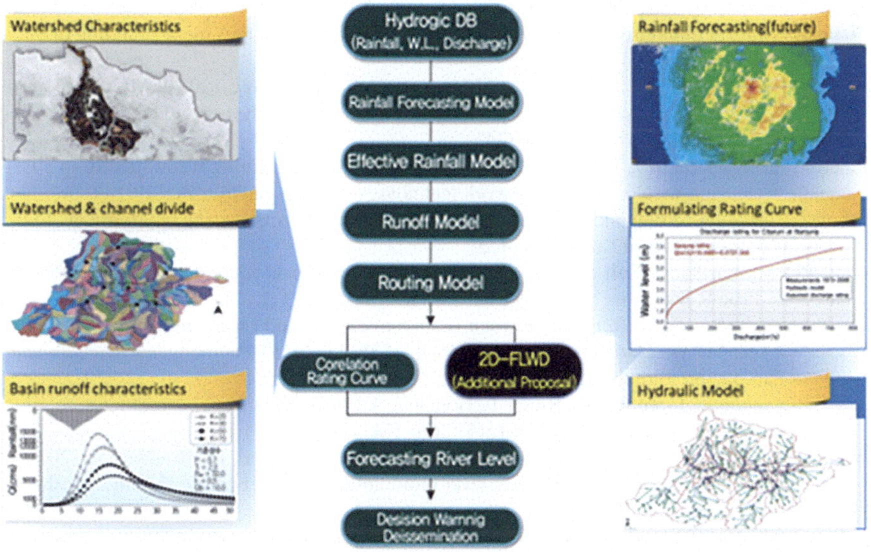

Fig. 12.18 Developed Method of Forecasting and Warning System in FFWS Phase I. *Source* Presentation FFWS File from Dr. Ko, Ick Hwan. Reproduced with permission

(MCWQM) Project, which innovates in real-time water quality monitoring of rivers and lakes is very useful and gives benefits for environmental monitoring (Blanco and Fernández 2018).

In monitoring the water quality of Laguna Lake (the Philippines), microbial source-tracking was carried out to identify the sources of fecal contamination (e.g., sewage, human sources, poultry, swine, cattle) which used the combination of DNA fingerprinting of *E.coli* and machine learning algorithms (dela Peña et al. 2021). *Bacteroides* 16S rDNA-based markers were also developed for detecting fecal contamination from specific hosts such as humans, ducks, and pigs (Malajacan et al. 2023). Both methods are helpful to gain insights on determining the sources of microbial pollution and to strengthen water quality management.

References

Ali H, Khan E (2018) Trophic transfer, bioaccumulation, and biomagnification of non-essential hazardous heavy metals and metalloids in food chains/webs—concepts and implications for wildlife and human health. Hum Ecol Risk Assess Int J 25(6):1353–1376

Blanco AC, Fernández M (2018) Data assimilation techniques for environmental monitoring and prediction. Environ Model Softw 106:10–23

Blanco AC (2015) Integrating real-time data from environmental monitoring systems: a case study. J Environ Manage 156:1–10

Coissac E, Riaz T, Puillandre N (2012) Bioinformatic challenges for DNA metabarcoding of plants and animals. Mol Ecol 21:1834–1847

dela Peña LBRO, Labrador KL, Nacario MAG, Bolo NR, Rivera WL (2021) Microbial source tracking of fecal contamination in Laguna Lake, Philippines using the library-dependent method, rep-PCR. J Water Health 19(5):762–774

Demetillo AT, Japitana MV, Taboada EB (2019) A system for monitoring water quality in a large aquatic area using wireless sensor network technology. Sustain Environ Res 29:1–9

Ficetola GF, Miaud C, Pompanon F, Taberlet P (2008) Species detection using environmental DNA from water samples. Biol Lett 4:423–425

Hebert PD, Cywinska A, Ball SL, DeWaard JR (2003) Biological identifications through DNA barcodes. Proc R Soc Lond Ser B Biol Sci 270(1512):313–321

https://byjus.com/biology/difference-between-biomagnification-and-bioaccumulation/

https://www.kompasiana.com/henkthomas/62f46e823555e45b16751404/mengenal-wilayah-sungai-ws-dan-daerah-aliran-sungai

Keck F, Vasselon V, Tapolczai K, Rimet F, Bouchez A (2017) Freshwater biomonitoring in the Information Age. Front Ecol Environ 15(5):266–274

Kermarrec L, Franc A, Rimet F, Chaumeil P, Frigerio JM, Humbert JF, Bouchez A (2014) A next-generation sequencing approach to river biomonitoring using benthic diatoms. Freshw Sci 33(1):349–363

Ko IC (2024) Development of the FFWS in the Citarum River Basin, Indonesia, 2nd Phase (Presentation File)

Leese F, Altermatt F, Bouchez A, Ekrem T, Hering D, Meissner K, Mergen P, Pawlowski J, Piggott J, Rimet F, Steinke D (2016) DNAqua-Net: developing new genetic tools for bioassessment and monitoring of aquatic ecosystems in Europe. Res Ideas Outcomes 2:e11321

Lefrançois E, Apothéloz-Perret-Gentil L, Blancher P, Botreau S, Chardon C, Crepin L, Cordier T, Cordonier A, Domaizon I, Ferrari BJ, Guéguen J (2018) Development and implementation of

eco-genomic tools for aquatic ecosystem biomonitoring: the SYNAQUA French-Swiss program. Environ Sci Pollut Res 25:33858–33866
Li L, Zheng B, Liu L (2010) Biomonitoring and bioindicators used for river ecosystems: definitions, approaches and trends. Procedia Environ Sci 2:1510–1524
Mächler E, Deiner K, Steinmann P, Altermatt F (2014) Utility of environmental DNA for monitoring rare and indicator macroinvertebrate species. Freshwater Science 33(4):1174–1183
Madheshiya P, Gupta GS, Sahoo A, Tiwari S (2022) Biomonitoring tools and bioprogramming: an overview. In: New paradigms in environmental biomonitoring using plants, pp 341–366
Malajacan GT, Nacario MAG, Obusan MCM, Rivera WL (2023) Host-associated bacteroides 16S rDNA-based markers for source tracking of fecal pollution in Laguna Lake, Philippines. Microorganisms 11(5):1142
Markert BA, Breure AM, Zechmeister HG (2003) Definitions, strategies and principles for bioindication/biomonitoring of the environment. In: Trace metals and other contaminants in the environment, vol 6, pp 3–39. Elsevier
Peraturan Menteri Pekerjaan Umum dan Perumahan Rakyat Nomor 04/PRT/M/2015 tentang Kriteria dan Penetapan Wilayah Sungai.
REMA (Rwanda Environment Management Authority) (2010) Tool and guideline # 10 practical tools on water monitoring methods and instrumentation. P.O. Box 7436 Kigali, Rwanda
Shokralla S, Spall JL, Gibson JF, Hajibabaei M (2012) Next-generation sequencing technologies for environmental DNA research. Mol Ecol 21(8):1794–1805
Taberlet P, Coissac E, Hajibabaei M, Rieseberg LH (2012) Environmental DNA. Mol Ecol 21:1789–1793
Undang-undang Sumber Daya Air Nomor 17 Tahun 2019 tentang Sumber Daya Air
Visco JA, Apothéloz-Perret-Gentil L, Cordonier A, Esling P, Pillet L, Pawlowski J (2015) Environmental monitoring: inferring the diatom index from next-generation sequencing data. Environ Sci Technol 49(13):7597–7605
Yarsan E, Yipel M (2013) The important terms of marine pollution "biomarkers and biomonitoring, bioaccumulation, bioconcentration, biomagnification". J Mol Biomark Diagn S1(2)

Part V
Lesson Learned

Chapter 13
Local Wisdom: Values and Perspectives

K. L. Sandoval, D. M. M. Marundan, N. A. Silva, P. A. Permatasari, Sri S. Tumuyu, and Luki Subehi

In this section, the perspectives of local wisdom are characterized and exemplified clearly within two different countries, the Philippines and Indonesia. Through observation of the situation in Calamba City, Laguna—the Philippines, the discussion examines the validity of indigenous knowledge (IK), questions the present-day application of local wisdom to tropical ecosystem problems in academic research, and suggests the need to investigate the dynamism, interaction, and transformation of knowledge beyond the framework of local areas in order to better understand local realities and build a broader network of cooperation. Later, the subsequent discussion explores the potential of the local wisdom in freshwater management of several urban areas in Indonesia. Various cases show the connection between local wisdom and indigenous, religious, and cultural values whereas numerous local wisdom applications can provide guidance for stakeholders to adopt the concept of local wisdom in managing sustainable freshwater ecosystems.

K. L. Sandoval (✉) · D. M. M. Marundan · N. A. Silva
School of Arts and Sciences, National University, Philippines, Manila, Philippines
e-mail: klsandoval@national-u.edu.ph

P. A. Permatasari (✉)
Department of Landscape Architecture, Faculty of Agriculture, IPB University, Bogor, Indonesia
e-mail: pritaa07@apps.ipb.ac.id

Center for Environmental Research, IPB University, Bogor, Indonesia

S. S. Tumuyu
School of Environmental Science, Universitas Indonesia, Kota Depok, Indonesia

L. Subehi
Research Center for Limnology and Water Resources, National Research and Innovation Agency (BRIN), Jakarta, Indonesia

© The Author(s), under exclusive license to Springer Nature Singapore Pte Ltd. 2025
A. C. Santos-Borja et al. (eds.), *Progress on Ecosystem Restoration of Tropical Inland Waters*, https://doi.org/10.1007/978-981-96-2284-9_13

13.1 Rethinking the Local Wisdom Approach in Ecosystem Restoration Through Local Knowledge from Fishermen of Calamba City, Laguna: The Concept of Yankaw

Exploration of each community has various discoveries in every revisitation. We might end up finding more than our prior objectives as we begin to come across each social groups. Level by level, communities tend to reveal more cultures, practices, and/or relationships which surface on primary interactions and intercourses with the members of the community.

Similarly, *Calamba City, Laguna, Philippines*, specifically *Sitio Ronggot*, was observed to have its unique ways of life on lakesides. Throughout the several years, the said community has developed local practices/knowledge to adapt in each circumstance and challenge they have encountered to sustain the livelihood of its people and to contribute on the restoration of the lake which is the *Laguna de Bay*.

Laguna de Bay is known to be the largest lake in the Philippines and is rich in resources which support the villagers near its perimeters; hence, lake fishing becomes one of the major sources of earnings in *Sitio Ronggot*. However, as seasons change, the lake has also experienced different calamities, such as fish kills, and degradation, due to climate change and waste mismanagement, which affects the people who are dependent on it. As a normal response of humans to changes, the people of the community have learned their way out of this catastrophe. They have used their nurtured local knowledge on creating an efficient pathway to contribute to the lake's restoration.

Yankaw (Fig. 13.1) is one of the major documented innovation derived from local knowledge/wisdom that the people of the community have utilized over a number of generations. As described by the fisherfolks in *Calamba City*, it is a natural fishing gear composed of clustered camachile tree branches (Fig. 13.2).

Per its structure, it is considered to be an alternative mangrove (in saltwater) in *Laguna de Bay* (in freshwater is *Yankaw*) that serves as a breeding ground for the aquatic species in the lake. This has been their solution to efficiently cultivating the fishes in the lake and the fishing itself. Over the past years, this has been seemingly projecting positive impact on the everyday fishing of the fisherfolks. The harnessed local knowledge prior and after the utilization of *Yankaw* represents the ability of the locals to observe and response to their current and future needs and the needs of the lake.

To support this claim, according to the local fisherfolks, series of observations were conducted to justify the effectiveness of *camachile* as the medium of *Yankaw*. Prior to the utilization of *camachile* in *Yankaw*, Bamboo and *Sampaloc* tree branches were also tested but it was observed that camachile projects more effectivity and compatibility to the profile of the lake and its species. Using *camachile* tree branches has been sustained until this present time.

However, studies on *Yankaw* have focused heavily on its benefits to the fishing communities and its sustainability and adaptability to the environment. These have

Fig. 13.1 *Yankaw* (own work)

Fig. 13.2 Different tree branches used in utilization of the *Yankaw*. **a** Camachile tree; **b** Bamboo tree; and **c** Sampaloc tree (own work)

accounted experiences on Yankaw in its relation to the fishing activities. Research gap remains on the holistic understanding of Yankaw providing different aspects (physical, chemical, ecological, and biological) to the topic.

It is set to believe that local knowledge, wisdom, and practices can play a huge role in efforts to restore lake ecosystems which can also give opportunities to sustain the livelihood of the communities. By definition, local wisdom is a local or indigenous knowledge/practice acquired by the inhabitants of a community through their life experience which in this case, creating such innovation like *Yankaw* that aids their daily fishing activity and lake ecosystem restoration.

Local wisdom can also mean practices on (1) protecting and cultivating the camachile trees, (2) gathering different institutions to support different causes on the lake, or (3) looking at alternative sources of income aside from fishing during calamities such as extreme weather in the lake instead of desperately reverting to illegal fishing; such efforts are just some of the practices that can be correlated to effective mechanism on lake ecosystem usage.

Deciding which of which could be called as local wisdom can be considered as ambiguous, formulating such could possibly not be comprehensible by local cultural and social contexts of the local villagers. Over the past decades, local wisdom and indigenous knowledge have been debated by sociologists and anthropologists and have its concept be critiqued.

Similarly, indigenous knowledge (IK) is a variety of local wisdom that considers the origin and longevity of a knowledge in a certain area. It supports wide variation of programs that both support practical and ethical goals of a community. IK proponents believe that it does not only provide practitioners primary information on previously documented and new known facts but also empowers native inhabitants entitled to their own local wisdom. However, sociologists and anthropologists argued differently and have questioned it value and validity which they have claimed that it could result to disadvantageous effects as well.

Numerous studies may have considered local wisdom to be relevant in lake restoration but extensive analysis on this matter and its applications shows uncertain results on its contribution to lake ecosystem restoration efforts and goals.

This chapter ventures in providing insights and perspectives on the value and validity of local wisdom on lake ecosystem restoration goals and objectives, together with its related academic approaches. This would examine the concept of indigenous knowledge on conservation practices and provide an overview of *Yankaw*'s current status as a local knowledge. Subsequently, this chapter also outlines the progression over several years of the local wisdom utilization in the Philippines to demonstrate the process on how it was formed as an ideology and its application on lake ecosystem challenges. Also, this chapter highlights the limitations of the local wisdom in *Calamba City*, *Laguna*, and provides recommendations and suggestions on how it could impact the lake ecosystem restoration of *Laguna de Bay*. Local wisdom in this chapter may project different approaches that can be attributed to the sole efforts of local residents of the community of *Calamba City, Laguna*, which could confine sources of knowledge into one main source rather being open and dynamic to various communications.

13.1.1 The Emergence of Indigenous Knowledge (IK)

It is undeniable that local wisdom had traveled to different regions and communities since ancient times and the interaction between many cultures and concepts like "Western," "Asian," "European," "Eastern," "Centralized," or "Scientific" knowledge is the primary means of comprehending the idea of IK.

Through colonialism, European nations had taken the lead in various parts of the world by the middle of the seventeenth century. European countries have influenced colonized areas with knowledge and technologies which developed scientific knowledge in those communities. For instance, a special emphasis was placed on local plants and herbs that were beneficial for medicinal treatment throughout the sixteenth and seventeenth centuries (Ellen and Harris 2000).

Despite the fact that this process relied on local knowledge, it eventually integrated it into centralized, Western, scientific knowledge systems, thereby muting its original integration into the local community and culture. Local knowledge in the colonies was seen as inferior to Western scientific understanding as colonialism spread, a perception that persisted until the latter half of the twentieth century. However, during the 1960s, the relationship of the prevailed Western knowledge and local knowledge that were initially introduced was reevaluated and reconsidered.

Similarly, to this instance, *Yankaw*'s origin became questionable due to prior documentation of similar tool in Northern Luzon Philippines. Agustin Umali's 1950 Guide to the Classification of Fishing Gears in the Philippines included information on *Rama* (Fig. 13.3). According to one of the oldest fishermen in *Sitio Ronggot*, who reportedly came from the aforementioned location in Northern Luzon, *Rama* is very similar to *Yankaw* which also makes use of grouped tree branches that are submerged in waist-deep river water (National Museum of the Philippines Ilocos 2020). This only shows that every indigenous knowledge also travels with the inhabitant's relocation from one community to another which becomes local knowledge to each new location.

Numerous social and cultural anthropologists, human geographers, and development sociologists have recognized the benefits of IK and emphasized the significance of recording and utilizing "indigenous knowledge systems" (Warren et al. 1995; Slikkerveer and Dechering 1995). IK is important and useful in development projects, according to Warren et al. (1995), because it "forms the information base for a society which facilitates communication and decision-making" (Warren et al. 1995).

It was also stated that "once the methodologies for documenting these [indigenous knowledge] systems are introduced into training institutes in a given country, the recorded systems can be systematically deposited and stored for use by development practitioners" (Warren et al. 1995).

In the 1980s, the idea of IK was also used to ecological conservation initiatives as a response to unsuccessful initiatives that did not include local residents (Berkes 2004; Lanzano 2013; Slikkerveer and Dechering 1995). IK has been idealized in

Fig. 13.3 *Rama* (own work)

the decades that have followed as the result of generations of residents coexisting peacefully and symbiotically with their natural environment.

IK's practical effectiveness is defined by UNESCO as "a foundation for locally appropriate sustainable development" and it is ethically valued as "important facets of cultural diversity" (including language, ritual, spirituality, etc.).

These have shown that the emergence of indigenous knowledge has begun several centuries prior this present times and exhibited transpositions in each exchange of information from a local community to another.

13.1.2 Local Wisdom in Philippines and Laguna Lake Restoration

Compared to research conducted in other nations, the Philippines exhibits a less consistent use of the term "IK." To convey different facets of the knowledge of the locals, multiple expressions are used.

The phrase "indigenous knowledge" in English combines the definition of "local knowledge" as defined by Clifford Geertz in relation to cultural patterns with the definition provided by UNESCO. Slikkerveer et al. (2019) use the phrase "indigenous knowledge" in a particular sense, referring to social networks and institutions that have the potential to alleviate poverty. Using the phrase "local knowledge" instead of "scientific" or "professional" knowledge, Nugroho et al. (2018) investigate how "local knowledge" could influence policy.

The phrase "local wisdom" has been more prevalent in recent studies on Yankaw which examines the interactions between local populations and their natural settings, as well as the generation-to-generation transmission of resource management methods.

During the Philippines' period of decentralization and democratization, the phrase "local wisdom" gained widespread usage. As previously stated, the term "indigenous" is inextricably tied to the transnationally defined notion of "indigenous peoples." Nonetheless, the Philippine government declares that all ethnic groups in the country are "indigenous" or "native" and does not use the term "indigenous" to refer to any particular group of people in national programs (Ellen and Harris 2000; Osawa 2022). Furthermore, it is not unexpected that the term "local" has been used in this context instead of "indigenous" because it is more acceptable and helps reduce the possibility of becoming involved in discussions about indigeneity.

Moving forward, fish kill in *Laguna de Bay* in 2012 claimed 14 metric tons of tilapia (*Oreochromis niloticus*) and bangus (*Chanos chanos*), which harmed the main beneficiaries of the freshwater resources in Laguna de Bay. Residents living next to the lake suffered casualties totaling one million pesos due to the intense southwest monsoon that occurred a month after the fish kill. This led to the creation of the Laguna Lake Health, Environment, and Agricultural Diversity (LakeHEAD) Project in September 2012 through a partnership between the local administration, Japanese scientists, and the Calamba Fisheries and Aquatic Resources Management Council (CFARMC).

The *Yaman ng Lawa* Project was established by the Research Institute for Humanity and Nature and the University of the Philippines—Los Baños in collaboration with Community-Based Researchers in Action (CoBRA) as a result of the initiative, which looked into the accumulation of local knowledge from the local fishermen (Buen 2016).

In July 2013, the Yankaw Fish Garden Project was launched in response to natural calamities that occurred in *Laguna de Bay*, and *Yankaw* showed effectiveness in a variety of fishing endeavors inside the Bay. With their combined efforts, the Yankaw Fish Garden was built at the Fish Protected Areas in *Calamba City, Laguna* municipality. This initiative aims to effectively restore fish reproduction in Laguna de Bay, with a particular focus on tilapia, which is the predominant species in the lake (Buen 2022).

Many initiatives were undertaken to maintain *Yankaw*'s visibility to nearby communities in order to execute the initial measures to reestablish the fishing activities in *Laguna de Bay*. The Toyota Foundation International funded the creation of the MATH: Yaman ng Lawa booklet in 2018, which was then used as the basis for a conference on Social Participation and Co-Management of Natural Resources for Sustainable Development in *Calamba City, Laguna*. The conference focused on highlighting a community-oriented and watershed-based approach for balancing environmental conservation and regional economies in Indonesia, the Philippines, and Japan (Buen et al. 2018).

Hence, IK is usually described as "local wisdom" in reference to interactions between humans and their surroundings. Local wisdom is understood as a multi-generational relationship between individuals and their natural surroundings. As a result, it indicates that the people living there are capable of managing the resources and environment in a sustainable manner. Thus, appreciating and honoring local knowledge benefits the community in three ways: (1) it helps create "models" for sustainable development; (2) it gives residents social, cultural, and economic advantages; and (3) it makes it easier for people to "participate" in policymaking. The idea of "local wisdom" enables us to construct an idealized picture of harmony between individuals and the environment that is based on enduring and significant aspects of the local way of life.

While some researchers suggest that exchanges with foreign cultures, laws, and policies dynamically generate and transform elements of local wisdom, it is still a minority opinion. This perspective is still generally subdued in the latest studies on the lake ecosystem restoration, which places a strong emphasis on the "local" in "local wisdom."

13.1.3 Practical Limitations of Local Wisdom in Solving the Freshwater Ecosystem Problem

Even before the formal establishment of the local wisdom concept, several studies have researched on the effectiveness of using local knowledge to resolve lake ecosystem issues. According to Hidayat (2000) and Noor and Jumberi (2007), "cultural wisdom" or "local cultural wisdom" encompasses the traditional customs, institutions, and beliefs of the area as well as the agricultural techniques used to address the fish kill problem in the 1990s. An example of such knowledge includes *Yankaw*, *Rama*, and all other freshwater tools or fishing gears built by the local residents of a community.

Case studies indicate that this knowledge—which is considered local wisdom—is critical to freshwater restoration. The studies do, however, have a propensity to assume historical continuity and sustainability, which highlights the impact of an idealized conception and language of local wisdom that limits the viewpoints of the researchers.

In order to support freshwater ecosystems in a results-driven manner, the study strategy concentrates on extracting local knowledge or practices; however, it ignores the process and motivations behind the adoption of these practices or knowledge within the local context. The historical background of the knowledge held by the inhabitants and its relevance to day-to-day activities are likewise concealed.

The institutions, materiality, environmental features, aspirations, relational reciprocity, cosmology, and other factors that maintain the lives of the locals are all

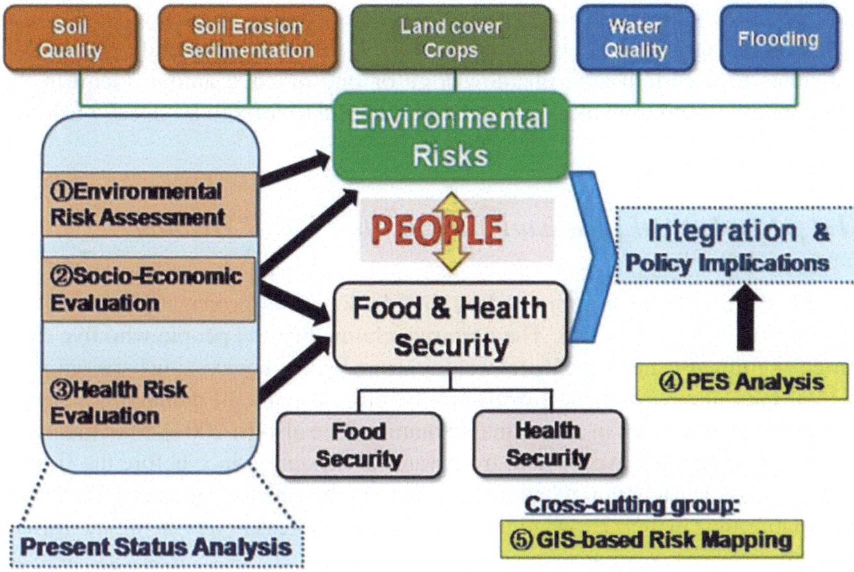

Fig. 13.4 General research framework of the Lake HEAD Project. Courtesy of Prof. Ryohei Kada, Ph.D. *Note* This model was produced by Kada in 2014 summarizing the trans-disciplinary research. From "The *Yaman ng Lawa* (Blessings of the Lake) Initiative in Santa Rosa City, Laguna, Philippines: An inclusive, participatory approach to public policy development" by Paterno et al. n.d.

interlinked with their knowledge and practices. The researchers compress this knowledge and practice, only emphasizing its effectiveness in relation to a certain external goal, by condensing this intricate interconnection to a single idea of "local wisdom" (Fig. 13.4).

In simple terms, the contexts, dynamism, and diversity of local knowledge and practices are not sufficiently considered by the dominant academic approach to local wisdom in freshwater studies. It is therefore uncertain to describe one particular practice or body of knowledge as "local wisdom" since it may oversimplify its history, misrepresent its institutions, ignore its varied applications within a local community, and overlook the various ways it is shared and modified in interactions with non-locals.

Through scientific methods or by connecting it to a disciplinary paradigm, researchers document local wisdom—a component of knowledge and practice that they have arbitrarily chosen as relevant to solving the freshwater problem. Local wisdom becomes a researcher's perspective of the world through this procedure, taking it out of the disorganized, dynamic setting of ordinary life in a location. The relationship between knowledge from many cultures and local or indigenous knowledge is thus not relativized by the local wisdom method (Kada 2013, 2015).

Rather, it supports the distinctions between the two bodies of knowledge and affirms the superiority of the former. The idea of "local wisdom" might be useful in getting scientists, particularly those studying nature, to pay more attention to the local way of life. Nevertheless, it also bears the risk of restricting the study perspective of scientists to a certain area of knowledge or one practice among many in the community at a given time that is thought appropriate to refer to as "local wisdom."

13.1.4 Ideological Idealism: Yankaw and Local Wisdom

This limited study focus is supported by the ideology that propagates an idealized representation of local wisdom. The narrative claims that the people who live there acquired an immense amount of local wisdom from years of coexisting harmoniously with their natural environment. Locals are therefore capable of managing the environment and its resources in a sustainable manner (see also Li 2001). Specifically, it is assumed that people have lived in freshwater ecosystems since before the significant increase in degradation, which means that (1) their historical use of freshwater is sustainable and (2) their knowledge can be used to address issues facing freshwater ecosystems.

The use of local knowledge, *Yankaw*, is said to be sufficient. Its application to particular development and conservation initiatives could redefine the dominance of centralized or global data by empowering communities that have struggled with the "top-down" imposition of such knowledge. Locals' engagement may also be encouraged in certain situations. As long as "local wisdom" is empirically verified (ecological, biological, even historical), it can be utilized as a useful analytical concept in academic research.

Academic research can examine the issues with freshwater ecosystems, customs and knowledge, and community contributions. This can help to improve understanding and alleviate the issues at hand. The concept's strong ideological idealism runs the risk of focusing inadequately and obstructing proper evaluation of local realities, which could cause researchers to misinterpret local circumstances and communicate incorrectly with locals. As a result, conservation efforts and development plans are created with less effectiveness.

13.1.5 Limits of Tradition

It is said to be challenging to find local wisdom in *Sitio Ronggot, Calamba City, Laguna* due to the theoretical limits of the local wisdom approach and its disparity from the reality of the field scenario. Yet again, the majority of academic studies adopt the local wisdom approach based on the uncertain presumption that (1) people have lived in and around lake environments continuously since before the increase in lake degradation, (2) their traditional use of *Rama* is sustainable in Northern Luzon, and (3) their knowledge can be applied to identifying an approach to the issues associated with lake degradation. These presumptions' shortcomings are immediately apparent in the field.

Multiple field workers who examined *Laguna de Bay* in Laguna prior to or shortly after the fish kills report that although the lake is still acceptable for living, it has changed after the calamity, causing people to rely on a lake or river that connects upstream and downstream regions in order to collect fish products.

13.1.6 The Trans-locality of the Problem

Ultimately, the most significant drawback of using a local wisdom approach to the freshwater or lake environment crisis is that local communities cannot find a solution on their own for a problem they did not create. It is necessary for various groups outside of the local community to interact and take action. The residents of *Laguna de Bay* have used other freshwater ecosystems besides their own, which have been harmed and degraded by numerous natural disasters like typhoons. They supply water to other nearby settlements as well. This indicates that the villagers' contribution to their degradation has been minimal at most.

The process of identifying and showcasing local wisdom in scholarly works and media essentially involves confining the cause—or, at worst, the solution—of this trans-local issue to the local community. This provides people who live in cities with the idea that since the issue with the freshwater ecosystem is occurring in rural areas, the locals have to handle it. If raising the local standard of living was the primary aim, the freshwater degradation problem is an environmental issue with which many stakeholders are fundamentally concerned. Freshwater ecosystems typically occur outside of the villagers' daily lives, while companies and private owners in urban areas are strongly linked to the reasons. By emphasizing local knowledge, we risk absolving ourselves of the primary duty for identifying the problems' causes and placing the responsibility of solving it back on the locals.

Therefore, encouraging collaboration and information exchange between the various sectors involved in and impacted by the problem is more important in solving it than emphasizing local expertise, which could alienate locals from non-local sectors. Making the connection between the issue and outsiders is especially important. The establishment and promotion of a knowledge network and, if feasible, collaboration between local and non-local sectors must be the primary focus when attempting to find answers.

13.2 Local Wisdom: Potential for Preserving Freshwater Ecosystem in Indonesian Urban Areas

Urban areas have a variety of unique characteristics compared to suburban and rural areas from the landscape and social conditions. The high population rate makes urban more vulnerable to environmental degradation than other areas (Feinberg et al. 2020). Nearly, 70% of the world's population is predicted to live in urban areas by 2050 (Veerkamp et al. 2021). High population leads to the massive use of natural resources that cause environmental quality deterioration. Environmental degradation in urban areas has closely related to population and economic growth (Wen et al. 2021). On the other hand, to be livable, a city needs to provide optimal services for its residents and reduce the dependence on various resources and services from other regions. To provide optimal ecosystem resources and services, environmental degradation in various ecosystems in urban areas needs to be minimized.

Freshwater is one of the features in urban areas that has a high potential to provide various ecosystem services for the community (Chen et al. 2017). Natural (lakes) or artificial (reservoirs), both are ecosystems that have various functions to support human life such as food supply, agricultural production, fisheries, water quality, recreation, etc. (Ye and Qiu 2021). It is estimated that as much as 23.2% of ecosystem services worldwide came from lakes and wetlands (Xu et al. 2018). This ecosystem can also provide water and nutrition for the life of living things and function as a habitat for various species. The existence of water resources in the ecosystem is related to its function as a flood control and drought.

The quality of fresh water in urban areas in Indonesia has shown varying conditions in recent years (Permatasari et al. 2017). Most rivers and lakes in urban areas do not meet the water quality standards set by the government. Various human activities have a high risk to produce pollutants that flow into lakes and reservoir waterbodies (Spears et al. 2022). Water purification service has a major effect on other ecosystem services and determines what functions can be given by a lake ecosystem (Piaggio and Siikamäki 2021). Bad water purification ecosystem services could have various consequences on other ecosystem services. Poor conditions of water quality will limit the lake and reservoir to provide their function to organisms, especially humans (Kritzberg et al. 2020).

Poor water quality will eliminate various other ecosystem services such as water provision, fish production, and recreational activities (Ye and Qiu 2021). In the worst case, the lake will lose its ability to provide optimal functions for the population. This condition makes the ecosystem lose its ability to provide ecosystem services and makes urban areas highly dependent on the surrounding area. On the other hand, the degradation of water quality makes urban areas vulnerable to environmental disasters and public health problems (Veerkamp et al. 2021).

Basically, people in urban areas have a strong desire to make their surrounding area comfortable and livable (Dzhambov et al. 2018; Garrett et al. 2019). However, various pressures such as hustle and bustle, economic pressure, self-centered and individualistic personality, and the loss of cultural values are factors inhibiting society's contribution to freshwater ecosystems sustainability (Wei et al. 2020). On the other hand, in some regions, local culture is still maintained for urban ecosystem management. Indigenous people use local wisdom as a middle way in ecological management and social sovereignty. This condition is like a ray of hope for other urban communities amidst the degradation of urban environment (Widaningsih et al. 2020).

Several indigenous communities in Indonesia, including those in urban areas, basically still hold the values of Eastern culture and have high tolerance values with the principle of unity in diversity. By forming collaboration between stakeholders, management of freshwater ecosystems in urban areas can be realized through local values adoption in each region. However, adopting local wisdom to become an ecosystem management program is not easy. In urban areas, adopting local wisdom requires adaptation and strong commitment from all stakeholders. The objective is to investigate and identify lessons learned from local wisdom implementation on freshwater ecosystem that can be applied to developing the community-based management system through traditional value.

13.2.1 Case Study: Situ Babakan Lake, Jakarta

Jakarta is the capital city and one of the biggest provinces in Indonesia which has a high lake and reservoir potential. No less than 108 lakes and reservoirs from 5 administrative cities are available in this province. The lake and reservoir in Jakarta have a vulnerability to ecological values decline caused by urban development. In the last five years, the condition of the lake and reservoir water quality in DKI Jakarta has decreased (Fig. 13.5), as indicated by the increasing number of high polluted lakes. Most of lake water in Jakarta was polluted by organic material, nutrients, heavy metals, and bacterial contamination (Henny and Meutia 2014).

Most of local community has water-adjacent or waterfront culture due to high dependence of water (Yodsurang et al. 2022). Nowadays, due to the degraded ecosystem, most of them only use lakes as tourist attractions or sources of livelihood without implementing special programs to protect lake ecosystems (Permatasari et al. 2023). Some of local communities who do not live adjacent to the lake ecosystem are also reluctant to contribute to lake management or lake cleaning program. Lake management efforts are still dominated by the provincial government. On the other hand, local communities contribute to pollutants flowing into lake water bodies originating from domestic waste.

Situ Babakan Lake (SBL) is located in the Betawi (Jakarta traditional tribe) cultural heritage area. The high sense of belonging of the people of the cultural heritage area toward the lake makes the community participation in this location in good condition (Mierdhani et al. 2022). Local communities realize the important value of lakes as a source of food, recreation, and livelihood. A degraded lake ecosystem will have an impact on reducing the value of economic benefits for local

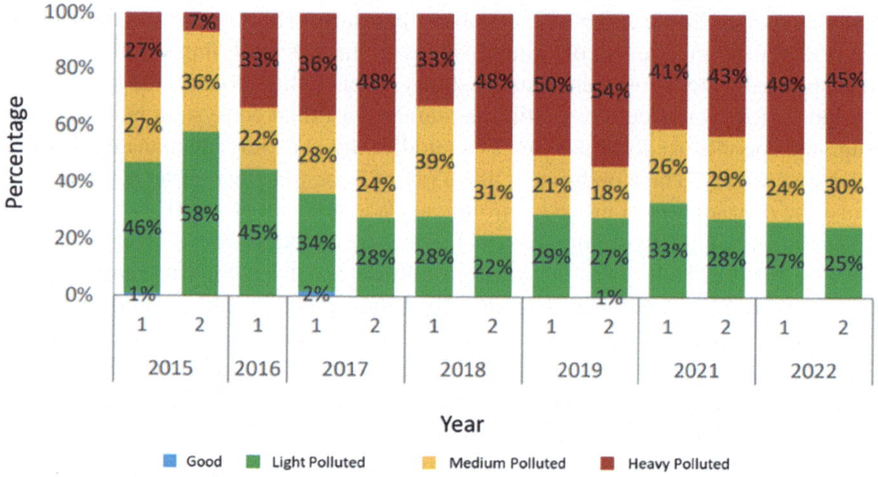

Fig. 13.5 Water quality in Jakarta Lakes and Reservoirs (Jakarta Provincial Environmental Service 2021)

communities. By understanding the relationship between anthropogenic pressure and ecosystem services, lake ecosystem management can be carried out well (Desta 2021).

A high level of concern for the freshwater ecosystems management can have a positive influence on ecosystem sustainability. The awareness to protect local heritage and water resources became the key in environmental movement (Santini and Taji 2019). Some of the programs carried out related to lake management include lakeshore commercial area decoration, Lebaran Betawi, and other lake festivals related to Betawi culture. Even though these activities are not directly aimed at maintaining lake water quality, various cultural festivals (Fig. 13.6) make local communities work together to maintain the cleanliness of the lake ecosystem because of the high potential for tourists and profits from the tourism sector.

The application of local wisdom in SBL is more focused on cultural values. Even though the majority of Betawi people are Moslems, not all cultural programs carried out are related to religion. Moreover, Jakarta as a melting pot of cultures and ethnicities has experienced cultural, racial, and religious assimilation and acculturation for many centuries (Martinez and Masron 2020). The various local wisdoms found at SBL in preserving the lake have a big influence from inherited religion and culture. As a cultural area that has blue open space elements, the SBL community is not only required to maintain traditional art but also a cultural heritage area that shows Betawi identity.

Based on the water quality values in SBL in 2022, several water quality parameters meet the quality standards set by the government except BOD (Table 13.1). This parameter reached 7.26 mg/L. The high number of settlements around the lake gives high risk of pollutants flow originating from domestic waste indicated by high BOD. SBL also gets water sources from the West Kalibaru River. The high level of river

Fig. 13.6 Horse-drawn carriage in lake festivals. Private collection. *Photo credit* Prita Ayu Permatasari

Table 13.1 Water quality in Situ Babakan Lake, Jakarta

No.	Parameter (ppm)	Quality standard	Value
1	BOD	3	7.26
2	COD	25	24.05
3	DO	4	5.17
4	TDS	1000	143.38

Source Jakarta Provincial Environmental Service (2021)

pollutants entering the lake also affected water quality. Point source flow and the potential for runoff when heavy rain enters the lake are factors that are difficult to control in managing water quality (Ruan et al. 2021). Fortunately, since the beginning of 2023, the local government has provided a wastewater treatment plant (WWTP) at the lake inlet point to prevent the entry of point source pollutants, especially those originating from domestic waste.

13.2.2 Case Study: Ayung River, Bali

Famous for its natural beauty and high cultural values, Bali becomes an island with highest tourism potential in Indonesia (Arismayanti et al. 2019). The strong bond between local communities and God makes that community activities are related to local cultural values and beliefs. These values are applied in all aspects of life, including environmental management. Local communities believe that achieving ecosystem sustainability requires implementing concepts that balance the relationship between God, humans, and nature or *Tri Hita Karana.*

Tri Hita Karana is a concept that Balinese use in daily life, not only in worship activities. Consisting of the elements of God (Parahyangan), humans (Pawongan), and the environment (Palemahan), this concept regulates harmony so that these three elements achieve balance without destroying each other. By implementing this concept, each generation can play a role in utilizing nature without damaging environmental sustainability. The application of this concept has led traditional Balinese communities to become the main actor in their own region without having to be marginalized in their own hometown.

The high tourism value of Bali creates the high globalization on this island. However, as long as the *Tri Hita Kirana* concept is firmly adhered to the community and other stakeholders, environmental degradation could be minimized (Roth and Sedana 2015). Due to the high tourism, property development is increasing every year. Many investors from local and overseas countries join the investment due to the high yields of property. Various hotels, resorts, apartments, restaurants, man-made tourism sites, and buildings supporting tourism activities have experienced a rapid increase in the last few decades. This condition contributes a negative influence on the river water quality in Bali.

Ayung (Fig. 13.7) is one of the rivers that flow in Bali and passes through various administrative areas. Most of the land use in Ayung Watershed is agricultural areas, followed by residential areas and forests. The high number of tourism properties such as hotels and restaurants in the Ayung watershed makes this river vulnerable to the pollutants flow through domestic waste release. Ayung River water is not only used as a source of irrigation and drinking water but also as a location for religious rituals which are closely related to the water element. This condition requires the participation of various stakeholders to maintain water quality so that it is safe to use for various human activities.

In general, the water quality of the Ayung River meets river water quality standards as a water recreation area (Table 13.2). However, the TSS parameters in downstream exceeded the quality standards. This condition shows the high level of urban development along the watershed. High TSS is also associated with seasonality, especially in the rainy season. Increasing TSS can cause turbidity in aquatic ecosystem. High turbidity results in low oxygen levels in the water (hypoxia), thereby triggering aquatic biota mortality (Permatasari et al. 2023). Human health can also be threatened by high TSS considering the suspended solid content which often contains algae, bacteria, sand, and silt. Moreover, the Ayung River is used by the Balinese people for ritual activities that require contact between water and human skin.

As time goes by, the *Tri Hita Karana* concept experiences challenges in its implementation. Land and water are seen as commodities since property values rise. Massive use of natural resources is carried out to support economic growth. This has an impact on the quantity and quality of river water flowing in Bali. Since the development era in the 1980s, several rivers in Bali are unsuitable for drinking, recreation purposes, or religious rituals such as melasti, mapag toya, banyu pinaruh,

Fig. 13.7 Rafting in Ayung River

Table 13.2 Water quality in Ayung River, Bali

No.	Parameter (ppm)	Quality standard	Upstream	Downstream
1	BOD	3	< 2	< 2
2	COD	25	< 6	< 6
3	DO	4	6.4	5.2
4	TSS	50	9	114

Source Widyasari and Ardi Putra (2022)

and nganyut due to the decline of river water quality. However, these rituals are still carried out in several Hindu temples with springs.

The importance of providing clean water sources in Bali is not only related to religious rituals or tourism business needs. The high demand for water in Bali due to increasing population and tourism activities makes it important to provide fresh water infrastructure and management due to the threat of water scarcity (Yamamoto et al. 2021). Based on the decree of the Minister of Environment and Forestry number 146/2023, Bali has become a province with exceeded water utilization status. This shows the low quality and quantity of water resources due to pressure from environmental changes.

Since 2000, Bali local government created the *Tri Hita Karana* Award for entrepreneurs in the tourism sector who are committed to run a business while maintaining environmental sustainability values based on this concept. This realization is not only important to apply to the tourism business sector but also other exploitative business sectors such as mining and residential areas as the main contributors to domestic waste. This program received high enthusiasm from entrepreneurs as proof of nature-responsible business. However, the government needs to balance rewards and punishment for environmental protection efforts. The existence of awards and punishment will encourage stakeholders to participate in river management.

13.2.3 Case Study: Code River, Yogyakarta

Apart from inherited traditions, community-based freshwater ecosystem management activities can also be newly initiated due to encouragement from the community to create a clean and livable place. *Merti Kali* is one of the community activities in Yogyakarta Province to clean the river ecosystem from the dengue fever outbreak that stalks many people. Although this activity began to be implemented in 2015, the local community expressed great enthusiasm in this program because of the strong encouragement to create a healthy environment, especially on river banks.

In contrast to other areas which focusing on certain freshwater ecosystems, this program is implemented in some rivers such as Code, Gajah Wong, Winongo, Kuning, Bedog, Kuncig, Tambak Bayan, Oyo, Belik, and Bulus. Initially, only a few rivers were included in this activity. As the year progresses, several rivers are also included. The wide coverage area of this activity is motivated by the government's important role. The existence of environmental laws encourages local governments to create ideas for this program and involve various stakeholders such as NGOs and local communities.

The Code River is one of the most highlighted *Merti Kali* (Fig. 13.8) ritual locations. This river flows from Mount Merapi and passes through urban areas. In the upstream area, water quality is pretty good and is characterized by a fairly high diversity of benthic species. Water quality decreases in urban areas, characterized by high levels of organic pollutants and low abundance of macrobenthos in these locations (Arumsari and Adharini 2021). This condition triggers high urgency from the surrounding community to carry out community-based environmental management.

Even though *Merti Kali* was only initiated in the last decade, this activity still pays attention to traditional elements in its procedures and rituals. The community is not only invited to clean the river together but understands the philosophy of the river as a source of life. On the Code River, the ritual begins by collecting water from seven different sources along the river as a symbol of social harmony. Some activities also include tree planting as an effort to protect springs. Furthermore, local communities will be invited to clean the river together as a symbol of community service. This ritual became an annual event and packaged as a tourist attraction.

The communities around the river are dominated by non-indigenous people. However, to increase loyalty toward this program, several main actors were selected, such as the community elders, religious leaders, village heads, and mosque administrators. Yogyakarta's close relationship with traditional culture makes the adoption of activities with cultural concepts easy for the community to comply with. The high influence of Islam and mosques on people's lives makes people obedient in implementing this program. With a religious approach, the local committee provides

Fig. 13.8 *Merti Kali* ceremony (own work)

outreach regarding the important value of rivers and the function of environmental management.

Based on the results of water quality monitoring in the Code River in 2022 (Table 13.3), BOD and COD show a high concentration. High BOD is associated with high levels of microbes which have an impact on the spread of waterborne disease such as cholera, dysentery, and gastroenteritis (Some et al. 2021).This condition indicates a high flow of pollutants originating from domestic and industrial waste. However, in the last three years, water quality in Code River has improved and met the quality standard for water recreation.

Even though *Merti Kali* is one of the leading programs in river management efforts, this program is not sufficient to improve river water quality. *Merti Kali* focuses on river cleaning activities by removing domestic solid waste in water bodies and riverbanks. Meanwhile, high BOD and COD are often sourced from liquid waste both industrial and domestic waste (Singkran et al. 2019). Therefore, combining *Merti Kali* with various other programs is very necessary to make river revitalization

Table 13.3 Water quality in Code River, Yogyakarta

No.	Parameter (ppm)	Quality standard	2020	2021	2022	2023
1	BOD	3	22.81	2.90	1.95	1.72
2	COD	25	65.62	13.43	10.24	9.83
3	DO	4	7.58	7.02	6.49	7.44
4	TSS	50	9.96	15.80	49.07	27.11

Source Yogyakarta Provincial Environmental and Forestry Service (2022)

efforts more optimal. Collaboration between stakeholders such as the community and government as decision makers is also needed to carry out this program.

Regulating the disposal of liquid waste and providing WWTP in both residential and industrial areas is highly recommended to reduce the flow of pollutants entering water bodies (Brontowiyono et al. 2022). Planting riparian buffer strips can also be the solution for pollutant absorbing since the local communities already have tree planting programs. Riparian buffer strips have many benefits in improving landscape aesthetics, protecting biodiversity, and mitigating diffuse pollution (Cole et al. 2020). However, the *Merti Kali* program still has a number of benefits. Increased environmental awareness and public health are the benefits felt by the local community after *Merti Kali* was carried out.

13.2.4 Case Study: Rawa Pening Lake, Central Java

A different approach in managing lake ecosystems occurs in Rawa Pening, Ambarawa, Central Java. The local community uses fairy tales to mobilize the community to manage the lake ecosystem. *Baru Klinting* is a popular fairy tale that tells the history of Rawa Pening Lake. The story gave the message regarding the human greed that lead into disaster. This fairy tale is used by local people to remind them of the importance of protecting the environment and sharing it wisely with others (Retnowati et al. 2020). By adopting the philosophy, *Baru Klinting* became an annual festival in Rawa Pening. By holding festivals regularly, the local community will try to keep the lake ecosystem clean. As a result, a high number of tourists come to visit and improve the economy of the surrounding community.

The main environmental problem that occurs in Rawa Pening is the high growth of water hyacinth (Prasetyo et al. 2022). The high growth of water hyacinth can threaten the sustainability of aquatic biota because of the high oxygen absorbed by plants. This invasive plants also impact the endemic organism loss by blocking the sunlight in waterbody and causing the ecosystem balance disorder (Maulidyna et al. 2021). Aesthetically, water hyacinth can reduce tourist value. Tourist boats that should be able to be used by tourists can also be hindered by the presence of water hyacinth (Kimijima et al. 2020). With the *Baru Klinting* festival (Fig. 13.9), the local community is trying to clean the water hyacinth so that the lake can be used as a tourist attraction.

Based on water quality monitoring in 2018, several water quality parameters in Rawa Pening exceed the class 2 water quality standard. High BOD and COD values indicate domestic and industrial waste flows that exceed quality standards. The TP value indicates high levels of phosphate in a water body which originates from fertilizer or floating net cages above the water body. High intensivity of fisheries sector in this lake caused the high feed waste and fisheries residue. The high agricultural area around the lake also has the potential to flow excess nutrients into the water body (Piranti et al 2018). The high flow of pollutants entering water bodies also affects the growth of water hyacinth in this lake.

Fig. 13.9 *Baru Klinting* festival (own work)

The poor water quality in Rawa Pening shows the complexity of water quality problems. This problem can not only be overcome through local wisdom or management programs that prioritize clean lake surfaces. The stakeholders should take a big concern on land use regulation, fertilizer use, and floating net cages management considering the ecosystem conditions that can receive pollutant flows from various sources (Vasconcelos et al. 2018). Lakes as water systems that are integrated with each other and other water bodies require an integrated management system. Thus, programs on a local scale are difficult to overcome these complicated problems.

The fairy tale approach to community-based environmental management has had a positive influence on tourism sector and the economy of the surrounding community but this system has not been effective in maintaining water quality and controlling the growth of water hyacinth in Rawa Pening. Local governments need to create a more comprehensive management system that takes into account land use and pollutant flows. Involvement of various stakeholders around Rawapening is also needed to realize harmonization of ecosystem management.

13.3 Remarks

1. There is no doubt that addressing and mitigating the degradation of lake ecosystems requires an investigation of local knowledge and practices. Such research gives academics a deeper comprehension of the reality of the community, which

in turn improves meaningful engagement with them and the creation of successful restoration programs.

Yet, a local knowledge strategy has been used in much recent academic research to address the issue of the lake environment in general as well as this particular investigation. By designating particular practices and information as "local wisdom," they aim to find unique or "grassroots" approaches to problem-solving. Like the IK technique, this one has a lot of drawbacks. "Local wisdom" permits outsiders to recognize something as local wisdom without giving local circumstances enough thought, and it leaves open problems regarding historical continuity and interactions with outsiders. As a result, the strategy may cause misconceptions about local expertise and cause locals to be disregarded by non-local sectors. When it comes to solving the lake ecosystem, it limits our comprehension of the communities that surround the freshwater ecosystem and the lake issue itself, which runs the danger of assigning only locals and organizations accountability for the issue. The intellectual idealism that this concept expresses shows these shortcomings. Despite the dynamic nature of local knowledge and practice resulting from interaction and conversation with others, local wisdom provides researchers with the impression of a static existence among locals. In order to address these issues, researchers should concentrate on the interaction and information transmission among various groups relevant to the problem, rather than local wisdom. Practically speaking, what is required is a strategy that allows stakeholders to engage in long-term engagements while also enabling ongoing communication and collaboration between them—primarily across local and non-local divisions. On a practical level, better comprehension of the circumstances faced by the people will enable ongoing, interactive conversation and collaboration. Lastly, it is critical to emphasize the value of knowledge exchange. Researchers must be informed by locals about the intricate circumstances and varied realities of each case in addition to imparting their knowledge to them based on scientific methods during the knowledge-sharing process. These exchanges could lead to productive collaboration and communication among scholars as well as a shift in their viewpoints. When it comes to solving the issue of freshwater ecosystem restoration, the possibility for information exchange clearly demonstrates the error of confining expertise or experience to locals alone.

2. Local wisdom can be influenced by various aspects. From the various examples presented, each case study has unique conditions. Inherited traditions, belief in the concept of the universe, obedience to religious leaders, the spread of disease outbreaks, and belief in fairy tales are the factors behind community-based environmental management in freshwater ecosystems. Local knowledge possessed by a person can be disseminated through the process of communication and interaction (Chaiphar et al. 2013). The availability of common goods in a location can be the main trigger for the formation of community action. Meanwhile, community trust and reciprocity also strengthen collective action in protecting the common good in a place. Even though community-based environmental management programs have ideas and applications that are rooted in local

communities, program consistency is still difficult to implement. Various challenges such as the modernization of urban communities, increasing economic needs, regional globalization, and the loss of interest in programs can prevent communities from consistently implementing community-based environmental management programs. Social movement change and transformation can be affected by several factors: political opportunity availability, movement group's preparedness, movement framing availability, and movement innovation availability (Roth and Sedana 2015). Not only the community, local government, and traditional/religious leaders also have their own challenges in implementing the program. Limited funds, weak political will of the central government, low participation from other stakeholders are some of the inhibiting factors (Ali et al. 2013). The effectiveness of local wisdom in encouraging people to carry out environmental management is also determined by the applicable sanctions. There are punishments from God and social sanctions given by society for violators. The strong belief and obedience will lead into good collective action. In several areas, local wisdom has been successfully legalized into regional regulations so that it is more formal and valid from the legal point of view. However, to make local wisdom into a legal written regulation requires many challenges. Freshwater ecosystems are connected to each other by nutrient flow. As a result, ecosystems in the lowlands are highly susceptible to the accumulation of pollutants and nutrients which leads to a decrease in water quality (Heino et al. 2021). Local-scale management will not produce optimal impacts because of the high level of interconnectedness. The flow of pollutants originating from other freshwater and terrestrial ecosystems carried by runoff makes managing freshwater ecosystems a complex matter. Coordination with stakeholders in other areas is highly recommended especially in the area that does not apply the same local wisdom. Therefore, the application of local wisdom should be adopted as a program that can be implemented in larger areas, not in traditional boundaries.

The application of local wisdom in freshwater ecosystem management is not only limited to maintaining water quality below quality standards, but also maintaining the sustainability of the ecosystem so that it can provide optimal ecosystem services for future generations. Therefore, it is not only limited to cleaning solid waste from water bodies but also ensuring that the flow of pollutants is acceptable by water purification carrying capacity. Natural source extraction should avoid overharvesting that caused depletion. Land use regulation in watershed scale is very needed to control pollutant flow, shoreline transformations, and hydrological changes (Furgała-Selezniow et al. 2021).

Cultural services can be seen from various points of view by society such as environmental spaces, cultural practices, cultural benefits, and cultural goods (UNESCO 2021). High awareness of the surrounding community regarding the high value of cultural services in an ecosystem will encourage the community to protect the ecosystem so that its value can be maintained or increased. The value or benefits produced by an ecosystem are not always seen as material or monetary gain but sometimes in intangible values that are important for society such as spirituality, sense of belonging, health, knowledge, etc. When awareness of protecting the environment meets local knowledge, local wisdom will be formed.

Several local wisdoms were processed into legal systems due to the importance of local values in influencing people's social life. In Eastern Indonesia, *Sasi* program is implemented as a program to maintain marine ecosystem sustainability, as well as respect for nature as the source of livelihood. In urban areas, legalization of local wisdom is seen as something that is not crucial due to the high abundance of natural resources from within and outside the area. The top-down approached governance tends to reduce the role of the community, including indigenous communities in urban areas. The individual character and modern life of urban communities also weaken awareness of developing local wisdom. However, local wisdom can be adopted in urban society as a resolution to many problems related to culture, social, and economics (Diab et al. 2022). Apart from improving environmental conditions, some local traditions have a negative impact on the environment. Floating the agricultural harvest into water bodies for expressing gratitude to water god has a bad influence by increasing pollutants. A similar tradition also occurred in Kyoto where traditional cloth was dyed and the dye residue flowed into the river (Santini and Taji 2019). Therefore, before being adopted, local wisdom needs to be accompanied by the latest knowledge to select what can be adopted and what cannot. Modifications can also be done as a middle way between indigenous knowledge and modern knowledge. Both can complement each other and synergize in supporting freshwater ecosystem management programs.

References

Ali DM, Islamy MI, Supriyono B, Muluk MRK (2013) Bringing local wisdom-based local government into practice: a quadrant strategy analysis. Public Policy Adm Res 3(5):21–34

Arismayanti NK, Sendra IM, Suwena IK, Budiarsa M, Bakta IM, Pitana IG (2019) Tourism villages' development in Bali, mass or alternative tourism? J Tour Hosp Manag 7(2):117–139. https://doi.org/10.15640/jthm.v7n2a11

Arumsari PL, Adharini RI (2021) Structure and distribution of macrobenthos community in code river, Yogyakarta, Indonesia. Indones J Limnol 2(2):21–27. https://doi.org/10.51264/inajl.v2i2.17

Berkes F (2004) Rethinking community-based conservation. Conserv Biol 18(3):62–630. https://doi.org/10.1111/j.1523-1739.2004.00077.x

Brontowiyono W, Boving T, Asmara AA, Rahmawati S, Yulianto A, Wantoputri NI, Lathifah AN, Andriansyah Y (2022) Communal wastewater treatment plants' effectiveness, management, and quality of groundwater: a case study in Indonesia. Water (Switz) 14(19):1–23. https://doi.org/10.3390/w14193047

Buen CC (2016) Yaman ng Lawa: adaptive co-management of lake resources: the experience of Calamba City, Laguna, Philippines. Presentation at the special seminar of food-risk CR Project, RIHN, Kyoto, Japan

Buen CC (2022) AfterMATH: Yaman ng Lawa. A GAD special edition 2022 by the City Government of Calamba City in partnership with City Agricultural Services Department and Gender and Development Division (GAD)
Buen CC, Guerrero EJ, Espino TML (2018) MATH for Yaman ng Lawa: strategies for sustainable development of the wealth of the lake the case of Laguna de Bay, Calamba city, Laguna, Philippines. In: 17th World lake conference, Ibaraki, Japan
Chaiphar W, Promsaka Na Sakolnakorn T, Naipinit A (2013) Local wisdom in the environmental management of a community: analysis of local knowledge in Tha Pong village, Thailand. J Sustain Dev 6(8):16–25. https://doi.org/10.5539/jsd.v6n8p16
Chen X, Chen Y, Shimizu T, Niu J, Nakagami K, Qian X, Jia B, Nakajima J, Han J, Li J (2017) Water resources management in the urban agglomeration of the Lake Biwa region, Japan: an ecosystem services-based sustainability assessment. Sci Total Environ 586(3):174–187. https://doi.org/10.1016/j.scitotenv.2017.01.197
Cole LJ, Stockan J, Helliwell R (2020) Managing riparian buffer strips to optimise ecosystem services: a review. Agric Ecosyst Environ 296(Feb):106891. https://doi.org/10.1016/j.agee.2020.106891
Desta H (2021) Local perceptions of ecosystem services and human-induced degradation of Lake Ziway in the Rift Valley region of Ethiopia. Ecol Indic 127. https://doi.org/10.1016/j.ecolind.2021.107786
Diab AL, Pabbajah M, Nurina Widyanti R, Muthalib LM, Fajar Widyatmoko W (2022) Accommodation of local wisdom in conflict resolution of Indonesia's urban society. Cogent Soc Sci 8(1). https://doi.org/10.1080/23311886.2022.2153413
Dzhambov AM, Markevych I, Hartig T, Tilov B, Arabadzhiev Z, Stoyanov D, Gatseva P, Dimitrova DD (2018) Multiple pathways link urban green- and bluespace to mental health in young adults. Environ Res 166(June):223–233. https://doi.org/10.1016/j.envres.2018.06.004
Ellen R, Harris H (2000) Introduction. In: Ellen R, Parkes P, Bicker A (eds) Indigenous environmental knowledge and its transformations: critical anthropological perspectives. Harwood Academic Publishers, Amsterdam, pp 1–33
Feinberg A, Ghorbani A, Herder PM (2020) Commoning toward urban resilience: the role of trust, social cohesion, and involvement in a simulated urban commons setting. J Urban Aff 00(00):1–26. https://doi.org/10.1080/07352166.2020.1851139
Furgała-Selezniow G, Jankun-Woźnicka M, Kruk M, Omelan AA (2021) Land use and land cover pattern as a measure of tourism impact on a lakeshore zone. Land 10(8). https://doi.org/10.3390/land10080787
Garrett JK, White MP, Huang J, Ng S, Hui Z, Leung C, Tse LA, Fung F, Elliott LR, Depledge MH, Wong MCS (2019) Urban blue space and health and wellbeing in Hong Kong: Results from a survey of older adults. Heal Place 55(Apr 2018):100–110. https://doi.org/10.1016/j.healthplace.2018.11.003
Heino J, Alahuhta J, Bini LM, Cai Y, Heiskanen AS, Hellsten S, Kortelainen P, Kotamäki N, Tolonen KT, Vihervaara P, Vilmi A, Angeler DG (2021) Lakes in the era of global change: moving beyond single-lake thinking in maintaining biodiversity and ecosystem services. Biol Rev 96(1):89–106. https://doi.org/10.1111/brv.12647
Henny C, Meutia AA (2014) Urban lakes in Megacity Jakarta: risk and management plan for future sustainability. Procedia Environ Sci 20:737–746. https://doi.org/10.1016/j.proenv.2014.03.088
Hidayat T (2000) Studi kearifan budaya petani Banjar dalam pengelolaan lahan rawa pasang surut. J Kalimantaan Agrikultura 7(3):105–111
Jakarta Province Environmental Service (2021) Final Report Jakarta Lake Water Monitoring 2021
Kada R (2013) Yaman ng Lawa: a platform for sustainable management and utilization of the Laguna lake resources. In: RIHN-LakeHEAD community forum 2013 (proceedings), pp 183–190
Kada R (2015) Managing environmental risks to food and health security in Asian watersheds. Research Institute for Humanity and Nature (RIHN), Shijonawate Gakuen University. https://www.chikyu.ac.jp/rihn_e/project/R-06.html

Kimijima S, Sakakibara M, Amin AKMA, Nagai M, Arifin YI (2020) Mechanism of the rapid shrinkage of Limboto lake in Gorontalo, Indonesia. Sustainability 12(22):1–14. https://doi.org/10.3390/su12229598

Kritzberg ES, Hasselquist EM, Škerlep M, Löfgren S, Olsson O, Stadmark J, Valinia S, Hansson LA, Laudon H (2020) Browning of freshwaters: consequences to ecosystem services, underlying drivers, and potential mitigation measures. Ambio 49(2):375–390. https://doi.org/10.1007/s13280-019-01227-5

Lanzano C (2013) What kind of knowledge is 'indigenous knowledge'? Critical insights from a case study in Burkina Faso. Transcience 4(2):3–18

Li TM (2001) Masyarakat adat, difference, and the limits of recognition in Indonesia's forest zone. Mod Asian Stud 35(3):645–676

Martinez R, Masron IN (2020) Jakarta: a city of cities. Cities 106(Jan)

Maulidyna A, Alicia F, Agustin HN, Dewi IR, Nurhidayah I, Dewangga A, Kusumaningrum L, Nugroho GD, Jumari J, Setyawan AD (2021) Review: economic impacts of the invasive species water hyacinth (*Eichhornia crassipes*): case study of Rawapening Lake, Central Java, Indonesia. Int J Bonorowo Wetl 11(1):18–31. https://doi.org/10.13057/bonorowo/w110103

Mierdhani MIR, Dewi L, Mularsari A (2022) Community participation in recovering cultural tourism in Betawi Cultural Village Setu Babakan. Int J Econ Manage Bus Soc Sci (IJEMBIS) 2(2):282–291

National Museum of the Philippines Ilocos (2020) May is also the month of the ocean. [Facebook]. Published date: 27 May 2020. Retrieved from: https://www.facebook.com/nationalmuseumilocosregion/photos/a.470961193382801/1123871044758476/?type=3&paipv=0&eav=AfbMPcrGHMzkzCewtrbfy177ivOCxuozCP5LZ1vy4Z-N019T49xrlpHQ_uJ3UoqzMuE&_rdr, 12 Oct 2023

Noor M, Jumberi A (2007) Kearifan local dalam perspektif pengembangan pertanian di lahan rawa. https://repository.pertanian.go.id.handle/123456789/6293

Nugroho K, Carden F, Antlov H (2018) Local knowledge matters: power, context and policy making in Indonesia. Policy Press, Bristol. https://doi.org/10.51952/9781443348085

Osawa T (2022) At the edge of mangrove forest: the Suku Asli and the quest for indigeneity, ethnicity and development, Kyoto area studies on Asia, vol 29. Kyoto University Press; Trans Pacific Press, Kyoto; Tokyo

Paterno RPP, Galvez Tan JZ, Palapar LA, Bermudez ANC, Guiang AJB (n.d.) The Yaman ng Lawa (Blessings of the Lake) initiative in Santa Rosa City, Laguna, Philippines: an inclusive, participatory approach to public policy development. https://www.pref.ibaraki.jp/soshiki/seikatsukankyo/kasumigauraesc/04_kenkyu/kaigi/docments/kosyou/15/2014wlc_ramon_pedrop_p_paterno.pdf. Accessed 26 June 2024

Permatasari PA, Setiawan Y, Khairiah RN, Effendi H (2017) The effect of land use change on water quality: a case study in Ciliwung Watershed. IOP Conf Ser Earth Environ Sci Conf Ser 1–7

Permatasari PA, Setiawati S, Subehi L (2023) Fisheries' sustainability of three lakes in Jakarta, Indonesia. AACL Bioflux 16(3):1245–1253

Piaggio M, Siikamäki J (2021) The value of forest water purification ecosystem services in Costa Rica. Sci Total Environ 789. https://doi.org/10.1016/j.scitotenv.2021.147952

Piranti AS, Rahayu DRUS, Waluyo G (2018) Evaluasi status mutu air Danau Rawapening. J Pengelolaan Sumberd Alam dan Lingkung (J Nat Resour Environ Manag) 8(2):151–160. https://doi.org/10.29244/jpsl.8.2.151-160

Prasetyo S, Anggoro S, Soeprobowati TR (2022) Water hyacinth *Eichhornia crassipes* (Mart) Solms management in Rawapening Lake, Central Java. AACL Bioflux 15(1)

Retnowati R, Purwanto P, Suryanto H (2020) The empowerment of Rawa Pening communities in enhancing cleanliness and environmental sustainability. J Int Conf Proc 3(1):206–220. https://doi.org/10.32535/jicp.v2i4.796

Roth D, Sedana G (2015) Reframing Tri Hita Karana: from 'Balinese Culture' to politics. Asia Pacific J Anthropol 16(2):157–175. https://doi.org/10.1080/14442213.2014.994674

Ruan S, Hong Y, Zhuang Y (2021) Evolution and restoration of water quality in the process of urban development: a case study in urban lake, China. Environ Monit Assess 193(7). https://doi.org/10.1007/s10661-021-09182-z

Santini T, Taji T (2019) Natural urban heritage and preservation policies: the case of Kyoto's waterways. Int J Environ Sci Sustain Dev 4(2):95–106. https://doi.org/10.21625/essd.v4i2.559

Singkran N, Anantawong P, Intharawichian N, Kunta K (2019) The Chao Phraya river basin: water quality and anthropogenic influences. Water Sci Technol Water Supply 19(5):1287–1294. https://doi.org/10.2166/ws.2018.167

Slikkerveer LJ, Dechering WHJC (1995) LEAD: the Leiden ethnosystems and development programme. In: Warren DM, Slikkerveer LJ, Brokensha D (eds) The cultural dimension of development: indigenous knowledge systems. Intermediate Technology Publications, London, pp 435–440

Slikkerveer LJ, Baourakis G, Saefullah K (eds) (2019) Integrated community-managed development: strategizing indigenous knowledge and institutions for poverty reduction and sustainable community development in Indonesia. Cooperative management. Springer, Cham. https://doi.org/10.1007/978-3-030-05423-6

Some S, Mondal R, Mitra D, Jain D, Verma D, Das S (2021) Microbial pollution of water with special reference to coliform bacteria and their nexus with environment. Energy Nexus 1(Aug):100008. https://doi.org/10.1016/j.nexus.2021.100008

Spears BM, Hamilton DP, Pan Y, Zhaosheng C, May L (2022) Lake management: is prevention better than cure? Inl Waters 12(1):173–186. https://doi.org/10.1080/20442041.2021.1895646

Umali A (1950) Guide to the classification of fishing gears in the Philippines. United States Government Printing Office

UNESCO (n.d.) Local and indigenous knowledge systems: what is local and indigenous knowledge? Retrieved from: http://www.unesco.org/new/en/natural-sciences/priority-areas/links/related-information/what-is-local-and-indigenous-knowledge, 04 Mar 2024

UNESCO (2021) The United Nations World Water Development Report 2021: valuing water, Paris

Vasconcelos FR, Menezes RF, Attayde JL (2018) Effects of the *Nile tilapia* (*Oreochromis niloticus* L.) on the plankton community of a tropical reservoir during and after an algal bloom. Hydrobiologia 817(1):393–401. https://doi.org/10.1007/s10750-018-3591-2

Veerkamp CJ, Schipper AM, Hedlund K, Lazarova T, Nordin A, Hanson HI (2021) A review of studies assessing ecosystem services provided by urban green and blue infrastructure. Ecosyst Serv 52(Apr 2020):101367. https://doi.org/10.1016/j.ecoser.2021.101367

Warren DM, Slikkerveer LJ, Brokensha D (1995) Introduction. In: Warren DM, Slikkerveer LJ, Brokensha D (eds) The cultural dimension of development: indigenous knowledge systems. Intermediate Technology Publications, London, pp xv–xviii

Wei W, Gao Y, Huang J, Gao J (2020) Exploring the effect of basin land degradation on lake and reservoir water quality in China. J Clean Prod 268:122249. https://doi.org/10.1016/j.jclepro.2020.122249

Wen C, Zhan Q, Zhan D, Zhao H, Yang C (2021) Spatiotemporal evolution of lakes under rapid urbanization: a case study in Wuhan, China. Water (Switz) 13(9):1–20. https://doi.org/10.3390/w13091171

Widaningsih TT, Diana R, Rahayunianto A (2020) Community based cultural tourism development Setu Babakan, Jakarta. J Environ Manag Tour 11(2):486–495. https://doi.org/10.14505/jemt.v11.2(42).26

Widyasari NL, Ardi Putra IK (2022) Water quality analysis of Ayung river using Storet method. J Ecocentrism 2(2):51–59. https://doi.org/10.36733/jeco.v2i2.4931

Xu X, Jiang B, Tan Y, Costanza R, Yang G (2018) Lake-wetland ecosystem services modeling and valuation: progress, gaps and future directions. Ecosyst Serv 33(Aug):19–28. https://doi.org/10.1016/j.ecoser.2018.08.001

Yamamoto EMS, Sayama T, Takara K (2021) Impact of rapid tourism growth on water scarcity in Bali, Indonesia. Indones J Limnol 2(1):1–16. https://doi.org/10.51264/inajl.v2i1.14

Ye Y, Qiu H (2021) Environmental and social benefits, and their coupling coordination in urban wetland parks. Urban Urban Green 60(Feb):127043. https://doi.org/10.1016/j.ufug.2021.127043

Yodsurang P, Uekita Y, Shimizu I (2022) Water-based settlement and the loss of community water resilience. Interiority 5(2):179–196. https://doi.org/10.7454/in.v5i2.210

Yogyakarta Province Environmental and Forestry Service (2022) Final Report River Water Monitoring 2022

The manufacturer's authorised representative in the EU is Springer Nature Customer Service Centre GmbH, Europaplatz 3, 69115 Heidelberg, Germany. If you have any concerns regarding our products, please contact ProductSafety@springernature.com

Printed and bound by CPI Group (UK) Ltd, Croydon, CR0 4YY
01/10/2026
02225929-0010